Target Organ Toxicity in Marine and Freshwater Teleosts

New Perspectives: Toxicology and the Environment

Target Organ Toxicity in Marine and Freshwater Teleosts

Volume 2 – Systems

Edited by

Daniel Schlenk

Department of Environmental Sciences
University of California
Riverside, California
USA

William H. Benson

Gulf Ecology Division, National Health and Environmental Effects
Research Laboratory
US Environmental Protection Agency
Gulf Breeze, Florida
USA

London and New York

First published 2001
by Taylor & Francis
11 New Fetter Lane, London EC4P 4EE

Simultaneously published in the USA and Canada
by Taylor & Francis
29 West 35th Street, New York, NY 10001

Taylor & Francis is an imprint of the Taylor & Francis Group

Typeset in Times by
Prepress Projects Ltd, Perth, Scotland
Printed and bound in Great Britain by
St Edmundsbury Press, Bury St Edmunds

British Library Cataloguing in Publication Data
A catalogue record for this book is available
from the British Library

Library of Congress Cataloging in Publication Data

ISBN 0–415–24839–6

Contents

Contributors

Lisa D. Arcand-Hoy is at Blasland, Bouck & Lee, Inc., Syracuse, NY, USA

William H. Benson is at the Gulf Ecology Division, National Health and Environmental Effects Research Laboratory, US Environmental Protection Agency, Gulf Breeze, FL, USA

Sandra K. Brewer is at the Department of Biological Sciences, University of Missouri, Columbia, MO, USA

Russell L. Carr is at the Center for Environmental Health Sciences, Mississippi State University, MS, USA

Janice E. Chambers is at the Center for Environmental Health Sciences, Mississippi State University, MS, USA

Edward E. Little is at the US Geologic Survey, Columbia Environmental Research Center, Columbia, MO, USA

Charles D. Rice is at the Department of Environmental Toxicology, Clemson University, Pendleton, SC, USA

Daniel Schlenk is at the Department of Environmental Sciences, University of California, Riverside, CA, USA

Foreword

The field of aquatic toxicology is increasing in importance because it is now commonly accepted that humans cannot continue to discharge their wastes and byproducts into streams, rivers, and oceans without having impacts on those ecosystems. As a result, efforts are focusing on reducing wastestreams from both point and non-point sources, as well as reducing the impacts of historic discharges through remediation and cleanup. This has led to considerable contention around the question "How clean is clean enough?" To answer this question aquatic toxicologists have conducted extensive research and written many books, looking at the effects of toxicants both at comparatively simple levels of biological organization (e.g. the biomarker approach) and less commonly at highly complex levels of organization (population and community impacts).

The challenge has been to link these two approaches. It is comparatively easy to determine the effects and mechanisms by which chemical contaminants affect cellular function, largely through laboratory investigations. Field investigations can demonstrate causal relationships between certain classes of contaminants and effects on individual organisms, but it is extremely rare for effects of chemical contaminants to be specifically discernible at the population or community level, especially in open marine and estuarine ecosystems. Increasingly, we are realizing that fitness of individual organisms may need to be the level of biological function which we need to protect in aquatic ecosystems, much as we currently do for human health. This requires assessing organ function, and it is surprising that there are not good reference texts for assessing organ system function in fish species exposed to chemical contaminants. This second volume addresses that void and will be a valuable reference to the practicing aquatic toxicologist.

Tracy K. Collier
Northwest Fisheries Science Center

Copyright acknowledgments

The authors and publishers would like to thank the following for granting permission to reproduce material in this work.

W.B. Saunders Co., Philadelphia, for permission to reproduce Figure 3.1, originally published as Figure 3.1 in Tizard, I.R. 1996. *Veterinary Immunology – An Introduction.*

W.B. Saunders Co., Philadelphia, for permission to reproduce Figure 3.2, originally published as Figure 11.1 in Stoskopf 1996. *Fish Medicine.*

W.H. Freeman and Co., New York, for permission to reproduce Figure 3.6, originally published as Figure 10.8 in Kuby, J. 1994. *Immunology*, 2nd edn.

The Company of Biologists Ltd, Cambridge, UK, for permission to reproduce Figure 5.2, originally published as Figure 1 in Satoh, N. and Egami, N. 1972. Sex differentiation of germ cells in the teleost, *Oryzias latipes*, during normal embryonic development. *Journal of Embryology and Experimental Morphology* 28(2): 385–395.

1 General mechanisms of toxicity

Daniel Schlenk

Introduction

The overall purpose of this book is to provide a reference describing specific toxicities in various organs of teleost fish. It is important to note that any given agent may have multiple targets, and that the concentration of the agent to which the fish is exposed can also have dramatic effects on the target(s) for the xenobiotic in question. As many of the mechanisms eliciting tissue damage at a particular target are shared, this introductory chapter is intended to focus upon general cellular mechanisms of toxicity. Most of the mechanisms that are to be discussed have been identified primarily in mammalian systems (see Gregus and Klaassen, 1996) with limited studies performed in teleosts. Consequently, the purpose of this chapter is to discuss potential mechanisms of cellular toxicity that have been elucidated primarily from mammalian studies and to identify information gaps or mechanisms that have yet to be explored in fish.

Disposition of xenobiotics

As demonstrated in Figure 1.1, xenobiotic disposition as well as interaction with a target involves several processes, including absorption, distribution, biotransformation, and elimination. Involvement of these processes depends not only upon the physiochemical parameters of the xenobiotic but also upon the location and molecular constitution of the target (i.e. receptor/protein, lipid, DNA, etc.). There are primarily three mechanisms by which a toxicant gains access to a cellular target: direct interaction with the target; interaction with the target following absorption; and interaction with the target following absorption and biotransformation.

Direct interaction with target

Because fish are in direct contact with the aqueous environment, exposure occurs over the entire body surface. Many of the mechanisms involved in direct target interactions involve disruption of plasma membranes or direct inhibition or destruction of specific receptors located in the plasma membrane. Putative targets may include: (1) the dermal surface of the fish which is susceptible to non-receptor-

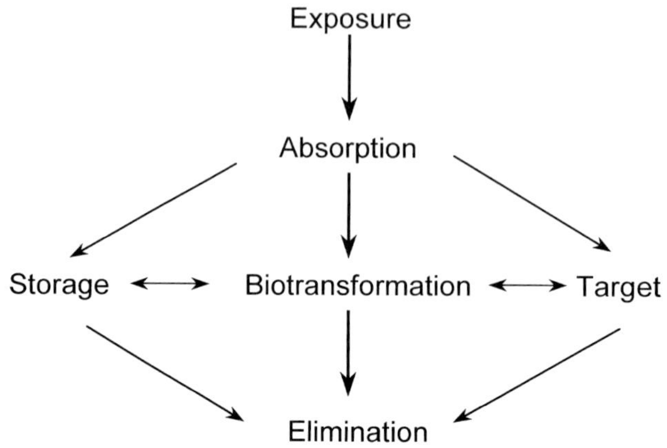

Figure 1.1 Dispositional processes of xenobiotic fate.

mediated plasma membrane disruption by caustic agents (i.e. acids/bases), leading to a variety of lesions such as loss of sensory tubricles or ulcerations of the epidermis and/or erosion of fins; (2) extracellular receptors involved in osmoregulation and/or gas exchange found primarily in the gill (i.e. Na^+,K^+-ATPases, H^+ transporters, Cl^- channels) – numerous agents, including organic surfactants or metals, may have significant interactions with these targets; (3) extracellular receptors involved in nutrient uptake or nitrogenous waste excretion (ammonia) in the gut or gill respectively; (4) epithelial tissue of gut or eye which may undergo general ulceration following interaction with caustic agents and/or metal oxides (gut).

Interaction with target following absorption

Absorption of xenobiotics may occur either passively or actively. Passive mechanisms of absorption are controlled by a number of factors, including size, lipophilicity and/or polarity of the compound as it moves down a concentration gradient, and the thickness of the plasma membrane. Active processes of xenobiotic absorption usually involve the substitution, or 'hijacking' by a structurally similar xenobiotic, with an endogenous ligand for a specific receptor/transporter. Such receptor-mediated processes may or may not require energy.

Intracellular damage may occur to the tissue which directly absorbs the xenobiotic following passage through the plasma membrane. Such intracellular damage is common in the gill. Alternatively, the compound may be transported or distributed to other organs, where it may cause more significant damage. For example, the absorption of a cholinesterase-inhibiting pesticide may not cause significant damage at the site of absorption (i.e. gill) but will affect the central nervous system following transport to the appropriate target(s) (neuron or neuromuscular junction). Transport and distribution of a particular agent to a target is typically implemented by the circulatory system of the organism and is influenced by several factors:

Binding and subsequent sequestration of the agent at the site of absorption

One of the best examples of this phenomenon is the binding of organic and inorganic agents by melanin in the skin of fish (Van Leeuwen *et al.*, 1986). The binding to melanin efficiently sequesters the compound in the skin, preventing its distribution to other organs.

Binding of the agent in the circulatory system by plasma proteins (i.e. albumin and lipoproteins)

Several studies with therapeutic agents used in fish culture have indicated that plasma protein binding significantly alters bioavailability and distribution in certain species such as catfish and trout (Plakas *et al.*, 1992; Jarboe *et al.*, 1993). Metals are also primarily bound to blood proteins (Roesijadi and Robinson, 1994). The specific form of plasma protein may also be critical as some tissues have albumin-specific receptors and may preferentially accumulate albumin-bound xenobiotics (Heath, 1995).

Specialized barriers

As in mammals, blood–brain barriers have been identified in several species of fish through dye uptake studies and histologically (Castejon, 1983; Lane, 1991). In addition, evidence for a blood–testes barrier has also been observed. Consequently, distribution to these particular tissues would be diminished unless the xenobiotic is highly lipophilic or can mimic endogenous ligands taken up by active transport mechanisms within those particular systems.

Storage sites

In contrast to intracellular sequestration at the site of absorption, organic lipophilic agents may be transported to specific storage sites that are abundant in lipid, such as fat bodies. Alternatively, certain metals have the propensity to accumulate in bone. Although lipid content in fish has been shown to have considerable impact on the accumulation of organic agents, the role of bone metabolism has not been extensively examined.

Blood flow to and from potential target systems

Because most gut-absorbed material is transported via the portal venous system to the liver, this tissue is generally considered to be the primary target for gut-absorbed chemicals. Chemicals that are absorbed through the gill or skin would have a different primary circulatory route from dietary chemicals. Blood transport from the gill would suggest that the brain or, possibly, the kidney (depending on the species) may receive a more significant blood flow and thus may be more susceptible targets.

Elimination away from the target

As mentioned above, cells possess numerous membrane transporters that are not only important for cellular absorption but also involved in cellular excretion of absorbed compounds. P-Glycoprotein homologs, also known as the multi-xenobiotic resistance mechanism (MXR), are often found in cells undergoing neoplastic transformation (Juliano and Ling, 1976). The MXR non-specifically exports a host of organic and inorganic toxicants out of cells following absorption. The MXR has been observed in tumors and in damaged tissues from several fish species (Kleinow *et al.*, 1996; Kohler *et al.*, 1998; Cooper *et al.*, 1999). Additional mechanisms to enhance elimination from the animal through the urine or feces involve various biotransformation processes discussed below.

Interaction with target following biotransformation

Bioactivation versus detoxification

Most organic xenobiotics are absorbed in fish and other organisms as relatively non-polar lipophilic chemicals. In an attempt to enhance the elimination of chemicals of this nature, the cell utilizes a series of enzymes to modify radically the structures to polar, water-soluble compounds. The enzymatic process of altering organic chemical structure (biotransformation) consists of two phases. Phase I biotransformation reactions typically involve the insertion of a polar atom (i.e. oxygen) or fragmentation of the molecule, leading to access of polar atoms (i.e. hydrolysis). Such radical alterations of structure often lead to the formation of reactive intermediates, necessitating a second phase of biotransformation which conjugates the intermediates to endogenous macromolecules which are in large cellular supply (i.e. glucose, water, peptides). If certain phase II processes are compromised or overwhelmed, the reactive intermediates may adversely affect proteins, lipids or nucleic acids which serve critical cellular functions, resulting in toxicity. Although uncommon, reactive intermediates may also be formed through phase II pathways. For example, in mammals, glutathione conjugates of halogenated aliphatic hydrocarbons may eventually be hydrolyzed to thiols or methyl thioether metabolites which may rearrange to electrophilic episulfonium species or become oxygenated by phase I monooxygenases to electrophilic sulfoxides respectively. Certain glucuronides may even become hydrolyzed by bacteria after reaching the gut, allowing the aglycone to be reabsorbed; a process known as enterohepatic circulation. Although phase I bioactivation has been observed in fish, bioactivation through phase II processes has yet to observed.

Ultimate toxicants and reactive intermediates

Biotransformation of non-polar lipophilic compounds to polar hydrophilic products typically involves radical structural modification which may result in the formation of reactive intermediates or ultimate toxicants. Although there are some exceptions, most reactive intermediates exist as either electron-deficient molecules (electrophiles) or unstable molecular fragments (free radicals).

ELECTROPHILES

Electron-deficient molecules often result from phase I biotransformation reactions (Table 1.1). Oxidation reactions catalyzed by cytochrome P450 or other enzymes can lead to the formation of numerous electrophilic metabolites including, but not limited to, epoxides, ketones, arene oxides, quinones, quinoneimines, or acyl halides. Heterolytic bond cleavage can also lead to the formation of positively charged electrophiles. For example, cleavage of the sulfoester conjugate of aromatic amines leads to the formation of the highly reactive aryl nitrenium ion. The oxidation of AsO_4^{3-} to $AsO_3^{2-}/(As^{3+})$ and elemental mercury to Hg^{2+} are examples of inorganic electrophiles. Electron-rich moieties found in amino acids and nucleic acids such as thiols and amines are generally thought to be the molecular targets for these intermediates.

FREE RADICALS

Containing one or more unpaired electrons in its outer orbital, free radicals are typically formed by accepting or losing single electrons during, respectively, reduction or oxidation reactions. Toxicants capable of accepting single electrons from reduction reactions can elicit toxicity by forming free radicals. The reduced toxicant typically transfers the extra electron to molecular oxygen, leading to the formation of a superoxide anion radical and the oxidized toxicant which can receive an additional electron. This process of the acceptance and donation of electrons and regeneration of the parent compound is known as 'redox cycling' (Figure 1.2) and can lead to the formation of a large number of superoxide anion molecules from a single molecule of toxicant.

Single electron reduction reactions may also lead to homolytic bond fission in certain xenobiotics, especially halogenated aliphatic hydrocarbons such as carbon tetrachloride. Such reactions lead to a dehalogenation of carbon tetrachloride to the trichloromethyl radical (Figure 1.3). Interaction of this reactive intermediate with molecular oxygen leads to the formation of another highly reactive metabolite known as the trichloromethylperoxy radical, which may abstract protons from lipids causing lipid peroxidation (see below). Additional one-electron processes that lead to the formation of free radicals are described by the Haber-Weiss and Fenton reactions (Figure 1.4). Transition metals such as iron and copper act as electron donors to various peroxides (which may be the dismutated products of superoxide anion), leading to the formation of the most reactive reduced oxygen species, the hydroxyl radical.

In addition to electron-accepting xenobiotics, nucleophilic toxicants (chemicals containing electron-donating atoms – aromatic amines, hydrazines, phenols, thiols) may form superoxide anion radicals through peroxidase-catalyzed transfer of one electron from the xenobiotic to molecular oxygen. Aromatic compounds having two *ortho* or *para* phenolic groups may undergo two sequential one-electron oxidations, ultimately leading to the formation of quinones. Depending upon the physicochemical properties of the specific toxicant, quinones not only may act as redox cyclers generating superoxide but also are classic electrophiles (see above)

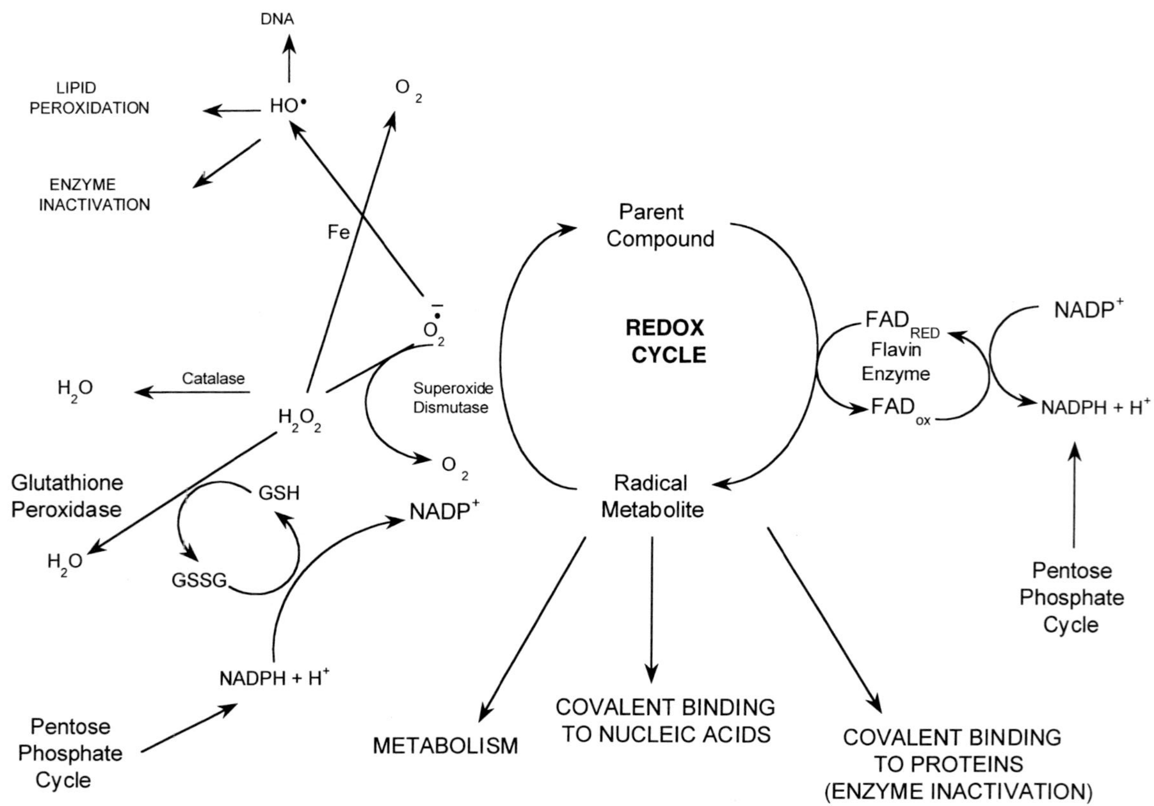

Figure 1.2 Potential pathways for oxidative stress (see Di Giulio *et al*., 1995).

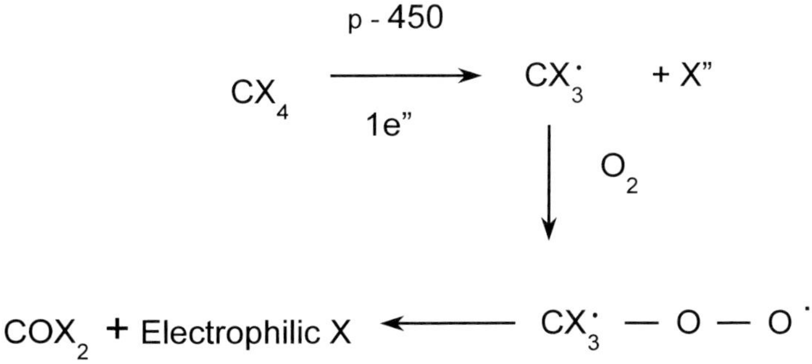

Figure 1.3 One-electron reduction of halogenated methanes.

$$O_2^{\dot{-}} + Fe^{3+}\text{ - chelate} \longrightarrow O_2 + Fe^{2+}\text{ - chelate}$$

$$+ Fe^{2+}\text{ - chelate} + H_2O_2 \longrightarrow {}^{\cdot}OH + OH^- + Fe^{3+}\text{ - chelate}$$

Figure 1.4 Haber–Weiss and Fenton reactions.

and undergo attack by nucleophilic molecules found in cellular proteins and nucleic acids.

Chemical reactions of reactive intermediates/ultimate toxicants

General cellular targets

Electron-deficient molecules formed through cellular biotransformation processes are prone to attack by nucleophilic molecules within any given cell, leading to the potential formation of toxicant–biomolecule adducts. Consequently, biomolecules having large electron densities, such as thiols and basic amine moieties, would be susceptible targets of electrophiles. As free radicals may donate or abstract electrons from targets, molecules containing atoms that are inclined toward electron acceptance or loss would be prime targets for free radical attack. Many biomolecules possess these general features, and it is likely that more than one type of reaction may occur with the cellular target. Discussion of general cellular targets will be limited to protein, lipid and nucleic acids.

Protein

As proteins contain numerous atoms that are excellent reducing agents (i.e. thiols), oxidation of protein thiols to disulfides or sulfenic acids is a common outcome following interactions with free radicals. Hydrogen abstraction of thiols by neutral free radicals may also form thiyl radicals, eventually leading to similar disulfides and sulfenic acid products. Hydroxyl radicals may also remove protons from methylene groups of amino acids, leading to the formation of carbonyls which tend to bind amines of proteins or DNA covalently, thus causing cross-linking. In addition, thiols are also excellent nucleophiles and provide potential sites for covalent binding of electrophilic toxicants. Non-covalent binding through hydrogen and ionic bonds may also occur with proteins having appropriate tertiary structure (i.e. helices) such as membrane receptors or enzymes. Covalent or non-covalent binding could significantly inhibit the catalytic activity of an enzyme or signal transduction pathway initiated by a receptor. For example, the mimicking of calcium by lead or 17β-estradiol by nonylphenols can have significant functional effects for their receptors or enzymes. Many enzymes and receptors require cofactors such as metals or organic moieties such as porphyrins or folic acid. If cofactors are modified (i.e. electron transfer to iron in hemoglobin) then protein function may be severely altered (i.e. methemoglobinemia); also, if the affected protein is critical for cell maintenance, it is likely that cell viability may be ultimately compromised.

Lipid

Possibly the most common reaction that occurs between toxicants and lipids is that of lipid peroxidation. The primary targets are unsaturated lipids found throughout cellular membranes. Bis-vinylic methylene protons are highly susceptible to abstraction by free radicals (Figure 1.5). Abstraction of either of these protons by free radicals such as hydroxyl radical leads to the formation of conjugated dienes, which may undergo Fenton chemistry and form lipid peroxyl radicals, and lipid hydroperoxides. These intermediates not only serve as endogenous toxicants but also result in significantly altered membrane integrity and function. As most cellular organelles as well as the plasma membrane is largely lipid in constitution, altered integrity could result in loss of osmotic gradients, leading to cell death if unrepaired.

Nucleic acids

Like thiol-containing proteins, nucleic acids are electron-rich molecules that can covalently bind electrophilic toxicants. In contrast to thiols, which tend to bind covalently soft electrophiles (i.e. quinones), purine-associated oxygens and nitrogens are more likely to attack hard electrophiles (i.e. alkyl carbonium ions). Covalent binding of DNA by various electrophiles disrupts replicative and transcriptional activities of polymerases, and, if unrepaired, may lead to genomic mutation. Because DNA exists as a double helix through significant hydrogen bonding, non-covalent intercalation by planar toxicants may also lead to mutation. DNA is also susceptible to attack by free radicals as hydroxyl radicals tend to bind the C-8 position of guanine and abstraction of protons on the pentose sugar

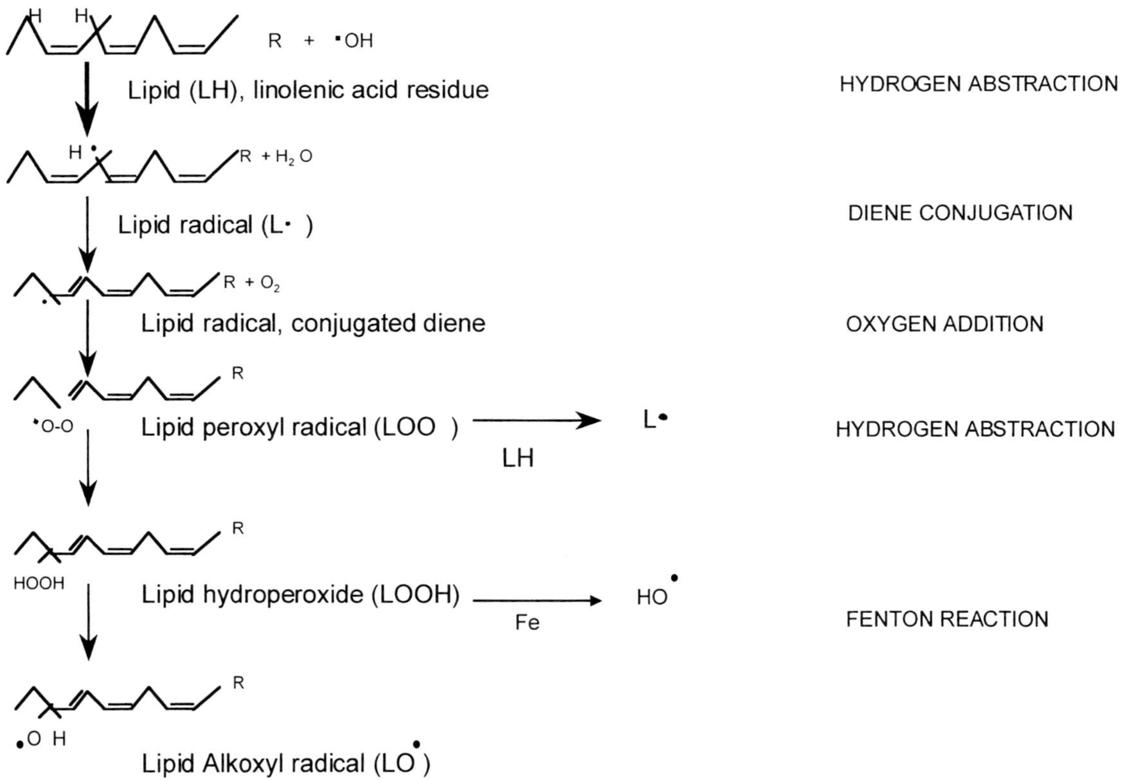

Figure 1.5 Initiation of lipid peroxidation.

of DNA by free radicals may lead to hydrolytic strand breaks. Alteration of the genome through any of these mechanisms may lead to modified protein function which could result in either cell death or carcinogenesis, depending upon the specific function of the affected protein or gene product (see below).

Exceeding capacity of cellular defense mechanisms

General types of cell defense

Many cellular toxicants elicit their adverse effects through formation of electrophilic reactive intermediates or free radical molecular fragments. Generally, cellular defense mechanisms (i.e. glutathione, antioxidants) are chemically similar to molecular targets (i.e. electron-rich molecules), but are in larger cellular concentrations and have greater binding capacities. As stated above, other mechanisms of defense rely on structural biotransformations (i.e. oxidation) to enhance polarity and cellular elimination. Since such defense mechanisms rely heavily upon oxidative processes, it is imperative for the cell to have adequate reducing equivalents (generally in the form of NADH or NADPH) to allow oxidative biotransformation reactions to occur. Regeneration of defending molecules as well as other vital cellular functions are also intimately related to the redox potential of the cell and the ability of the cell to produce reducing equivalents. Cellular toxicity generally occurs when either the scavenging capacity of defense mechanisms are exceeded or the regenerative capacity of the cell to produce reducing equivalents is overwhelmed.

Defense of nucleophiles

To prevent the donation of electrons from nucleophilic toxicants such as hydroxylated, thiol, or amine-containing chemicals, phase II conjugation reactions (i.e. glucuronidation, sulfation, acetylation; see Table 1.1) are primarily responsible for detoxification. For example, many amines may be acetylated through *N*-acetyl transferases to metabolites that do not donate electrons. If such phase II processes are compromised, chemicals containing these functional groups may undergo single electron reductions or oxidations through respective reductases or peroxidases, leading to the formation of free radicals. Some nucleophiles (amines, thiols, phosphines, selenides) are also detoxified by oxidation reactions catalyzed by flavin-containing monooxygenases to corresponding N, S, P, or Se oxides.

Defense of electrophiles

As stated above, many reactive intermediates from phase I processes are electrophilic and are typically conjugated with nucleophilic phase II molecules. For example, cellular concentrations of glutathione are very large and conjugate many reactive intermediates such as epoxides or acyl halides. Glutathione, in addition to the polypeptide metallothionein, also binds electrophilic metals such

Table 1.1 General pathways of xenobiotic biotransformation and their major subcellular locations.

Reaction	*Enzyme*	*Localization*
Phase I		
Hydrolysis	Carboxylesterase	Microsomes, cytosol
	Peptidase	Blood, lysosomes
	Epoxide hydrolase	Microsomes, cytosol
Reduction	Azo- and nitroreduction	Microflora, microsomes, cytosol
	Carbonyl reduction	Cytosol
	Disulfide reduction	Cytosol
	Sulfoxide reduction	Cytosol
	Quinone reduction	Cytosol, microsomes
	Reductive dehalogenation	Microsomes
Oxidation	Alcohol dehydrogenase	Cytosol
	Aldehyde dehydrogenase	Mitochondria, cytosol
	Aldehyde oxidase	Cytosol
	Xanthine oxidase	Cytosol
	Monoamine oxidase	Mitochondria
	Diamine oxidase	Cytosol
	Prostaglandin H synthase	Microsomes
	Flavin monooxygenases	Microsomes
	Cytochrome P450	Microsomes
Phase II	Glucuronide conjugation	Microsomes
	Sulfate conjugation	Cytosol
	Glutathione conjugation	Cytosol, microsomes
	Amino acid conjugation	Mitochondria, microsomes
	Acylation	Mitochondria, cytosol
	Methylation	Cytosol

as mercury and cadmium. Epoxide hydrolases also convert epoxides to the less reactive metabolites dihydrodiols. Reduction of hydroquiones by DT diaphorase may afford protection from covalent binding and prevent redox cycling. Covalent binding to cellular proteins by α,β-unsaturated aldehydes can be diminished through reduction by aldehyde dehydrogenase as well as by adduction to glutathione. Glutathione is a unique defense mechanism because not only does it have the capacity to bind many electrophiles non-enzymatically but also it may be enzymatically transferred to electrophilic substrates through cytosolic and microsomal glutathione transferase enzymes.

Defense of free radicals

Neutral free radicals such as lipid peroxyl or xenobiotic radicals are primarily detoxified by non-enzymatic reduction by glutathione, α-tocopherol (vitamin E) or ascorbic acid (vitamin C). Each of these systems can be regenerated through transfer of reducing equivalents ultimately from NADPH via specific reductases (glutathione reductase and glutaredoxin respectively) (Figure 1.6). Hydroxyl radicals, the most reactive endogenous radical (see Figure 1.4) cannot be scavenged by any of the antioxidant systems described above. Consequently, the cell's primary strategy is to prevent the formation of hydroxyl radical. Typically, superoxide is dismutated by cytosolic or mitochondrial superoxide dismutase to hydrogen peroxide which undergoes catalase or glutathione peroxidase-catalyzed peroxidation to water. Fenton reagents such as iron and copper are sequestered by specific metal-binding proteins that limit their participation in peroxide transformation. Metallothioneins have been shown not only to sequester potentially reactive metals but also to scavenge free radicals because of the large content of thiols. However, oxidation of thiols may also release metals such as copper, causing elevated oxidative damage. The overall mechanism of protection by metallothioneins from free radical damage is still unclear.

DNA repair

Although it may be argued that repair of DNA may not be classified as a defense mechanism preventing DNA damage, DNA repair processes do, in fact, defend against the transmission of damaged DNA to the next generation of cells, or, in other words, prevent the occurrence of mutation. There are primarily four types of DNA repair pathways that reverse DNA damage. The first of these is base-excision repair, in which apurinic endonucleases, DNA glycosylases, polymerases and ligases remove and replace minor lesions resulting from small molecule alkylation at specific bases. Second, if a bulky adduct is present (such as an aflatoxin adduct or pyrimidine dimer), several nucleotides surrounding the lesion may be removed by nucleotide excision, which utilizes an adenosine triphosphate (ATP)-dependent nuclease, and replaced by subsequent DNA polymerase and ligase activities. Third, direct repair mechanisms such as DNA photolyase and O_6-methylguanine-DNA methyltransferase specifically repair ultraviolet (UV)-induced pyrimidine dimers and methylguanine adducts respectively. The last type of DNA repair, recombinational repair, is reserved for massive structural alterations that are not corrected prior to DNA replication by any of the other three types of repair processes. Essentially, damaged DNA (i.e. double-strand breaks or massive single-stranded lesions) is replaced by a segment from a 'sister' or homologous strand of DNA that 'crosses over' to the damaged strand, leading to the formation of a complete single or double strand of DNA.

Specific cellular targets

Cellular regulation

Regulation of cell viability is ultimately controlled at the level of gene expression. Gene expression can be self-regulating especially in multitissue organisms which require extracellular signaling between tissues and intracellular signal transduction pathways for appropriate signals to reach specified genes. Alteration in any of

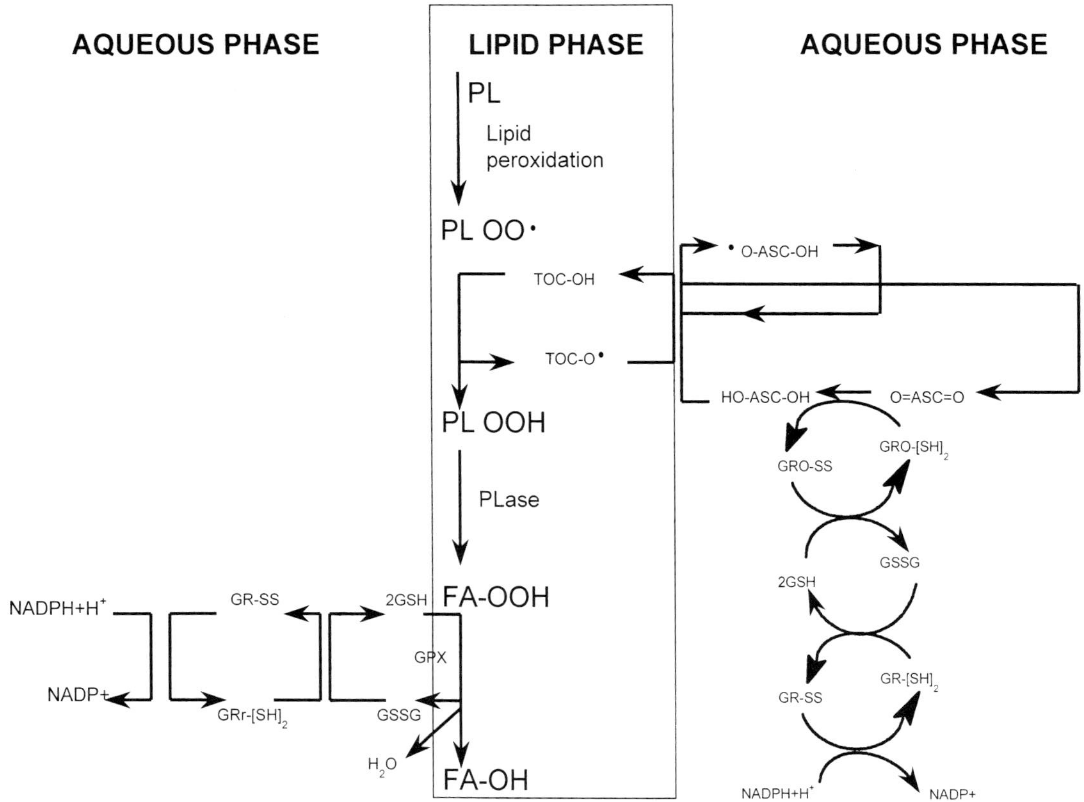

Figure 1.6 Activation and detoxification of lipid peroxidation reactions (see Gregus and Klaasen, 1996).

these pathways may dramatically affect processes that control viability and/or proliferation of any given cell type. As discussed above, modifications of gene expression may begin by a direct mutation of a gene which encodes a specific product that is critical for cell survival. Alternatively, the mutation may occur in a regulatory region of the gene, leading to enhanced or repressed expression of the gene product. A third possible scenario is disruption or enhancement of specific transcription factors that are necessary for gene expression. Many transcription factors also serve as intracellular receptors for various endogenous substances such as hormones and vitamins (Table 1.2). Consequently, if a xenobiotic competes for the binding site of the endogenous substance, the transcription factor may

Table 1.2 Toxicants acting on ligand-activated transcription factors.

Ligand-activated transcription factor	*Endogenous ligand*	*Exogenous ligand*	*Toxic effect*
Estrogen receptor	Estradiol	Ethynylestradiol Diethylstilbestrol DDT	Mammary and hepatic carcinogenesis
Glucocorticoid receptor	Cortisol	Dexamethasone	Apoptosis of lymphocytes Teratogenesis (cleft palate)
Retinoic acid receptor	All-*trans* retinoic acid	A metabolite of 13-*cis* retinoic acid (?)	Teratogenesis (craniofacial, cardiac, thymic malformations)
Ah receptor	Unknown	TCDD	Apoptosis of thymocytes Teratogenesis (cleft palate) Hepatocarcinogenesis in rats Altered protein synthesis (e.g. CYP1A1)
Peroxisome proliferator-activated receptor (PPAR)	Polyunsaturated fatty acids	Fibrate esters (e.g. clofibrate) Phthalate esters (e.g. DEHP)	Hepatocarcinogenesis in rats Peroxisome proliferation Altered protein synthesis (e.g. CYP4A1, acyl-CoA oxidase)
Metal-responsive element binding (MTF-1)	Zn^{2+}	Cd^{2+}	Synthesis of metallothionein

antagonize or potentiate gene expression depending on the affinity of the xenobiotic for the receptor and the specific activity or abundance of the transcription factor. Such is the case with xenoestrogens and their activation or inactivation of the estrogen receptor, which eventually serves as a transcription factor for estrogen-responsive genes (i.e. hepatic expression of vitellogenin). The acute and chronic effects of such interactions on reproduction and development can be significant depending on the dose of the xenobiotic and the stage of development when exposure occurs. Transcription factors may never interact directly with the xenobiotic, but may be activated through signal transduction pathways originating upstream from ligand–receptor interactions which may occur in another tissue. For example, extracellular signaling molecules such as macrophage-derived cytokines and growth factors eventually trigger the activation of parenchymal transcription factors, leading to gene expression. If any of the signal-transducing pathways become altered (i.e. phosphorylation, calcium metabolism), activation of transcription factors may be modified. In mammals, activation or increased expression of the transducing protein protein kinase C (PKC) by a xenobiotic (i.e. tetradecanoylphorbol acetate or lead) can cause stimulation of the transcription factor activation protein 1 (AP-1), which increases mitogenesis. Toxicants inducing oxidative stress activate other transcription factors such as nuclear factor kappa beta (NF-KB), which may alter cellular regulation.

Another potential mechanism affecting gene expression may involve disruption of an extracellular signal or hormone which initiates the cascade to gene expression. As mentioned above, preventing homeostatic function of circulating hormones either through antagonism at any given receptor or through inhibition of hormone synthesis may have serious effects on reproduction and development (i.e. alteration of sex steroids). Likewise, impeding the normal degradation and elimination of signaling hormones also impacts susceptible cellular systems through elevated gene expression.

Specialized cell functions

Tissues that are actively undergoing electrochemical modifications for specific functions [i.e. central nervous system (CNS), cardiovascular, somatic muscle] are extremely vulnerable to toxicant interference with ligand receptor and subsequent signal transduction pathways. Indeed, many pesticides mediate their toxic effects through disruption of one or more of the processes in these tissues. According to Gregus and Klaassen (1996), adverse effects in these systems primarily are relegated to modifications of (1) concentrations of neurotransmitters, (2) receptor function, (3) intracellular signal transduction, or (4) the signal-terminating processes (Figure 1.7).

Alteration in neurotransmitter concentration

Xenobiotics may alter concentrations of ligands in electrochemically sensitive cells by disrupting the ligand's synthesis, storage, release or removal from the

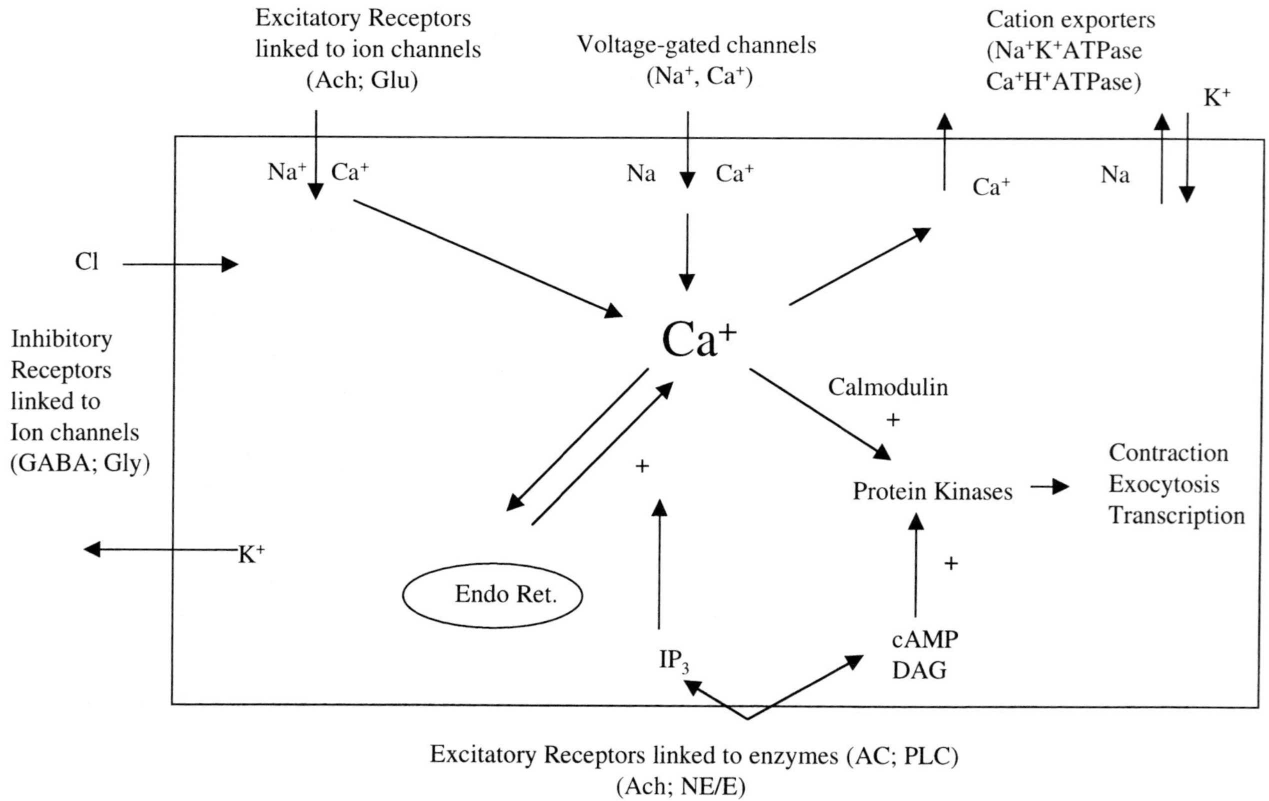

Figure 1.7 Examples of signal transduction pathways (see Gregus and Klaasen, 1996).

vicinity of the receptor. An example of the last mechanism would involve the organophosphate-induced inhibition of various acetylcholinesterases at the neuromuscular junction, brain or blood in fish leading to accumulation of acetylcholine, resulting in cholinergic overload and associated toxicities such as loss of opercular movement.

Toxicant–neurotransmitter receptor interactions

There are several means by which xenobiotics may interact directly with neurotransmitter receptors. Chemicals may act as direct or partial agonists, mimicking the endogenous ligand and activating the receptor. Alternatively, the compound may antagonistically bind to the receptor, thus failing to stimulate activity. If chemicals do not bind the active site of the receptor but stimulate activity, they are classified as activators. In contrast, inhibitors block receptor activation without binding directly to the active site of the receptor.

Toxicant–signal transducer interactions

Another mechanism of disrupting electrochemically active cells is that of interrupting intracellular signal transduction pathways. Red tide-produced saxitoxins are excellent examples as they block voltage-gated sodium channels. Opening of these channels is triggered by excitatory signals generated by ligand-gated cation channels at the neuromuscular junction. By blocking the voltage-gated channels, saxitoxin prevents transduction of the original excitatory signal, resulting in paralysis of muscular tissue. In contrast, DDT [1,1,1-trichloro-2,2-bis(*p*-chlorophenyl)ethane] causes an activation of voltage-gated channels, leading to overstimulation and muscular tremors and/or convulsions depending on the dose.

Toxicant–signal termination interactions

Cellular signals in electrochemically active cells are driven by cation flux into the cell. Consequently, termination of the signal is mediated by cation efflux via transporters or specific cationic channels. Disruption of cationic transporters may not be unique to neurons or myocytes as a mechanism of toxicity. The Na^+,K^+-ATPase transporter in gills is critical to osmoregulation in fish such that disruption of this transporter can have dramatic effects on gill function.

General cellular function

There are primarily two major pathways that contribute to cell death in most tissues: impairment of cellular energy and sustained increase in calcium. The former is brought about by agents that disrupt ATP synthesis, whereas the latter may be effected through several molecular pathways, including impairment of ATP synthesis.

Disruption of ATP synthesis

ATP is required for the maintenance of multiple cell functions, including cell division, protein, lipid, carbohydrate and nucleic acid synthesis, muscle contraction, polymerization of the cytoskeleton, regulation of cell volume, plasma membrane integrity, and transport of essential nutrients and ions (i.e. Ca^{2+}). The synthesis of ATP is accomplished through oxidative phosphorylation occurring in the mitochondria. As hydrogen is oxidized to water in the mitochondria, an electrochemical gradient along the mitochondrial membrane drives ATP synthase, eventually phosphorylating adenosine diphosphate (ADP) to ATP.

There are primarily four areas where agents may interrupt the synthesis of ATP (Gregus and Klaassen, 1996). Agents may diminish the delivery of reducing equivalents to the electron transport chain through disruption or inhibition of basic biochemical pathways that produce reducing equivalents (i.e. citric acid cycle, glycolysis, gluconeogenesis, or fatty acid oxidation). Reducing equivalents are critical in regenerating oxidized proteins or lipids during cellular repair. Second, xenobiotics may inhibit electron transport either by disrupting electron transport complexes (i.e. rotenone, diphenyl ether herbicides) or by acting as electron acceptors (carbon tetrachloride). Third, agents leading to hypoxia or reduction in oxygen transport to the electron transport chain (i.e. methemoglobinemia) will adversely affect ATP synthesis. Lastly, ATP synthesis may be impaired by agents that directly inhibit ADP phosphorylation either through a general inhibitor of protein synthesis or through specific inhibition of ATP synthase (DDT, chlordecone), the ADP/ATP antiporter (DDT), the phosphate transporter (*p*-benzoquinone), or through an uncoupling of the mitochondrial membrane potential from ATP synthase (pentachlorophenol, chlordecone) (Gregus and Klaassen, 1996).

When oxidative phosphorylation is diminished, ADP and its degradation products accumulate in the cell leading to reductions in pH and increased phosphoric acid, which is thought to bind excess calcium and accumulate because of dysfunctional ATP-driven Ca^{2+} exporters. The loss of ATP-dependent ion pumps eventually leads to an accumulation of sodium, resulting in cellular swelling or blebbing. If no repair occurs, the intracellular pH will begin to rise, stimulating phospholipase activity, rupturing blebs and forming free fatty acids which cannot be reacylated (ATP dependent). In addition, cellular concentrations of calcium will rise, leading to several other interrelated pathologic events.

Prolonged excess of intracellular calcium

In certain cell types, influx of calcium is a regular occurrence. However, its sustained increase and accumulation can significantly damage cells and lead to cell death. Cellular membranes are impermeable to calcium and either actively transport it out of the cell or sequester it in the endoplasmic reticulum and mitochondria (Figure 1.7). Thus, toxicants either cause elevated calcium by increasing influx or decreasing efflux from the cytoplasm. Mechanisms may vary from inhibiting specific organelle or plasma transporters (including ATP synthesis which drives the transporters) to general membrane disruption, allowing calcium to flow down its concentration gradient into the cytoplasm. There are three significantly harmful consequences of sustained intracellular calcium concentrations: depletion of energy reserves, dysfunction of microfilaments, and activation of hydrolytic enzymes.

Cellular energy reserves may be depleted by inhibiting the synthesis of ATP (see above). The resulting increase of intramitochondrial calcium activates mitochondrial dehydrogenases which, through augmented electron transport chain activity, may lead to the increased formation of partially reduced oxygen species (i.e. superoxide anion radical). As mentioned above, increased formation of free radicals may lead to damage of the inner mitochondrial membrane, further disrupting oxidative phosphorylation. Lastly, as calcium export is ATP dependent, increases of intracellular calcium may deplete ATP through consumption by specific calcium transporters. ATP depletion and increases of intracellular calcium concentrations are processes that are significantly linked and may potentiate cell damage through a cyclical interrelationship.

Sustained increases in intracellular calcium may also lead to dissociation of actin microfilaments from membrane-anchoring proteins, causing blebbing and eventually leading to membrane rupture. Cytoskeletal proteins may also be direct targets of xenobiotics. However, this has only been observed in mammalian systems.

In addition to exacerbating energy depletion and its cytoskeletal effects, calcium activates the endonuclease topoisomerase II, several proteases (i.e. calpains) and phospholipases, leading to degradation of DNA, proteins and lipids respectively. Hydrolytic enyzmes may also be released into the cytosol from general membrane disruption of lysosomes, which also cause influx of calcium through destruction or destabilization of plasma or mitochondrial or endoplasmic reticular membranes, thus creating other cyclical reactions.

Targets involving repair and cellular regeneration

Apoptosis and cell proliferation

As mentioned throughout this chapter, cellular damage may not be life-threatening because of numerous defense mechanisms that can limit or rapidly repair xenobiotic-induced damage. However, when defenses are overwhelmed or compromised, the cell may pass 'the point of no return' and either undergo apoptosis, which is a scheduled mechanism of cell death, or necrosis, which tends to be an unorganized rupturing of the damaged cell that leads to significant input by inflammation. The general characteristics of each pathway are described in Table 1.3.

Generally, apoptosis is an intra- and extracellular signaled pathway which contributes to cellular repair and is a means for tissue remodeling, which is critical in fetal development (Gregus and Klaassen, 1996). Apoptosis allows DNA-damaged cells to be placed in appropriate cell cycle phases for DNA repair to

Table 1.3 Features of necrosis versus apoptosis.

	Necrosis	*Apoptosis*
Histology	Cellular swelling Coagulation necrosis Disruption of organelles	Single cells Chromatin condensation Apoptotic bodies
DNA breakdown	Random, diffuse	Internucleosomal
Mechanisms	ATP depletion Membrane injury Free radical damage	Gene activation Endonuclease
Tissue reaction	Inflammation	No inflammation Phagocytosis of apoptotic bodies

occur. For example, following DNA damage, cells are cycled from the resting G_0 to the G_1 cell cycle phase by expression of various gene products, one of which in mammals appears to be c-*myc* (Martin *et al.*, 1994). Further progression to a proliferative phase is blocked by *p53* expression, allowing DNA repair to occur. If repair does not occur, the cell is diverted to the apoptotic pathway. Apoptosis may be blocked by expression of bcl-2 protein, which binds the transcriptional activator (bax) that initiates apoptosis. If the apoptotic signaling pathway is overwhelmed through massive cellular damage, cells typically undergo necrosis.

An additional repair mechanism is that of cellular proliferation. When cells experience damage, adjacent undamaged cells in the G_0 phase also rapidly (within 36 h) progress to the M (mitotic) phase and undergo proliferation. This is enhanced by the increased expression of immediate–early growth response genes such as c-*myc*, c-*fos* and c-*jun*. Expression of these transcription factors in addition to cytokines and other small molecules which are thought to be released by damaged cells allows DNA synthesis to gain top priority and override specialized cellular activities in adjacent cells. However, many of these signaling molecules stimulate macrophage and endothelial cells to release additional secondary signaling factors that not only aid in cellular proliferation but also carry out a host of various inflammatory reactions, including microcirculatory alterations and leukocyte recruitment which may engulf or destroy cells that could not be repaired. An example of a macrophage-derived proliferative signal in mammals is hepatocyte growth factor (HGF), which acts as a mitogen in many different tissues by stimulating parenchymal regeneration and proliferation of endothelial cells. Unlike HGF, which has yet to be observed in fish, the macrophage-derived transforming growth factor beta (TGF-β) has been observed in fish and stimulates non-parenchymal cells to synthesize extracellular matrix which integrates with newly proliferated cells. Regenerating cells also contribute to their own proliferation through the release of TGF-α. The function of TGF-α is the proliferation of parenchymal cells and macrophages. Termination of proliferation occurs as TGF-β accumulates and eventually overrides the mitogenic activity of TGF-α.

If apoptosis and cellular proliferation are overwhelmed by massive cell injury, macrophages and endothelial cells will release numerous other agents that can alter microcirculation (i.e. NO), oxidatively degrade cells (i.e. reduced oxygen radicals) and recruit peroxidative granulocytic lymphocytes (i.e. cytokines) which may potentially lead to inflammation and/or tissue necrosis. The efficiency of cellular repair is an important determinant of the dose–response relationship for toxicants that cause tissue necrosis. This is not only because the dose ensures sufficient concentration of the ultimate toxicant at the target site but also because that concentration of toxicant causes enough damage to overwhelm repair, resulting in progression of the injury (Gregus and Klaassen, 1996).

Consequences of overwhelming tissue repair

Fibrosis

As discussed above, acute tissue damage that overwhelms apoptosis and cellular proliferation typically leads to tissue necrosis. In an attempt to halt the spread of necrosis, inflammatory signals from non-parenchymal tissues continue to be released (i.e. TGF-β), resulting in protracted production of extracellular matrix that eventually compromises function through the formation of scar tissue. This 'non-functional' tissue causes the production of a diffusional barrier which may lead to malnourishment of cells, obliteration of blood vessels and parenchymal cells, and reduction in the flexibility of critical tissues such as the heart and gills.

Carcinogenesis

An alternative pathway for a cell that has undergone toxicant insult is that of carcinogenesis. Chemical carcinogenesis is a complex multistage process that can result from the failure of three basic repair mechanisms: DNA repair, leading to mutation and initiation; apoptosis, leading to uncontrolled cellular proliferation and promotion; and/or termination of cellular proliferation, leading to propagation or progression of the neoplastic tissue.

As mentioned earlier, DNA may be damaged through chemical adduction, oxidative addition, and breakage of strands. Each of these lesions can be repaired through several mechanisms (see above). However, if repair is compromised, a mutation may occur, with the lesion transmitted to the mitotic daughter cells. Adverse effects may still not be realized because of the large excess of non-protein-encoding DNA (99 percent of total genome) that presumably acts as a buffer protecting against potential alterations of critical regulatory proteins. If the lesion does occur in a coding or regulatory region, an additional protective mechanism lies in the degeneracy of the genetic code. Thus, even though mutations occur in the genome of an organism, they may still be 'silent' and not lead to an adverse effect. However, if the mutation leads to an alteration of the amino acid sequence which is important in the activation or function of the protein, two potential initiation pathways may occur. Most mutations typically reduce or eliminate protein function. If the protein is critical for cell viability then the mutation results in apoptotic cell death. Unfortunately, elimination of specific proteins involved in suppression of cell division (i.e. p53) may lead to uncontrolled proliferation of the cell. In mammalian systems, p53 acts as a transcription factor and induces the transcription of growth arrest and apoptotic genes through interaction with TATA-binding protein(s), inhibiting the transcription of many genes that do not have p53 regulatory elements. Tumor-suppressor genes have been isolated from fish (Van Beneden and Ostrander, 1994), but their functional homology with mammalian systems has yet to be verified. For example, Kraus *et al.* (1997) has identified the gene encoding for p53 in Japanese medaka, but sequencing of DNA obtained from tumor-bearing fish exposed to the carcinogen *N*-methyl-*N′*-nitro-*N*-nitrosoguanidine (MNNG) revealed no mutations within characteristic mutational hotspots. In addition, studies with *N*-diethylnitrosoamine-treated *Rivulus ocellatus marmoratus* indicated an increase of p53 protein in neoplasms rather than repression (Goodwin and Grizzle, 1997). Clearly, more studies in fish are needed to determine whether inhibition of tumor suppression is mechanistically similar to that observed in mammals.

An alternative pathway following mutation is the activation of one or more transducing proteins which may lead to uncontrolled cellular proliferation. The genes encoding such proteins are referred to as oncogenes because of their association with carcinogenesis. These gene products are generally growth factors, growth factor receptors, intracellular signal-transducing proteins/kinases, and nuclear transcription factors (see above). When these mutated cells undergo division, their descendants also have the propensity for proliferation, leading to further mutations that potentiate proliferation which, in turn, eventually override the apoptotic repair process. Although several oncogenic proteins have been observed in fish, as indicated above, few have been observed to correlate with tumors or cellular proliferation (for a review, see Van Beneden and Ostrander, 1994). The most consistent relationship in fish has been between the expression of *ras* in *N*-diethylnitrosamine-treated rainbow trout and *Rivulus ocellatus marmoratus* (Goodwin and Grizzle, 1997). Ras proteins are transduction proteins with GTPase activity that are normally triggered by various growth factor receptors and that initiate a kinase cascade which eventually leads to mitogenic activity. Specific mutations in the *ras* gene allow the expressed protein to remain in an activated state. The accumulation of genetic damage in the form of mutant oncogenes (which encode activated proteins) and mutant tumor-suppressor genes (which encode inactivated proteins) is the main driving force in the transformation of normal cells with controlled proliferative activity to malignant cells with uncontrolled proliferative activity in mammals (Figure 1.8). Whether this is the case in fish remains to be verified.

As stated above, even after mutational damage occurs, initiated cells may be removed by apoptosis. If tumor-suppressor protein expression is reduced, then the cell does not undergo DNA repair or apoptotic death, but rather promotion and clonal expansion. Thus, inhibition of apoptosis is detrimental because it facilitates both mutations and clonal expansion of preneoplastic cells (Gregus

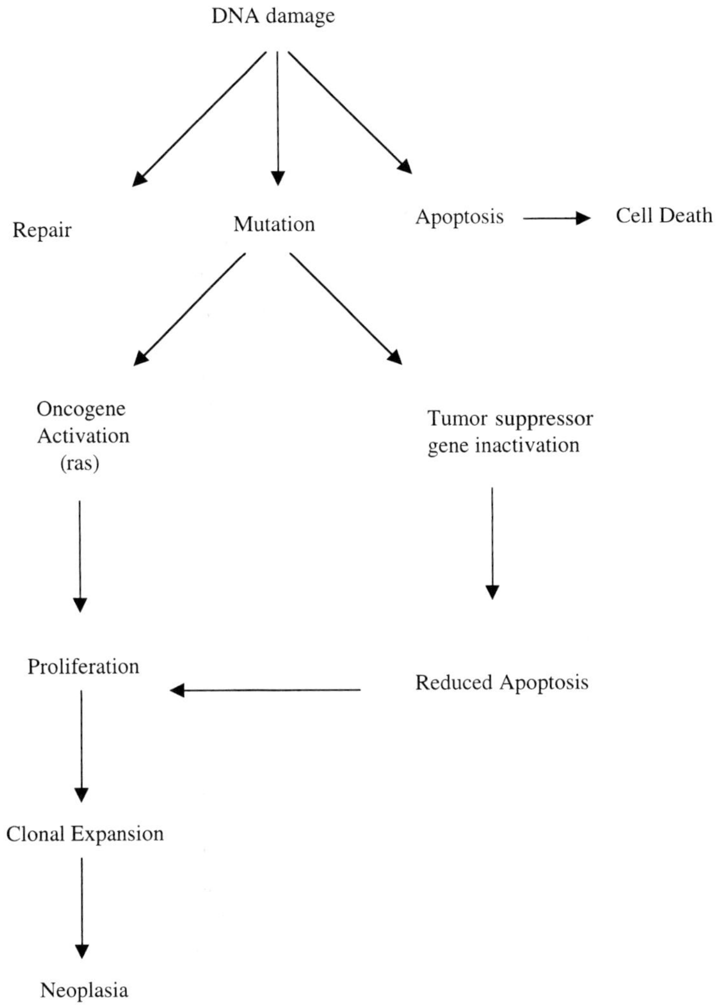

Figure 1.8 Simplified scheme of neoplasia pathogenesis.

and Klaassen, 1996). In addition to underexpression of tumor-suppressor proteins, overexpression of apoptotic inactivating proteins (i.e. bcl-2) have been shown to be associated with several mammalian cancers (McDonnell *et al.*, 1993). Consequently, enhancement of cellular proliferation through the activation of oncogenes or through xenobiotic activation of factors that prevent termination of cellular proliferation can result in non-genotoxic cancers. Thus, not all carcinogens directly damage DNA. It must be emphasized, however, that a non-genotoxic or epigenetic xenobiotic may actually cause cancer through a genotoxic mechanism by enhancing the potential for mutation from endogenous sources such as oxygen radicals resulting form respiratory bursts during chronic inflammation. Mitogenic activity can significantly enhance the possibility of mutation by reducing DNA repair time by shortening the G_1 phase which also tends to reduce the time for DNA methylation (transcriptionally active genes are typically hypomethylated). An additional effect of cellular proliferation that may lead to invasiveness (or be a result of expansion) is the loss in cell–cell communication by disruption of intercellular adhesion gaps. Although several promoting agents have been studied in fish (primarily rainbow trout), few epigenetic carcinogens have actually been identified (McDonnell *et al.*, 1993).

Summary

An understanding of the basic mechanisms of fundamental processes is considered essential in traditional mammalian toxicology. Although this objective is viewed as highly desirable in environmental toxicology and chemistry, often it is not considered essential in addressing immediate environmental concerns. Thus, compared with the wealth of information regarding cellular mechanisms of toxicity in mammalian systems, there have been relatively few mechanistic studies in fish. Although understanding the mechanistic basis of toxicant action is critical in assessing the risk of any given chemical to the environment, there are numerous gaps in our knowledge of how many toxicants elicit cellular and subsequent organelle damage. Fish share many common biochemical and physiologic pathways with mammals, but, obviously, are quite distinct from mammals with regard to life histories (i.e. reproductive strategies) and continual residence in aqueous environments. Thus, it is the goal of this text to provide a current comparative review of the unique interactions of various toxicants and their target organ systems in teleosts. It is hoped that this framework will aid risk assessors in determining appropriate hazard identifications for ecotoxicologic studies as well as biomedical researchers searching for unique model systems for bettering human health.

References

Castejon, O.J. 1983. Light, scanning and transmission electron microscopy study of fish cerebellar capillaries. *Scanning Electron Microscopy* 1: 151–160.

Cooper, P.S., Vogelbein, W.K. and Van Veld, P.A. 1999. Altered expression of the xenobiotic transporter P-glycoprotein in liver and liver tumors of mummichog (*Fundulus heteroclitus*) from a creosote-contaminated environment. *Biomarkers* 4: 48–58.

Di Giulio, R.T., Benson, W.H., Sanders, B.M. and Van Veld, P.A. 1995. Biochemical mechanisms: metabolism, adaptation, and toxicity. In *Fundamentals of Aquatic Toxicology: Effects, Environmental Fate and Risk Assessment*. Rand, G.M. (ed.), pp. 523–561. Taylor and Francis, Washington, DC.

Goodwin, A.E. and Grizzle, J.M. 1997. Oncogene expression in hepatic and biliary neoplasms of the fish *Rivulus ocellatus marmoratus*: Correlation with histologic changes. *Carcinogenesis* 15: 1993–2002.

Gregus, Z. and Klaassen, C.D. (1996). Mechanisms of Toxicity. In *Casarett and Doull's Toxicology: The Basic Study of Poisons*. Klaassen, C.D. (ed.), pp. 35–74. McGraw Hill, New York.

Heath, A.G. 1995. *Water Pollution and Fish Physiology*. Lewis Publishers, Boca Raton.

Jarboe, H., Toth, B.R., Shoemaker, K.E., Greenlees, K.J. and Kleinow, K.M. 1993. Pharmacokinetics, bioavailability, plasma protein binding and disposition of nalidixic acid in rainbow trout (*Oncorhynchus mykiss*). *Xenobiotica* 23: 961–972.

Juliano, R.L. and Ling, V. 1976. A surface glycoprotein modulating drug permeability in chinese hamster ovary cell mutants. *Biochimica et Biophysica Acta* 455: 152–162.

Kleinow, K., Venugopalan, C.S., Smith, A.A., Wiles, J.E. and Holmes, E. 1996. P-glycoprotein distribution and inducibility in the catfish intestine. *The Toxicologist* 30: 35.

Kohler, A., Lauritzen, B., Bahns, S. and Van Noorden, C.J.F. 1998. Clonal adaptation of liver tumor cells in flatfish to environmental contamination by multidrug resistance and metabolic changes (G6PDH, CYP450, GST). *Marine Environmental Research* 46: 191–195.

Krause, M.K., Rhodes, L.D. and Van Beneden, R.J. 1997. Cloning of the p53 tumor suppressor gene from the Japanese medaka (*Oryzias latipes*) and evaluation of mutational hotspots in MNNG-exposed fish. *Gene* 189: 101–106.

Lane, N. 1991. Morphology of glial blood brain barriers. *Annals of the New York Academy of Sciences* 633: 348–362.

McDonnell, T.J., Marin, M.C. and Hsu, B. 1993. The bcl-2 oncogene: Apoptosis and neoplasia. *Radiation Research* 136: 307–312.

Martin, S.J., Green, D.R. and Cotter, T.G. 1994. Dicing with death: Dissecting the components of the apoptosis machinery. *Trends in Biochemical Science* 19: 26–30.

Plakas, S.M., Stehly, G.R. and Khoo, L. 1992. Pharmacokinetics and excretion of phenol red in the channel catfish. *Xenobiotica* 22: 551–557.

Roesijadi, G. and Robinson, W.E. (1994). Metal regulation in aquatic animals: mechanisms of uptake, accumulation and release. In *Aquatic Toxicology: Molecular, Biochemical and Cellular Perspectives*. Malins, D.C. and Ostrander, G.K. (eds), pp. 387–420. Lewis Publishers, Boca Raton.

Van Beneden, R.J. and Ostrander, G.K. (1994). Expression of oncogenes and tumor suppressor genes in teleost fishes. In *Aquatic Toxicology: Molecular, Biochemical and Cellular Perspectives*. Malins, D.C. and Ostrander, G.K. (eds), pp. 295–325. Lewis Publishers, Boca Raton.

Van Leeuwen, C.J., Van Hameren, P., Bogers, M. and Griffioen, P.S. 1986. Uptake, distribution and retention of zineb and ziram in rainbow trout (*Salmo gairdneri*). *Toxicology* 42: 33–46.

2 Toxic responses of the nervous system

Russell L. Carr and Janice E. Chambers

Introduction

Following exposure to any potentially toxic chemical, modification of behavior in an organism frequently suggests that the nervous system is the toxicologic target of that chemical. Although this is quite true in situations concerning mammalian species, the same cannot be assumed for aquatic species. As the environments inhabited by the two groups are distinctly different, the interaction of each group with its environment is also distinctly different. Introduction of a contaminant into an aquatic system can produce behavioral responses in the inhabitants that are quite different from those behavioral responses observed in mammals exposed to a similar chemical. However, while modification of behavior in fish may not be indicative of the nervous system as a target, it is suggestive of neurotoxic activity.

In this chapter, the reported chemical-induced modifications of behavior in teleosts has been used to identify chemicals which may have a neurologic target (Smith, 1984; Heath, 1995). The goal of this chapter is to report on the mechanistic interactions of toxicants with neurologic systems present in teleosts. The majority of mechanistic work of neurotoxicants has been directed towards mammalian toxicity, despite the fact that the biochemical targets of many neurotoxic chemicals have been elucidated using non-mammalian systems. Therefore, in searching for the target of toxicants in teleost systems, one must use what is known concerning the effects of a chemical on targets in other systems as a possible target in teleosts. Here, a brief overview of neuron physiology will be provided and the similarity of these systems with teleost systems will be discussed. Historical references in which the biochemical targets of certain neurotoxic chemicals were first described and reports on whether or not teleosts and mammals share that target will also be provided.

As stated earlier, the chemicals discussed in this chapter were selected based on their reported disruption of behavior in fish. The organochlorine, pyrethroid, and anticholinesterase insecticides disrupt behavior but are also known neurotoxicants and have been responsible for many fish kills. Disruption of behavior in fish has been reported for several metals (mercury, lead, copper, cadmium, and zinc) which have neurologic targets. Other metals (i.e. chromium) disrupt behavior, but the literature suggests that the effects observed may not be mediated by affecting a neurologic mechanism. Several organic chemicals (solvents and polychlorinated biphenyls) have been reported to disrupt behavior. Additionally, several natural toxins which have been demonstrated to be toxic to fish (conotoxins), have been implicated in fish kills (cyanotoxins, saxitoxin, neosaxitoxin, brevetoxin, and *Pfiesteria piscicida*), or have the potential for toxicity in fish (kainic acid, domoic acid, and ciguatoxins) have been discussed.

Nervous system physiology

The teleost brain is largely similar to that of higher vertebrates. Many structures are the same, but some regions differ in size among the vertebrate classes. For example, the cerebellum and the optic and olfactory lobes are frequently quite large in teleosts, whereas the cerebrum is small and does not possess a cerebral cortex (Heath, 1995). Thus, many of the functions of the cerebrum in mammals are performed by other regions of the brain in teleosts. There can also be variability among teleost species. For example, the size of the optic and olfactory lobes is predicated on the method of feeding. Teleosts that primarily utilize sight for foraging have larger optic lobes whereas teleosts that primarily utilize olfaction for foraging have larger olfactory lobes. However, the biochemical processes within the nervous system in teleosts are very similar to those in mammals or other fish (i.e. elasmobranchs). In fact, much of the research which initially characterized the biochemical systems responsible for nervous system function in mammals and other higher vertebrates was performed using fish as an experimental model (e.g. *Torpedo*). However, most research on nervous system function in fish is oriented toward how that system is utilized to respond to natural changes in the environment rather than how it is impacted by a contaminant.

Nerve function

The neuron is the fundamental unit of the nervous system. The typical neuron consists of four regions, each of which carries out specialized functions of the cell. These include: (1) the dendrites, which are multiple projections extending from the cell body that receive signals from sense organs or other neurons, convert those signals to electrical impulses, and transmit them toward the cell body; (2) the cell body or soma, which contains the nucleus and the organelles for energy production and macromolecule synthesis; (3) the axon, which is a single fiber that transmits impulses, termed action potentials, away from the cell body; and (4) the axon (nerve) terminals, which are small branches of the axon that form a junction with another neuron (synapse) or with an effector, such as a muscle fiber (neuromuscular junction) (Figure 2.1). The signal passively spreads from the dendrites through the cell body to the axon hillock (the junction of the cell body and the axon). If the signal received by the axon hillock is sufficiently strong to change the membrane potential from its resting state to a certain level called the threshold potential, an action potential or larger electrical signal is generated at the axon hillock and conducted down the axon to the axon terminal synapse.

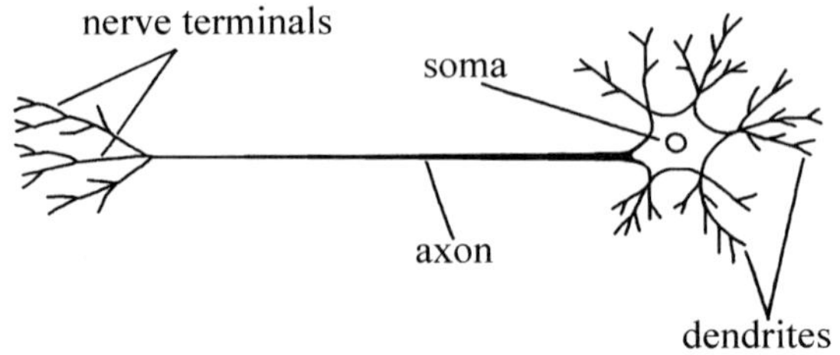

Figure 2.1 Structure of a typical neuron.

Once the action potential reaches the junction, the impulse is transmitted to the post-junctional cell. This transmission may be through either an electrical junction or a chemical junction. In an electrical junction, the impulse passes directly from the presynaptic cell to the post-synaptic cell through gap junctions. In a chemical junction, which is the most common type, the impulse stimulates the release of neurotransmitters stored in synaptic vesicles at the axon terminal into the space between the two cells, the synaptic cleft. The neurotransmitter release is stimulated by increases in calcium (Ca^{2+}), which enters the cell through voltage-gated Ca^{2+} channels. Once released, the neurotransmitter crosses the cleft and binds to receptors on the post-synaptic membrane, thereby transmitting the message to the effector cell. The chemical junction may be excitatory or inhibitory. For example, if the junction is an excitatory junction between two neurons (synapse), the neurotransmitter released would tend to induce an action potential in the effector cell. If the synapse is an inhibitory junction between two neurons (synapse), the neurotransmitter released would inhibit the induction of an action potential in the effector cell. The type of neurotransmitter released determines whether it is an excitatory or inhibitory junction.

The transmission of the action potential down the axon involves the use of electrochemical gradients. When the neuron is resting, a normal gradient is maintained in which the concentration of sodium (Na^+) is greater on the outside of the cell while the concentration of potassium (K^+) is greater inside the cell. The cell membrane itself is slightly negative and has a resting membrane potential of –60 mV. In the axon membrane, there are Na^+ and K^+ ionophores (channels) which are voltage dependent. When the action potential is generated at the axon hillock, thc electrical charge causes the closed Na^+ channels (closed but sensitive to a change in membrane potential) in that region to open, allowing Na^+ influx (Figure 2.2). This influx of Na^+ depolarizes the membrane in that area (i.e. increases the inside membrane potential toward the positive). This change in the membrane potential in that area then stimulates the opening of the Na^+ channels in the adjacent area, depolarizing that area, which then continues to stimulate the Na^+ channels to open in adjacent regions of the membrane. Thus, the action potential is propagated down the axon. After a slight delay, the opened Na^+ channels close and become inactivated (i.e. no longer sensitive to a change in membrane potential). The K^+ channels open, allowing K^+ to flow outward. The outflux of K^+ repolarizes the membrane (i.e. decreases the membrane potential). The K^+ channels close and

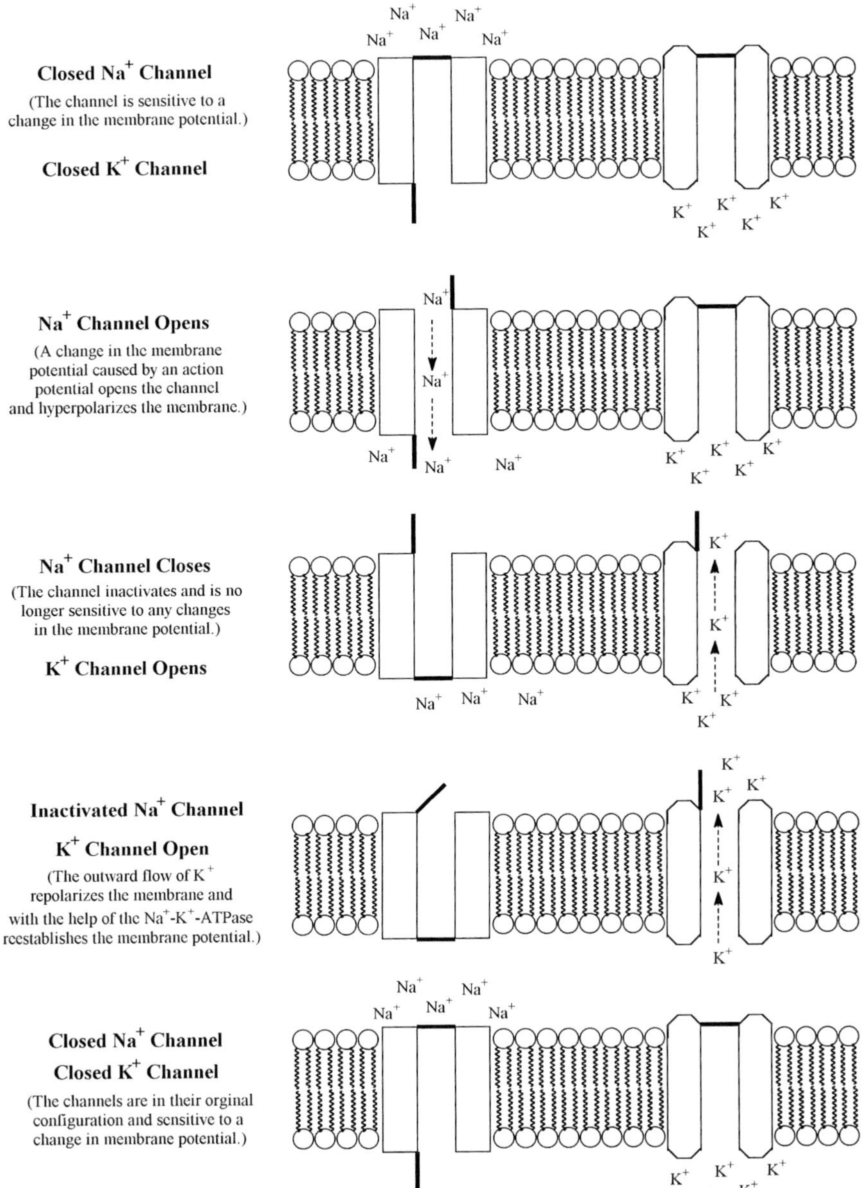

Figure 2.2 Simplistic representation of the process of sodium (Na^+) channel and potassium (K^+) channel opening and closing during an action potential.

the Na^+ channels return to the original closed state (i.e. sensitive to changes in membrane potential). The enzyme Na^+,K^+-ATPase functions to re-establish the proper gradient required for another action potential by pumping Na^+ out and K^+ in.

Sodium channels

Using pharmacologic agents, five distinct binding regions have been identified on the mammalian Na^+ channel (Catterall, 1988) (Figure 2.3). Some work has been carried out using these same agents to characterize the fish brain Na^+ channel. Although some binding sites on the Na^+ channel in fish are similar to those of mammals, some appear to be different.

Site 1 binds sodium channel blockers tetrodotoxin (TTX), saxitoxin (STX) (see Figure 2.11) and the mu-conotoxins which block Na^+ permeability. Site 1 in fish seems similar to that in mammals. TTX effectively blocks Na^+ permeability in rainbow trout (*Oncorhynchus mykiss*) synaptoneurosomes as it does in mammals (Stuart *et al.*, 1987; Rubin and Soderlund, 1992). However, although it appears that the binding characteristics of [^{3}H]-STX are similar in rainbow trout brain synaptoneurosomes (Rubin and Soderlund, 1993) and in garfish nerve membrane (Henderson *et al.*, 1973), the binding characteristics are slightly different from those in rat brain (Catterall *et al.*, 1979).

Site 2 binds the lipophilic activators aconitine (ACN), veratridine (VTD), grayanotoxin, and batrachotoxin (BTX), which cause persistent spontaneous opening of the Na^+ channel. Site 2 in fish seems similar to that in mammals, but is not as responsive to activation. ACN, VTD, and BTX will activate the opening of Na^+ channels and allow Na^+ uptake in both rainbow trout and mouse synaptoneurosomes (Stuart *et al.*, 1987; Rubin and Soderlund, 1992). Also, the ranking of potencies (BTX > VTD > ACN) in rainbow trout is identical to that in mouse synaptosomes (Rubin and Soderlund, 1992). However, the magnitude of

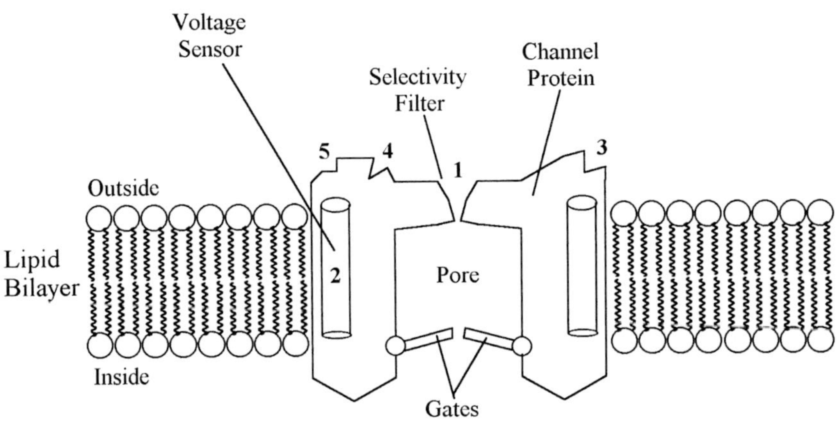

Figure 2.3 Theoretical diagram of the structure of a sodium channel and the hypothesized binding sites for ligands. The actual shape and structure of the sodium channel are not known. Binding sites: (1) tetrodotoxin (TTX), saxitoxin (STX), and mu-O-conotoxins; (2) aconitine (ACN), veratridine (VTD), grayanotoxin, and batrachotoxin (BTX); (3) North African scorpion venom α-polypeptide toxin (α-ScV) and sea anemone nematocyst venom (ATX); (4) American scorpion venom β-polypeptide toxin (β-ScV); and (5) ciguatoxins and brevetoxins.

Na^+ uptake in response to these activators is much lower in rainbow trout brain than in mice (Rubin and Soderlund, 1992) or rats (Eells *et al.*, 1993).

Site 3 binds the α-polypeptide toxins isolated from the venom of North African scorpions (α-ScV) (e.g. *Leiurus quinquestriatus*) and sea anemone nematocysts (ATX) (e.g. *Anthopleura xanthogrammica* and *Anemonia sulcata*) which interfere with the Na^+ channel closing such that, once activated, the Na^+ channel will remain open longer. Site 3 is similar in some aspects but not in others. α-ScV and ATX enhance the activation of Na^+ uptake by ACN, VTD, and BTX in mouse and rat preparations, but rainbow trout brain preparations are less sensitive to this enhancement (Rubin and Soderlund, 1992; Eells *et al.*, 1993). While ACN-activated Na^+ uptake is inhibited by the pyrethroid insecticide deltamethrin in the mouse (Rubin and Soderlund, 1992) and by permethrin, deltamethrin, and cypermethrin in the rat (Eells *et al.*, 1993), these pyrethroids increase ACN-activated Na^+ uptake in rainbow trout.

Site 4 binds the β-polypeptide toxins isolated from the venom of American scorpions (β-ScV) (i.e. *Centruroides noxius* and *Tityus serrulatus*) which also prolong the time that the Na^+ channel is open once activated. Little pharmacologic work has been done comparing site 4 in fish and mammals. However, an antibody which effectively inhibits the binding of *Tityus serrulatus* gamma scorpion toxin to site 4 in rats also antagonized the binding of the toxin to electric eel (*Electrophorus electricus*) Na^+ channels (Barhanin *et al.*, 1985).

Site 5 binds the ciguatoxins produced by benthic dinoflagellates (*Gambierdiscus toxicus*) and the brevetoxins produced by red tide dinoflagellates (*Ptychodiscus brevis*) which cause persistent repetitive spontaneous opening of the Na^+ channel. Site 5 is similar in mammals and fish in some aspects but not in others. Although both ciguatoxins and brevetoxins bind to the Na^+ channel in fish (Capra *et al.*, 1987; Stuart and Baden, 1988), *in vitro* the brevetoxins are more potent activators of the Na^+ channel in fish brain than the ciguatoxins, whereas the opposite is true for mouse brain Na^+ channels (Lewis, 1992).

Overall, a toxicant which binds to the Na^+ channels in mammals should theoretically bind to Na^+ channels in fish. However, the species differences in the composition of the channel may increase or decrease the efficacy of a toxicant in fish compared with mammals.

Cholinergic system

Acetylcholine (ACh) is the neurotransmitter associated with the cholinergic system of all developed organisms. ACh acts on two types of receptors, nicotinic and muscarinic (Table 2.1). The nicotinic receptors were originally distinguished by the action of the agonist nicotine and the muscarinic receptors by the action of the agonist muscarine, a mushroom alkaloid. Subtypes of both nicotinic and muscarinic receptors have been identified. Nicotinic receptors are associated with ligand-dependent ion channels (usually Ca^{2+} and Na^+ channels) and can be found in the neuromuscular junction of skeletal muscles in the peripheral nervous system (PNS) and in the autonomic ganglia, adrenal medulla, and other areas of the central nervous system (CNS). The nicotinic receptor has been well characterized in mammals and consists of five subunits. In the neuromuscular junction, the receptor consists of two α subunits, one β subunit, one γ subunit, and one δ subunit. In the CNS, there are eight α subtypes and four β subtypes of nicotinic receptor, but these receptors do not contain γ or δ subunits. Muscarinic receptors can be found in the heart, smooth muscle, secretory glands, autonomic ganglia, and other areas of the CNS. Muscarinic receptors are associated with guanine nucleotide-binding proteins (G proteins) which are activated by the binding of the neurotransmitter to the receptor and, in turn, activate (G_s) or inhibit (G_i) second messenger systems to produce the appropriate cellular response. The type of second messenger system associated with the receptor depends on the subtype of the receptor. For example, activation of mammalian M_1, M_3, and M_5 muscarinic receptor subtypes causes stimulation of phospholipase C to hydrolyze phosphatidylinositol 4,5-bisphosphate to the second messengers inositol 1,4,5-trisphosphate (which increases intracellular Ca^{2+}, thus activating other Ca^{2+}-dependent processes) and diacylglycerol (which, along with Ca^{2+}, activates protein kinase C). Activation of mammalian M_2 and M_4 muscarinic receptor subtypes causes inhibition of adenyl cyclase, activation of K^+ channels, and suppression of the activity of the voltage-gated Ca^{2+} channels. The muscarinic subtypes are found in many different tissues (see Ashkenazi and Peralta, 1994) with several subtypes present in the same tissue.

Teleosts possess nicotinic receptors which appear to be similar to mammalian nicotinic receptors. The neuromuscular nicotinic receptor in electric eel has been purified and sequenced and found to be highly homologous to that of mammals (Numa, 1986) and to other non-teleosts (e.g. *Torpedo californica*) (Conti-Tronconi *et al*., 1984). Functionally, the alpha-conotoxins from the marine snails of the genus *Conus* are effective antagonists for nicotinic receptors in fish and other species (Hopkins *et al*., 1995), causing paralysis of the skeletal muscles in all species. The nicotinic antagonist α-bungarotoxin will bind as effectively to nicotinic receptors in teleosts as it does in other species (Deplano, 1988). Additionally, the differential sensitivities of fish nicotinic receptors to nicotinic agonists (e.g. nicotine) and antagonists (e.g. tubocurarine) suggest that different subtypes of nicotinic receptors with differing functions are present in teleost nervous tissue (Deplano, 1988).

Little work has been carried out in identifying the presence of different subtypes of muscarinic receptors or on the pharmacologic characterization of these subtypes in fish. However, some muscarinic subtypes have been reported in teleosts. Using specific M_1 and M_2 antagonists to block carbachol-stimulated contractions, the M_2 subtype has been suggested to be present in the smooth muscle of rainbow trout but not the M_1 subtype (Burka *et al*., 1989). The M_3 subtype has been suggested to mediate the cholinergically induced aggregation of pigment in the melanophores of glass crystal catfish (*Kryptopterus bicirrhis*) and mailed catfish (*Corydoras paleatus*) (Hayashi and Fujii, 1994). With respect to muscarinic receptors in the brain, few binding studies have been carried out, with descriptive studies available only for goldfish (*Carassius auratus*) (Francis and Schechter, 1980), carp (*Cyprinus carpio*) (Szabó *et al*., 1989), and brook trout (*Salvelinus fontinalis*) (Jones and King, 1995). In the last two, the reported numbers of the M_1 subtype were significantly greater in teleost brain than the M_2 subtype, similar to mammals.

Once ACh has stimulated a cholinergic receptor, it is rapidly degraded by the enzyme acetylcholinesterase (AChE). AChE is closely associated with the target receptors on the post-synaptic membrane and functions to hydrolyze ACh to choline and acetate, thus eliminating any further stimulation. In teleost brain, the number of molecular forms of AChE can vary from one to five, and these mainly have a substrate preference for ACh compared with other choline analogues such as propyl-, butyryl-, and methylcholine (Kozlovskaya *et al.*, 1993). The active site of AChE contains a serine hydroxyl group to which ACh binds during hydrolysis. However, there are species differences in the configuration of the active site of AChE. The active site of rainbow trout brain AChE is smaller than that of rat brain AChE and has a weaker nucleophilic strength (Kemp and Wallace, 1990). However, the active site of channel catfish (*Ictalurus punctatus*) brain AChE appears to have a strong nucleophilic site similar to that of rat (Carr and Chambers, 1996). It is the characteristic of the binding site that influences its affinity for various inhibitors of the enzyme (e.g. for different organophosphates and carbamates) and that can lead to species differences in toxicity among teleosts for the same compound. Thus, species differences in the characteristics of the enzyme AChE in teleosts could play a role in the varying susceptibility to toxicants which bind to this enzyme. A discussion of anticholinesterase insecticides follows in the section on acetylcholinesterase insecticides.

GABAergic system

γ-Aminobutyric acid (GABA) is a major inhibitory neurotransmitter which is widespread in the vertebrate CNS (DeLory and Olsen, 1994). The inhibitory neurons which utilize GABA as a neurotransmitter form a synapse on the cell body of other neurons. The depolarization of the presynaptic inhibitory neurons stimulates the release of GABA into the synaptic cleft which diffuses across the cleft to the target receptors on the post-synaptic surface. In the brain, there are two major subtypes of GABA target receptors, $GABA_A$ and $GABA_B$ (Table 2.1).

The $GABA_A$ receptor is directly coupled to a chloride (Cl^-) ionophore or channel. When GABA binds to the $GABA_A$ receptor, the permeability of the post-synaptic membrane to Cl^- increases and Cl^- flows into the post-synaptic cell through the Cl^- channel. This influx of Cl^- drives the membrane potential toward the equilibrium potential of Cl^-, which is more negative than the resting potential. This hyperpolarization of the post-synaptic membrane by Cl^- makes the neuron less sensitive to the depolarization resulting from the opening of voltage-gated Na^+ channels. This reduces the probability that an action potential can be initiated by the action of excitatory neurotransmitters. The permeability of the membrane to Cl^- induced by GABA is long lasting (seconds) compared with the rapid depolarizing effects of Na^+ (milliseconds), so GABA will induce a slow, inhibitory post-synaptic response.

Table 2.1 Nervous system receptors and their associated neurotransmitters, agonists, and antagonists.

System	*Receptor*	*Neurotransmitter*	*Agonists*	*Antagonists*
Cholinergic	Muscarinic	Acetylcholine	Muscarine Pilocarpine	Atropine
	Nicotinic	Acetylcholine	Nicotine	α-Bungarotoxin *d*-Tubocurarine
GABAergic	$GABA_A$	γ-Aminobutyric acid	Muscimol Isoguvacine	Bicuculline Picrotoxin
	$GABA_B$	γ-Aminobutyric acid	Baclofen	Saclofen
	$GABA_C$	γ-Aminobutyric acid	–	–
Catecholaminergic	α1	Norepinephrine Epinephrine	Methoxamine Phenylephrine	Indoramin Benoxathian
	α2	Norepinephrine Epinephrine	Clonidine Oxymetazoline	Piperoxan Yohimbine
	β	Norepinephrine Epinephrine	Isoproterenol	Propanolol Timolol
Dopaminergic	D_1-like	Dopamine	Ergot alkaloids SKF-38393	SCH23390
	D_2-like	Dopamine	*N*-Propylnorapomorphine Aminotetralines	Sulpiride Doperidone Butyrophenone
Glutamatergic Ionotropic	NMDA	Glutamate/aspartate and glycine (simultaneously)	Cysteinesulfinate Quinolinate	Dizocilpine (MK801) D-AP5
	AMPA	Glutamate	Quisqualate	NBQX
	Kainate	Glutamate	Kainic acid Domoic acid	GAMS
Metabotropic	L-AP4	Glutamate	L-AP4	–
	ACPD	Glutamate	*trans*-ACPD Ibotenate	L-AP3

GABA, γ-aminobutyric acid; NMDA, *N*-methyl-D-aspartate; D-AP5, D-amino-5-phosphonopentanoiacid; AMPA, α-amino-3-hydroxy-5-methyl-4-isoxazole propionic acid; NBQX, 6-nitro-7-sulfamobenzo[f]quinoxaline-2,3-dione; GAMS, γ-D-glutamylaminomethylsulfonate; L-AP4, 1,2-amino-4-phosphonopropionic acid; ACPD, aminocyclopentyldicarboxylic acid; L-AP3, L-2-amino-3-phosphonopropionic acid.

The $GABA_A$ receptor–ionophore complex is believed to be made up of five similar subunits (Burt and Kamatchi, 1991). Five binding domains have been identified pharmacologically: (1) the GABA binding site to which both agonists, such as muscimol from the mushroom *Amanita muscaria*, and antagonists, such as the convulsant bicuculline, can bind; (2) the benzodiazepine site to which benzodiazepines such as the drugs diazepam (Valium) and chlordiazepoxide (Librium) bind as well as several other non-benzodiazepines, including β-carbolines, cyclopyrrolones, and imidazopyridines; (3) the barbiturate site to which barbiturates such as phenobarbital and pentobarbital bind; (4) the neuroactive steroid site to which steroids bind; and (5) the Cl^- binding site to which channel blockers such as the plant alkaloid metabolite picrotoxin or the compound *t*-butylbicyclophosphorothioate (TBPS) can bind. Binding of the benzodiazepines, barbiturates, and neuroactive steroids to the GABA site potentiate GABA action, resulting in depression of the CNS. Binding of a channel blocker to the Cl^- site results in a state of uncontrolled excitation.

It has been suggested that the $GABA_A$ receptor has been conserved over vertebrate evolution (Cole *et al.*, 1984; Eshleman and Murray, 1990). However, such sweeping statements about the receptor as a single unit cannot be made. Although some of the binding sites on the $GABA_A$ receptor in fish brain are similar to those in other species, there are insufficient data on the physiologic effects of pharmacologic agents in fish to compare them with other species.

The subunits of the GABA binding site in goldfish demonstrate about 75 percent homology with those of rodent and bird (Bahn *et al.*, 1996). Two binding sites for γ-aminobutyric acid have been demonstrated in channel catfish brain (Mathis and Tunnicliff, 1984), and similar dual binding systems for γ-aminobutyric acid have been demonstrated in mammals (Olsen *et al.*, 1981; Tunnicliff and Smith, 1981). The potencies of several inhibitors of $[^3H]$-GABA binding were similar in the channel catfish to that reported or rats (Mathis and Tunnicliff, 1984). The receptor density, affinity constants, and specific activity for $[^3H]$-muscimol binding in codfish (*Gadus morhua morhua*) brain are similar to those in mammalian brain (Deng *et al.*, 1991). Additionally, the effect of different free fatty acids on the binding of $[^3H]$-muscimol was similar among seventeen fish species and various mammals and amphibians (Witt and Nielsen, 1994). However, there may be differences in the functional aspects of the site. The stimulation of $^{36}Cl^-$ uptake into brain membrane vesicles by the agonists GABA and muscimol was significantly lower in freshwater eel (*Anguilla anguilla*) brain than in rat brain (Corda *et al.*, 1989). Additionally, the antagonist bicuculline inhibited $^{36}Cl^-$ uptake into brain membrane vesicles in freshwater eel brain to a greater extent than in rat brain.

The binding site for benzodiazepines has been identified in fish brain as well. This site is suggested to be highly conserved, but only with higher bony fish and tetrapods. For example, it has been suggested that cyclostomes and elasmobranchs do not possess benzodiazepine binding sites (Fernholm *et al.*, 1979), whereas teleosts, mammals, and birds possess these sites that have similar affinities (Friedl *et al.*, 1988; Deng *et al.*, 1991) but are composed of different subunits (Hebebrand *et al.*, 1987). Additionally, there seems to be an age-dependent decrease in benzodiazepine binding sites in both rats and freshwater and saltwater teleosts (Corda *et al.*, 1991; Giannaccini *et al.*, 1997). At least three benzodiazepine binding sites have been postulated to exist in the brain of marine flatfish plaice (*Pleuronectes platessa*) and codfish (Nielsen *et al.*, 1978) but only one has been identified in the brain of saltwater mullet (*Mullus surmeltus*) (Giannaccini *et al.*, 1997). In the saltwater mullet, the benzodiazepine recognition site was similar to that in mammalian brain, but was not identical pharmacologically (Giannaccini *et al.*, 1997). Wilkinson *et al.* (1983) reported only one benzodiazepine binding site in rainbow trout brain but Eshleman and Murray (1989) utilized several different ligands and found two sites in rainbow trout brain which were similar pharmacologically in some aspects to mouse, calf, and human brain but which were different in others. Thus, multiple benzodiazepine binding sites may exist in both mammalian and fish brain but they are not pharmacologically identical. Additionally, differences in the benzodiazepine binding site exist between different species of fish. Witt and Nielsen (1994) demonstrated differences in the brain benzodiazepine binding site of several fish species by comparing the effect of different free fatty acids on the binding of [^{3}H]-diazepam to brain membranes. However, although different, it appears that the benzodiazepine binding sites in fish serve similar physiologic functions to those of mammalian systems (Rehnberg *et al.*, 1989).

Review of the literature found little information about the barbiturate site in fish. However, pentobarbital acted as an agonist and stimulated the influx of $^{36}Cl^-$ into trout brain synaptoneurosomes (Eshleman and Murray, 1991). This is consistent with the normal response of pentobarbital binding to the $GABA_A$ receptor. Likewise, review of the literature found little information about the neuroactive steroid site in fish. However, the $GABA_A$ receptor has been demonstrated to be involved in the release of gonadotropin from goldfish pituitary (Kah *et al.*, 1992; Trudeau *et al.*, 1993); thus a possible feedback binding site may exist for sex-related steroids.

The Cl^- binding site (picrotoxin/TBPS site) has been identified in several species of fish. The binding characteristics of [^{35}S]-TBPS in the brain of channel catfish (Carr *et al.*, 1998), Sacramento blackfish (*Orthodon microlepidotus*) (Cole *et al.*, 1984) and rainbow trout (Eshleman and Murray, 1990) are similar to reported values for human, cow, rat, and chicken brain. Thus, based on binding studies, it can be hypothesized that the picrotoxin/TBPS binding site in fish and mammals may be similar in many aspects. However, the functional interaction of this site with the other binding sites on the $GABA_A$ receptor differs between mammals and fish. For example, modulation of [^{35}S]-TBPS binding to the $GABA_A$ receptors by the GABA site ligands (GABA, muscimol, and bicuculline) was different in freshwater eel and rat brain (Corda *et al.*, 1989). Whereas GABA and muscimol completely inhibited [^{35}S]-TBPS binding in rat brain, less inhibition was observed in freshwater eel brain. Bicuculline significantly increased [^{35}S]-TBPS binding in freshwater eel brain but did not in rat brain. Additionally, there may be differences in the picrotoxin/TBPS site between fish species. The number of picrotoxin/TBPS binding sites in freshwater eel brain was similar to those reported for the species of fish listed above but the affinity of these sites was twofold lower (Corda *et al.*, 1989), whereas the affinity of the picrotoxin/TBPS binding sites in mosquitofish (*Gambusia affinis*) was similar to those reported for the species of fish listed above but the number of binding sites was lower (Bonner and Yarbrough, 1987).

Less is known about the $GABA_B$ receptor. The $GABA_B$ receptor is thought to be indirectly associated with a potassium (K^+) ionophore or channel (Bowery, 1989). When GABA binds to the post-synaptic $GABA_B$ receptor, cyclic adenosine 3′,5′-cyclic monophosphate (cAMP) production is decreased via intracellular mechanisms mediated by G proteins. This may result in the dephosphorylation of the K^+ channel, which opens the channel and increases the outward movement of K^+. The outward movement of K^+ is normally associated with repolarization of the membrane of a neuron after it has been depolarized during an action potential. If K^+ is flowing outward during a stimulus, the membrane potential will remain at resting potential or even be slightly hyperpolarized. As with the $GABA_A$ receptor, this resistance to the effect of the depolarization resulting from the opening of voltage-gated Na^+ channels reduces the probability that an action potential can be initiated by the binding of an excitatory neurotransmitter. It has also been proposed that $GABA_B$ can inhibit presynaptic release of the neurotransmitter by causing a decrease in Ca^{2+} influx, which is necessary for release, but most work suggests that the main response of the $GABA_B$ receptor is K^+ flux (Gähwiler and Brown, 1985). The $GABA_B$ receptor was initially distinguished from the $GABA_A$ receptor by its insensitivity to several $GABA_A$ agonists, THIP [4,5,6,7-tetrahydro-isoxazolo(5,4-c)pyridin-3-ol], APS (3-aminopropane sulfonic acid), and isoguvacine, and to the $GABA_A$ antagonist bicuclline. Also, $GABA_B$ can be stimulated by baclofen and blocked by 2-hydroxysaclofen or phaclofen, whereas $GABA_A$ is not. A large amount of work investigating the physiology of the $GABA_B$ receptor has been carried out using the spinal cord of the cyclostome lamprey eel as a study animal (Leonard and Wickelgren, 1986; Alford and Grillner, 1991; Christenson and Grillner, 1991; Dubuc *et al.*, 1993; Matsushima *et al.*, 1993; Tegnér *et al.*, 1993). In teleosts, however, little work has been done on $GABA_B$ receptors. The $GABA_B$ agonist baclofen has been used as a pharmacologic tool to eliminate the presence of $GABA_B$ receptors while investigating GABAergic functions in rainbow trout pineal gland (Meissl and Ekstrom, 1991), catfish retina (Dong *et al.*, 1994; Takahashi *et al.*, 1995), carp retina (Han *et al.*, 1997), and goldfish retina (Ishida and Cohen, 1988; Heidelberger and Matthews, 1991; Kah *et al.*, 1992; Trudeau *et al.*, 1993; Matthews *et al.*, 1994). Although the inhibition of Ca^{2+} currents by baclofen suggested the presence of $GABA_B$ receptors in the retinal ganglion cells of goldfish using patch clamp methodology, this inhibition of Ca^{2+} currents was not sensitive to the $GABA_B$ receptor antagonists 2-hydroxysaclofen or phaclofen (Bindokas and Ishida, 1991). As the $GABA_B$ receptor is found in the cyclostome spinal cord, it could be present in the teleost spinal cord as well.

More recently, a third type of GABA receptor, $GABA_C$, has been identified in vertebrate retina and to some extent in other parts of the CNS (Lukasiewicz, 1996).

As with $GABA_A$ receptors, $GABA_C$ receptors are closely associated with a Cl^- channel. However, using fish retina as a source of $GABA_C$ and $GABA_A$ receptors, these two receptors have been shown to be pharmacologically, molecularly, and functionally distinct. The retinal $GABA_C$ receptor has been shown to be insensitive to the $GABA_A$ agonist bicuculline in white perch (*Roccus americana*) (Qian *et al.*, 1997), carp (Han *et al.*, 1997), hybrid bass (*Morone chrysops* × *Morone saxitilis*) (Qian and Dowling, 1995), goldfish (Matthews *et al.*, 1994), and catfish (Dong *et al.*, 1994; Takahashi *et al.*, 1994, 1995; Dong and Werblin, 1996). Likewise, the $GABA_C$ receptor differs from the $GABA_B$ receptor by its insensitivity to the $GABA_B$ agonist baclofen in catfish retina (Dong *et al.*, 1994; Takahashi *et al.*, 1995). A binding site for the Cl^- channel blocker picrotoxin has also been demonstrated on the $GABA_C$ receptor in catfish retina (Takahashi *et al.*, 1994, 1995; Dong and Werblin, 1996). It appears that these receptors play an important role in the transmission and processing of visual information. Thus, if a toxicant alters functioning of the $GABA_C$ receptor, it could affect the teleost in many ways (e.g. detection of prey, detection of predators, and detection of changes in photoperiod necessary for controlling circadian rhythm).

Glutamate system

Glutamate and aspartate mediate the majority of the excitatory synaptic transmission in neurons. These amino acid transmitters act on ionotropic and metabotropic receptors (Table 2.1). The ionotropic receptors include the *N*-methyl-D-aspartate (NMDA), α-amino-3-hydroxy-5-methyl-4-isoxazole-propionic acid (AMPA), and kainate receptors. The metabotropic receptors include the aminocyclopentyl dicarboxylic acid (ACPD) and L-2-amino-4-phosphono-propionic acid (L-AP4) receptors.

The three ionotropic receptors have been identified in fish neurons (Henley and Oswald, 1988; Maler and Monaghan, 1991; Barnes and Henley, 1994; Peng *et al.*, 1995). In mammals, the activation of each of the ionotropic receptors opens an ion channel, allowing the influx of Ca^{2+} and/or Na^+, which depolarizes the neuronal membrane. The NMDA receptor is activated by the simultaneous binding of glycine and either glutamate or aspartate, whereas the AMPA and kainate receptors are activated only by glutamate (Dingledine and McBain, 1994). There is a great deal of similarity between these receptors in fish and other species. The cDNA sequence coding for a portion of the NMDA receptor in the electric fish *Apteronotus leptorhynchus* demonstrated 80 percent homology with DNA coding for the rat, human, and duck receptor (Bottai *et al.*, 1997). The cDNA sequence coding for the AMPA receptor in the brain of the tilapia *Oreochromis mossambicus* demonstrated 85 percent homology with the DNA coding for the rat brain receptor (Kung *et al.*, 1996; Chang *et al.*, 1998). Additionally, antibodies against specific domains of the AMPA receptor recognized similar proteins in rat, human, frog, chick, and goldfish brain samples (Bahr *et al.*, 1996). A purified kainate receptor polypeptide from goldfish brain containing the kainate binding site exhibited a similar size and an amino acid sequence homology of 40–60 percent compared with that of frog, chick, and pigeon brain (Ziegra *et al.*, 1992). However, the size of the kainate binding site in goldfish brain is very different from that in mammalian brain (Murphy *et al.*, 1993). Additionally, a kainate binding site, which is thought to be a metabotropic receptor rather than an ionotropic receptor, has been identified in goldfish brain but not in mammals (Willard *et al.*, 1991; Barnes *et al.*, 1993; Wo and Oswald, 1996).

The ionotropic receptors display a similar mechanism (mediate the influx of Ca^{2+} and/or Na^{+}) in fish to that in other species. Specific agonists for all three ionotropic receptors have been shown to increase Ca^{2+} influx into channel catfish retinal horizontal cells (Linn and Christensen, 1992). Additionally, the stimulated Ca^{2+} influx by each agonist was blocked by its specific receptor antagonist. This is similar to that observed in mammalian systems.

In mammals, the activation of metabotropic ACPD receptors by glutamate activates phospholipase C to convert phosphoinositides to inositol 1,4,5-trisphosphate (IP_3), which in turn stimulates the release of intracellular Ca^{2+} (Dingledine and McBain, 1994). In mammals, the activation of L-AP4 receptors by glutamate activates phosphodiesterase to hydrolyze a cyclic nucleotide (i.e. cAMP/cGMP to AMP/GMP) which causes the ionotropic Ca^{2+} ion channels to remain open, thus hyperpolarizing the neuron and leading to reduced synaptic transmission (Dingledine and McBain, 1994). Metabotropic glutamate receptors have been identified in goldfish olfactory epithelium (Cao *et al.*, 1998), and channel catfish olfactory neurons (Medler *et al.*, 1998). The ACPD receptor-mediated stimulation of IP_3 formation has been demonstrated in the olfactory rosettes of Atlantic salmon (Pang *et al.*, 1994), in salmon brain (Kubokawa *et al.*, 1996), and in carp retinal cells (Janssen-Bienhold *et al.*, 1994). However, little is known about the physiologic role of the ACPD metabotropic glutamate receptor in teleosts. A metabotropic receptor which exerted effects similar to those of a L-AP4 glutamate receptor has been identified in catfish retinal horizontal and bipolar cells (Dixon and Copenhagen, 1997), but very little other work has been carried out investigating teleost L-AP4 glutamate receptors.

There appears to be high homology between the glutamate receptors in teleost and other non-teleost species with regard to structure and mechanisms. In fact, the goldfish retina is commonly used as a model for investigating both ionotropic and metabotropic glutamate receptors (Ishida and Neyton, 1985; Tachibana, 1985; Yazejian and Fain, 1992; Schultz *et al.*, 1997; Villmann *et al.*, 1997). Glutamate receptors are involved in olfactory and optic tissue in teleosts. Additionally, they are involved the initial transmission of taste in teleosts (Smeraski *et al.*, 1998) and in the control of electric organ discharge in electric fish(Spiro, 1997). However, there has been little work investigating these receptors in fish brain.

Catecholaminergic system

The catecholamines norepinephrine, epinephrine and dopamine are neurotransmitters which exert actions in the CNS as well as in other tissues in the periphery. Norepinephrine and epinephrine bind to the adrenergic receptors and dopamine binds to the dopamine receptors (Table 2.1).

In mammals, there are three types of adrenergic receptors, α_1-, α_2-, and β-receptors. Regardless of tissue, these receptors are functionally coupled to second messenger systems. Once stimulated, the α_1-receptors activate phospholipase C and increase cellular Ca^{2+}, the α_2-receptors inhibit cAMP production, and the β-receptors increase cAMP (Weiner and Molinoff, 1994). In teleosts, the α_1-receptors (Zhang *et al.*, 1992; Moon *et al.*, 1993), α_2-receptors (Sammak *et al.*, 1992; Svensson *et al.*, 1997), and β-receptors (O'Connor *et al.*, 1989; Hubbard *et al.*, 1996) mediate their action through similar mechanisms. While there is some information available on adrenergic receptors in the CNS of teleosts, the majority of the work describing those receptors has been performed using peripheral tissue. However, these receptors mediate their actions by similar processes regardless of location.

All three types of adrenergic receptors have been identified in teleost tissues using pharmacologic agents (Iga, 1983; Morishita *et al.*, 1985; Garcia-Sainz *et al.*, 1995; Jozefowski and Plytycz, 1998). There are some similarities between the adrenergic receptors of teleosts and mammals. For example, the medaka α_1-receptor is similar (61 percent homology) to that of mammals (Yasuoka *et al.*, 1996) and the cuckoo wrasse (*Labrus ossifagus*) α_2-receptor is similar (47–57 percent homology) to that of humans (Svensson *et al.*, 1993). Likewise, the medaka α_1-receptor appears to be pharmacologically similar to that of mammals (Yasuoka *et al.*, 1996). However, the codfish brain α_1-receptor differs both structurally and pharmacologically from the rat brain α_1-receptor (Bergstrom and Wikberg, 1986) and the goldfish leukocyte β-receptor differs pharmacologically from that in mammalian leukocytes (Jozefowski and Plytycz, 1998). Additionally, although the α_2-receptor has been identified pharmacologically in several species of fish, there are differences in the receptors among fish species (Karlsson *et al.*, 1987).

In mammals, there are two families of dopamine receptors, D_1-like and D_2-like receptors. The D_1-like receptor family includes D_1 (D1A) and D_5 (D1B) which increase cAMP (Weiner and Molinoff, 1994). The D_2-like receptor family includes D_2 (D2A) (Weiner and Molinoff, 1994), D_3 (D2B) (Werner *et al.*, 1996; Robinson and Caron, 1997; Kuzhikandathil and Oxford, 1999), and D_4 (D2C) (Tang *et al.*, 1994; Werner *et al.*, 1996; Yan *et al.*, 1997) and inhibits cAMP production, activates K^+ channels, and blocks the opening of Ca^{2+} channels. Frequently, the literature on teleosts only identifies the receptors as D_1 and D_2 and does not distinguish between the subtypes in each family. In teleosts, the D_1-like receptors increase cAMP (Steffey *et al.*, 1991; Frail *et al.*, 1993; Cardinaud *et al.*, 1997) and the D_2-like receptors inhibit cAMP production (Dearry and Burnside, 1988; Wong *et al.*, 1996).

Using pharmacologic ligands, both D_1-like receptors (Frail *et al.*, 1993; Behrens and Wagner, 1995; Lamers *et al.*, 1997; Cardinaud *et al.*, 1997; Hirano *et al.*, 1998) and D_2-like receptors (Omeljaniuk and Peter, 1989; Rashid *et al.*, 1993; Hillman *et al.*, 1995; Omeljaniuk, 1995; Wang *et al.*, 1997) have been identified in fish tissue. As with the adrenergic receptors, there are some similarities in the adrenergic receptors of teleosts and mammals. For example, the D_1 receptor in goldfish is similar pharmacologically to that in mammals (Steffey *et al.*, 1991; Wong *et al.*, 1993), and its amino acid sequences have portions that are highly homologous to that of rat and human (Frail *et al.*, 1993). However, it lacks certain amino acid segments found in mammals (Frail *et al.*, 1993). Likewise, a ligand that is a full agonist of the D_1 receptor in rat brain is only a partial agonist of the D_1 receptor in carp (Kebabian *et al.*, 1992). There are also phylogenetic differences in the dopamine receptors. Although there are two types of receptors in the D_1-like family in mammals, additional D_1-like receptors have been observed in freshwater eel (four types) (Cardinaud *et al.*, 1997; Hirano *et al.*, 1998), in carp (five types) (Hirano *et al.*, 1998), and in tilapia (at least three types) (Lamers *et al.*, 1996). Three subtypes of the D_1-like family have been identified in carp, but none were present in freshwater eel (Hirano *et al.*, 1998).

Thus, although the mechanisms by which the catecholamine receptors mediate their actions are similar, there are differences in the structure, function, and sensitivity to pharmacologic agents of these receptors. These differences occur not only between teleosts and non-teleosts but also among different teleost species.

Chlorinated hydrocarbon insecticides

The chlorinated hydrocarbon (organochlorine) insecticides are a very diverse group of compounds. According to Matsumura (1985), these compounds are characterized by the presence of carbon, chlorine, hydrogen and, sometimes, oxygen atoms (including a number of C—Cl bonds), the presence of cyclic carbon chains, and the lack of any particular active intramolecular sites. They can be classified into three distinct groups: the dichlorophenylethanes, the hexachlorocyclohexanes, and the cyclodienes. These compounds are highly toxic to a wide range of insect pests and many are relatively safe to tetrapods. In addition, they are relatively inexpensive to manufacture, are highly lipophilic, have low volatility, and are chemically stable. These characteristics made them very effective insecticides and, from the mid-1940s to the mid-1960s, they were extensively used in the USA in both agricultural and domestic situations. However, these chemicals were very persistent and tended to bioaccumulate, resulting in many of these insecticides being banned in North America and Europe. Unfortunately, because they are effective, relatively safe, and inexpensive, they are still used in many developing nations.

Although many of these chemicals have been banned for years, the sheer volume at which they were manufactured and applied during their peak time of usage is astonishing. For example, in southern Arizona in 1965, 7.59 lb of toxaphene, 1.35–4.00 lb of DDT, 1.2 lb of endrin, 0.58–1.02 lb of endosulfan and undetermined quantities of aldrin, benzene hexachloride (BHC), heptachlor, and chlordane were applied to every acre of cotton (Matsumura, 1985). Given the chemical stability and patterns of usage of these compounds, it is not surprising that residues are still detectable in tissues of both freshwater and saltwater fish in areas where these chemicals have not been used for over 25 years.

Dichlorodiphenylethanes

The dichlorodiphenylethane compounds include the insecticide DDT [1,1,1-trichloro-2,2-bis(*p*-chlorophenyl)ethane] and its related analogues such as the insecticides 1,1-dichloro-2,2-bis(*p*-chlorophenyl)ethane (TDE or DDD or Rhothane®) and methoxychlor and the acaricides chlorfenethol (DMC or Dimite® or Qikron®), dicofol (Kelthane®), and Chlorobenzilate® (Figure 2.4). It was the discovery of the insecticidal properties of DDT by Peter Müller in 1939 that initiated the development of synthetic pesticides and the development of the DDT analogues and other chlorinated hydrocarbon insecticides. While DDT and many of its analogues are no longer manufactured for use in the USA, some members of this class, such as methoxychlor, dicofol, and Chlorobenzilate®, are still presently used.

Mechanism of action

The actual mechanism behind the toxic action of DDT is not totally clear. This is true for both tetrapods and fish. However, sufficient behavioral data in fish exposed to DDT (Anderson and Peterson, 1969; Anderson and Prins, 1970; Davy *et al.*, 1972; McNicholl and Mackay, 1975a,b; Bengtsson and Larsson, 1981) have been reported that suggest a neurochemical target for DDT in fish. There are four possible mechanisms of neurotoxicity: (1) interaction with Na^+ channels on nerve cell membranes; (2) disruption of K^+ membrane permeability in nerve cells; (3) inhibition of Na^+,K^+-ATPase, Mg^{2+}-ATPase, Ca^{2+},Mg^+-ATPase, and/or Ca^{2+}-ATPase; and (4) inhibition of the function of calmodulin.

In the case of Na^+ channels, DDT functions to block effectively the closing of the Na^+ channel once an action potential has occurred (Lund and Narahashi, 1981). Normally during an action potential, the channel opens, allowing Na^+ influx which depolarizes the membrane. The Na^+ channel then closes and the Na^+ gradient is re-established. In the presence of DDT, the closed Na^+ channel opens normally, but once open it becomes a modified open channel and inactivates (closes) slowly or, in other words, it remains open such that it interferes with the normal restoration of the Na^+ gradient. Thus, this disrupts the normal repolarization of the membrane. In addition to blocking closing of the Na^+ channel, DDT suppresses the permeability of the membrane to K^+ (Narahashi, 1979). As the efflux of K^+ is associated with repolarization of the membrane following an action potential, a decrease in K^+ permeability would further contribute to the disruption of the normal repolarization of the membrane. Thus, DDT causes a 'negative after potential' which makes the nerve much more sensitive to future stimuli such that the influx of lower amounts of Na^+ than normal will yield an action potential (Figure 2.5). This leads to a characteristic 'repetitive discharge' in nerve impulse patterns following a single stimulus. These effects were determined using squid and crustaceans as a source of nervous tissue. Using mammalian tissue, the action of DDT on the gating of the Na^+ channel has been further demonstrated. In mouse brain synaptosomes, DDT has been demonstrated to enhance the uptake of Na^+ into the synaptosomes that had been induced by the activators VTD and BTX

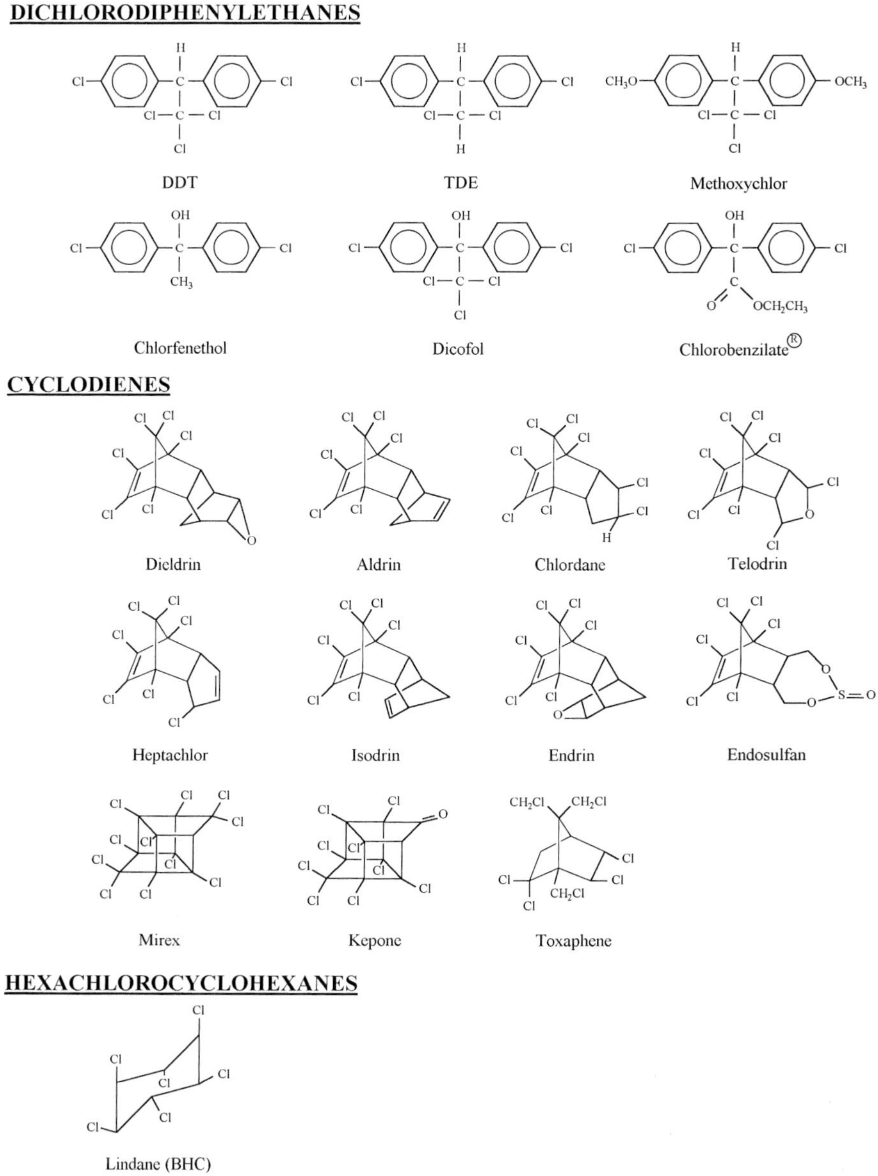

Figure 2.4 Chemical structures of some chlorinated hydrocarbon insecticides.

(Bloomquist and Soderlund, 1988). As α-ScV and ATX similarly enhance VTD- and BTX-induced Na^+ uptake, it suggests that the action of DDT or its analogues is on the gating mechanism of the Na^+ channel. Similar results have been demonstrated in rainbow trout brain synaptosomes, where DDT enhanced the uptake of Na^+ induced by VTD (Stuart *et al.*, 1987), suggesting an identical target

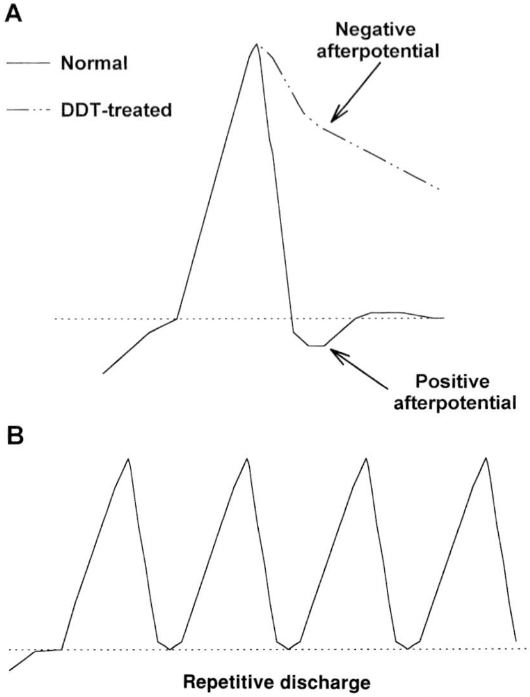

Figure 2.5 Effects of DDT on the action potential in neurons. (A) Demonstration of a negative afterpotential caused by DDT exposure. (B) Characteristic repetitive discharge caused by DDT as a result of a negative afterpotential.

in fish. Sublethal DDT exposure leads to observed hyperactivity in fish, for example flounder (*Platichthys flesus*) (Bengtsson and Larsson, 1981) and bluegill sunfish (*Lepomis macrochirus*) (Ellgaard *et al.*, 1977), and enhanced hyperactivity to stimuli (Heath, 1995). It is feasible to assume that the effects of DDT on Na^+ and K^+ channels play a role in the toxicity of DDT in fish.

DDT and its analogues have also been demonstrated to inhibit Na^+,K^+- and Mg^{2+}-ATPases in fish, tetrapods, and insects. The *in vitro* inhibition of brain ATPases by DDT or its analogues has been demonstrated in many species of teleosts, including lake trout (*Salvelinus namaycush*) (Koch, 1969), bluegill sunfish (Cutkomp *et al.*, 1971, 1972; Koch *et al.*, 1971), rainbow trout (Davis and Wedemeyer, 1971), and channel catfish (Desaiah and Koch, 1977). In addition, water column in combination with oral exposure to DDT *in vivo* has resulted in inhibition of both ATPases in fathead minnow (*Pimephales promelas*) (Desaiah *et al.*, 1975). It appears that Mg^{2+}-ATPase is more sensitive to inhibition by DDT in fish than Na^+,K^+-ATPase, whereas the reverse seems to be true in mammalian systems. As Na^+,K^+-ATPase plays an important role in the active transport of ions across the nerve membrane, it could be hypothesized that the inhibition of Na^+,K^+-ATPase in fish neurons could alter nerve transmission and contribute to the toxicity in this sense. In addition, mitochondrial Mg^{2+}-ATPase is an energy-regulating enzyme involved in oxidative phosphorylation. Inhibition of this enzyme in the brain would decrease the efficiency of aerobic metabolism in the brain and lead to an anaerobic state in the nerve cells. It was initially thought that inhibition of these enzymes in brain and other tissues (especially gill) was the target of action for DDT. However, there is no way to relate the inhibition of either of these enzymes directly to the characteristic signs of DDT intoxication or the action of DDT on the axonal membrane (Shankland, 1983).

Calcium is crucial to regulating the cellular processes of all types of cells. In the nerve cell, the Ca^{2+} gradient is important in neurotransmitter release and for stabilization of the axonal membrane. Ca^{2+} also plays an important role in second messenger systems and operation of certain ionic channels. Thus, disruption of the normal maintenance of Ca^{2+} in a nerve cell could result in altered function. It has been demonstrated that Ca^{2+}-ATPase from lobster nerve is very sensitive to inhibition by DDT (Matsumura and Ghiasuddin, 1979). Ca^{2+}-ATPase is thought to represent the ATP-dependent phosphorylation and dephosphorylation system associated with Na^{+}/Ca^{2+} exchange (Matsumura and Clark, 1982). DDT has also been shown to disrupt this Na^{+}/Ca^{2+} exchange system in squid nerve (Clark and Matsumura, 1982). Insufficient levels of Ca^{2+} on the external membrane of the axon induce membrane destabilization and can lower the threshold potential required to obtain an action potential. In addition, DDT can decrease the Ca^{2+}-carrying ability of calmodulin, thus increasing internal Ca^{2+} levels (Rashatwar and Matsumura, 1984), as demonstrated using a bovine calmodulin system. Inhibition of calmodulin function by DDT has been shown to have subsequent effects on calmodulin-stimulated Ca^{2+},Mg^{2+}-ATPase (Warngard *et al.*, 1988), which is thought to be identical to the Ca^{2+} ion pump that maintains proper low internal Ca^{2+} concentration and the gradient across membranes (Shankland, 1983). Furthermore, it has been demonstrated that systems which are regulated by Ca^{2+} in DDT-resistant strains of insects are less sensitive to the effects of Ca^{2+} than in susceptible strains. This suggests that the disruption of Ca^{2+} maintenance by DDT may be an important part of its toxicity. DDT has been shown to increase the amount of free acetylcholine (Tobias *et al.*, 1946), which could be the result of increased internal Ca^{2+} levels stimulating release of acetylcholine. There is some indication that DDT will inhibit Ca^{2+}-ATPase in the skeletal muscle of cod (*Gadus morhua morhua*) and wolffish (*Anarhikas lupus*) (Khokhryakova, 1981). However, it is not known whether DDT will disrupt Ca^{2+} maintenance in the nervous tissue of teleosts and, if it does, what role in toxicity this will play.

Cyclodienes

The cyclodiene compounds include dieldrin (the epoxide of aldrin), aldrin, chlordane, isobenzan (Telodrin®), heptachlor, isodrin (isomer of aldrin), endrin (the epoxide of isodrin), and endosulfan (Figure 2.4). Also included in this group, although they are not cyclodienes, are toxaphene, mirex, Kepone® (chlordecone), and Strobane®. The cyclodienes were developed in the mid-1940s following the introduction and success of DDT and the hexachlorocyclohexanes (discussed below).

As with DDT, inhibition of Na^+,K^+-ATPase, microsomal Mg^{2+}-ATPase, and mitochondrial Mg^{2+}-ATPase has been considered to be a possible mechanism of action for the cyclodiene insecticides in many species including fish. There is a voluminous amount of information on the effect of various cyclodienes on these enzymes. The *in vitro* inhibition of ATPases by various cyclodienes has been demonstrated in the brain of many species of teleosts, including lake trout (Koch, 1969), bluegill sunfish (Cutkomp *et al.*, 1971; Koch *et al.*, 1971; Yap *et al.*, 1975), channel catfish (Desaiah and Koch, 1975a,b,c, 1977), rohu (*Labeo rohita*) (Verma *et al.*, 1978a, 1979), and singii (*Saccobranchus fossilis*) (Verma *et al.*, 1979). The *in vivo* inhibition of ATPases has also been observed in freshwater murrel (*Channa gachua*) (Dalela *et al.*, 1978) and in rohu and singii (Verma *et al.*, 1978b). However, the epoxide forms of several of these cyclodienes are not as potent *in vitro* inhibitors as the parent compounds, even though they are equally toxic to fish (Cutkomp *et al.*, 1971; Koch *et al.*, 1971). Additionally, there is a large variation in the interaction of the compounds with the different ATPases. For example, not all cyclodienes inhibit ATPases and, when they do, the amount of inhibition varies among compounds and among species. Similar to that proposed for DDT and ATPases, Shankland (1983) also concluded that there is no way to relate the inhibition of ATPases directly to the characteristic signs of cyclodiene intoxication or to the action of these insecticides on the axonal membrane.

It was initially thought that the ultimate cause of toxicity for the cyclodiene insecticides was the direct induction of excessive and spontaneous release of acetylcholine into the synapse (Shankland and Schroeder, 1973; Shankland, 1979). Since the cyclodiene insecticide metabolite heptachlor epoxide inhibits Ca^{2+},Mg^{2+}-ATPase in the synapse of rat brain (Yamaguchi *et al.*, 1979), which could raise the concentration of free Ca^{2+} in the presynaptic region (Yamaguchi *et al.*, 1980), neurotransmitter release could be stimulated in this manner. However, while investigating the effects of cyclodiene insecticides on another neurotransmitter system, Ghiasuddin and Matsumura (1982) and Matsumura and Ghiasuddin (1983) demonstrated that heptachlor epoxide would bind to the picrotoxin/TBPS site on the GABA receptor. They proposed that this site is the mechanism of action for all cyclodienes. Furthermore, insects resistant to the cyclodienes are also resistant to picrotoxinin, the active principle of picrotoxin. Thus, it is now generally accepted that the toxicity of the cyclodiene insecticides results from their binding to the picrotoxin/TBPS site of the GABA receptor which blocks Cl^- flux into the cell.

In teleosts, the importance of the binding of cyclodiene insecticides to the GABA receptor has also been clarified. Bonner and Yarbrough (1987, 1988) identified that the mechanism behind the resistance to the cyclodiene endrin in mosquitofish (*Gambusia affinis*) was an alteration in the picrotoxin/TBPS binding site. This alteration in resistant mosquitofish included a decrease in the number of binding sites for TBPS as well as a decrease in the affinity of the binding site for endrin (Bonner and Yarbrough, 1989). In addition, Carr *et al.* (1998) determined the potency of several chlorinated insecticides to inhibit the binding of $[^3H]$-TBPS to the picrotoxin/TBPS site of the GABA receptor in channel catfish brain. When the inhibitory potency was compared with the reported 96-h LC_{50} values for channel catfish, a linear relationship (r^2 value of 0.98) was calculated. Although additional research is needed to characterize further the interactions between the cyclodienes and the picrotoxin/TBPS site, it can be assumed that the mechanism of action behind the acute toxicity of these chlorinated insecticides in fish is related to their interaction with that site on the GABA receptor.

Hexachlorocyclohexanes

The hexachlorocyclohexane compounds are commonly referred to as benzene hexachloride (BHC). BHC was synthesized by Michael Faraday in 1825, and four of the eight possible isomers of BHC were identified by Van der Linden in 1912. The insecticidal properties were identified in 1942, by Dupire and Rancourt in France and by Slade in the UK. The toxic γ-isomer was isolated by Slade and named lindane in honor of Van der Linden (Matsumura, 1985). Although BHC has been used as an insecticide, the toxicity of the mixture depends on the content of the γ-isomer, lindane (Figure 2.4).

Lindane has been shown to inhibit Na^+,K^+-ATPase in mouse brain (Magour *et al.*, 1984). and Na^+,K^+- and Mg^{2+}-ATPases in lake trout brain (Koch, 1969). However, lindane was not a very effective inhibitor of Na^+,K^+- or Mg^{2+}-ATPases in rainbow trout (Davis *et al.*, 1972) or in bluegill sunfish (Yap *et al.*, 1975).

As with the cyclodienes, it was initially thought that the ultimate cause of toxicity for lindane was due to the direct induction of excessive and spontaneous release of acetylcholine into the synapse (Shankland and Schroeder, 1973; Shankland, 1979). However, as the work with the cyclodienes progressed, it was proposed that the mechanism of action for lindane involved binding to the picrotoxin/TBPS site on the GABA receptor (Ghiasuddin and Matsumura, 1982; Matsumura and Ghiasuddin, 1983). As with the cyclodienes, insects resistant to lindane are also resistant to picrotoxinin, the active principle of picrotoxin. In channel catfish brain, lindane effectively competes with [^{3}H]-TBPS binding to the picrotoxin/TBPS site of the GABA receptor, and, in fact, it competes better than several cyclodiene insecticides (aldrin and heptachlor) (Carr *et al.*, 1998). Thus, it can be accepted that the toxicity of the lindane results from its binding to the picrotoxin/TBPS site of the GABA receptor which blocks Cl^- flux into the cell.

Pyrethroids

Historical perspective

The pyrethroids are very effective synthetic insecticides which were originally derived from the botanical insecticide pyrethrum (Matsumura, 1985). Pyrethrum was originally extracted from *Chrysanthemum* flowers, particularly the species *Chrysanthemum cinerariaefolium* and *Chrysanthemum coccineum*. Pyrethrum consists of six esters which make up the two major active ingredients: pyrethrin I from the chrysanthemic acid esters pyrethrin I, cinerin I, and jasmolin I, which possess the lethality properties of the insecticide, and pyrethrin II from the pyrethric acid esters pyrethrin II, cinerin II, and jasmolin II, which possess the knockdown properties of the insecticide (Figure 2.6). Pyrethrum is mainly active as a contact insecticide and is very susceptible to deactivation by metabolism.

The use of pyrethrum powder (ground flowers) was first reported around 1800 and, by 1851, its use was world wide (Matsumura, 1985). However, while the insecticidal powder form is fairly stable, it was wasteful and it was soon determined that it would be more efficient to extract the active ingredient and concentrate it for insecticide formulations. The first commercial production of pyrethrum was in Yugoslavia from flowers which contained about 0.7 percent active ingredient. Soon thereafter, the source of flowers shifted to Japan, where the flowers contained about 1.0 percent active ingredient. A strain of Kenyan flowers was found to have 1.3–3 percent of the active ingredient and is presently used as the primary source for pyrethrum. Unfortunately, the extracted pyrethrum is unstable in the presence of light, air, and moisture and this has led to synthesis of new analogues which are more persistent, less susceptible to deactivation by metabolism and are thus more toxic.

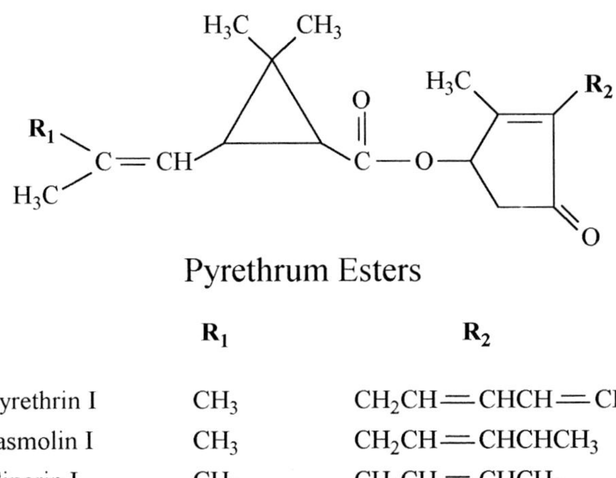

Pyrethrum Esters

	R1	**R2**
Pyrethrin I	CH_3	$CH_2CH{=}CHCH{=}CH_2$
Jasmolin I	CH_3	$CH_2CH{=}CHCHCH_3$
Cinerin I	CH_3	$CH_2CH{=}CHCH_3$
Pyrethrin II	$\underset{\underset{O}{\|}}{C}OCH_3$	$CH_2CH{=}CHCH{=}CH_2$
Jasmolin II	$\underset{\underset{O}{\|}}{C}OCH_3$	$CH_2CH{=}CHCHCH_3$
Cinerin II	$\underset{\underset{O}{\|}}{C}OCH_3$	$CH_2CH{=}CHCH_3$

Figure 2.6 Chemical structures of the six pyrethrum ester compounds with neurotoxic activity found in *Chrysanthemum* flowers.

Signs of toxicity

Initially after exposure to toxic levels of pyrethroids, there is an increase in the cough response and increased mucus secretion from the gills in species such as rainbow trout. In some species such as fathead minnows, there is a loss of schooling behavior and swimming near the water's surface. Soon after, fine tremors are observed which progress to violent whole body seizures, possibly including head shaking and twisting. During this time, the opercula become flared and ventilation is drastically increased. As time progresses, these seizures increase in intensity until a point is reached where the fish becomes inactive and unable to maintain its position in the water column, which is soon followed by death (Holcombe *et al.*, 1982; Bradbury *et al.*, 1987, 1991; Bradbury and Coats, 1989; Haya, 1989).

Mechanism of action

The mechanism of action of the pyrethroids was initially investigated using nerve preparations from insects, crayfish, and giant squid (Narahashi, 1971; Casida *et al.*, 1983). Based on the response of crayfish nerve cord exposed to the pyrethroids, Narahashi *et al.* (1977) reported that there are two types of pyrethroids. This description was further characterized based on differences in the signs of toxicity present in susceptible cockroaches exposed to various pyrethroids (Gammon *et al.*, 1981) and on differences in the toxicity of various pyrethroids in resistant strains of cockroaches (Scott and Matsumura, 1983). Generally, type I pyrethroids (Figure 2.7) block the closing of the Na^+ channel, causing excitation involving 'repetitive discharge' in axons, as with DDT (Lund and Narahashi, 1982). Type II pyrethroids (Figure 2.7) bind to open Na^+ channels, interfering with their closing (modified open channel), and bind to inactivated channels, changing them to 'modified' inactivated channels such that, to return to the normal closed state, they must go through an open channel configuration (Lund and Narahashi, 1981). There is no 'repetitive discharge' with type II pyrethroids as seen with the type I pyrethroids. Type I pyrethroids act on the PNS as well as the CNS, whereas type II pyrethroids act primarily on the CNS. The majority of the signs resulting from pyrethroid toxicity in mammals result from their action in the CNS. For example, it has been demonstrated that both type I and type II pyrethroids are more toxic to mice when injected intracerebrally than when administered peripherally (Ruzo *et al.*, 1979; Staatz *et al.*, 1980).

Pyrethroids are much more toxic to fish than to amphibians, mammals, and birds (Miyamato, 1976; Jolly *et al.*, 1978; Edwards *et al.*, 1986). Fish do not metabolically deactivate pyrethroids efficiently (Glickman *et al.*, 1982), leading to more of the insecticide reaching the brain after exposure (Glickman and Lech, 1982). A correlation between signs of toxicity in rainbow trout (onset of hyperactivity followed by seizures and loss of ability to maintain position in the water column) and concentration of permethrin or cypermethrin in rainbow trout brain has been demonstrated (Glickman and Lech, 1982; Edwards *et al.*, 1986). As permethrin (type I) and cypermethrin (type II) produce similar signs of toxicity, the type I/type II system of pyrethroid classification may not be relevant for fish (Edwards *et al.*, 1986).

Type I Pyrethroids

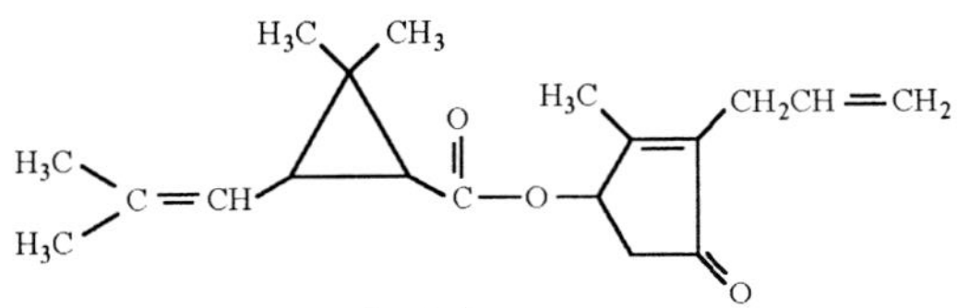

Allethrin

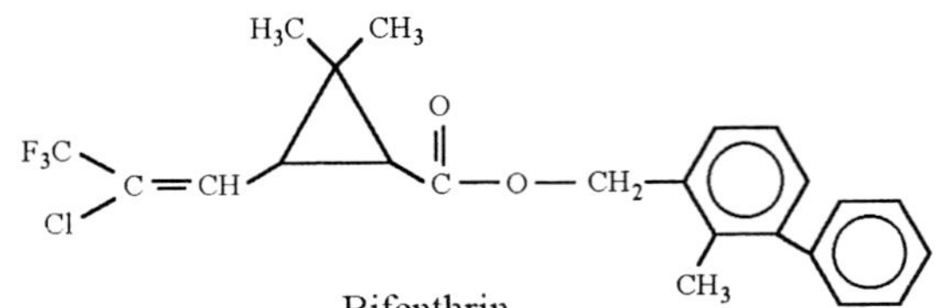

Bifenthrin

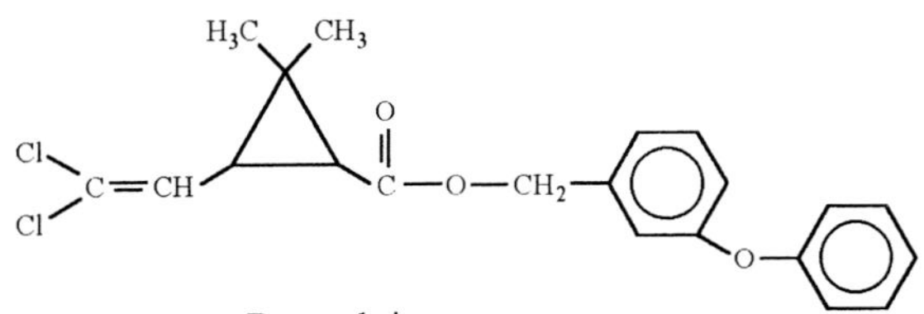

Permethrin

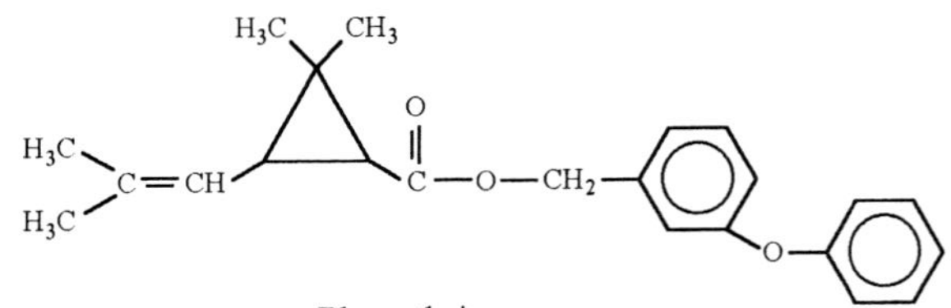

Phenothrin

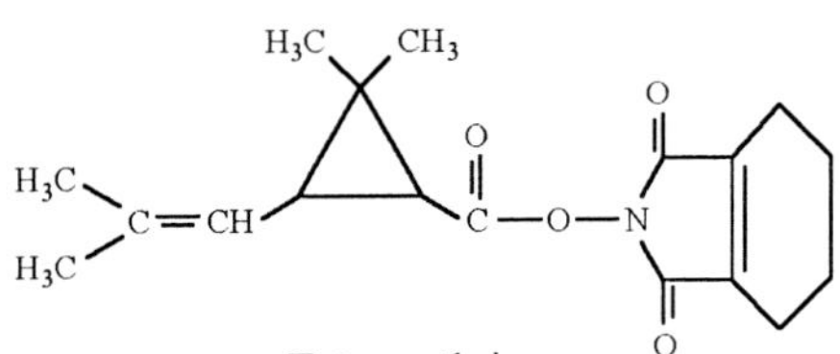

Tetramethrin

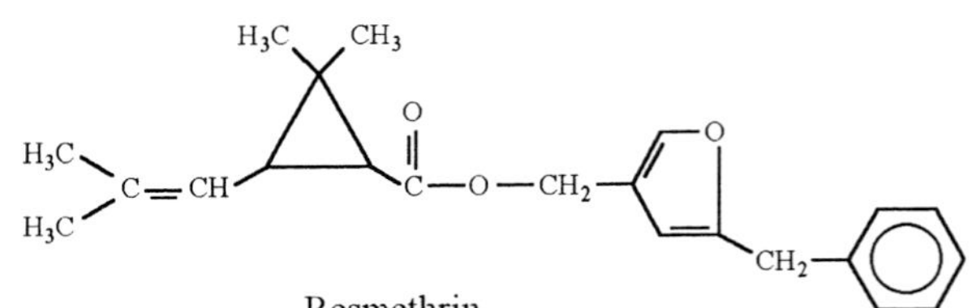

Resmethrin

Type II Pyrethroids

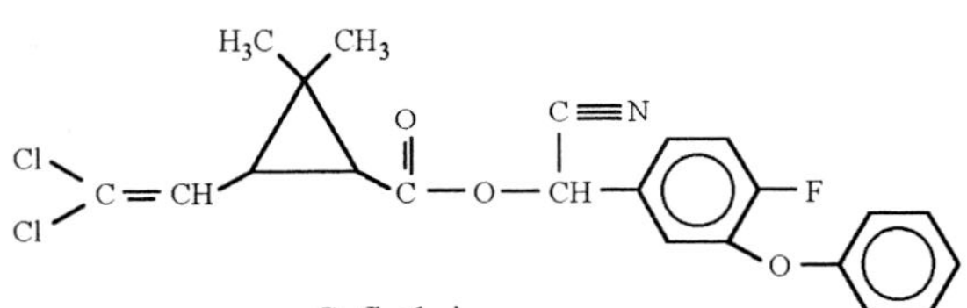

Cyfluthrin

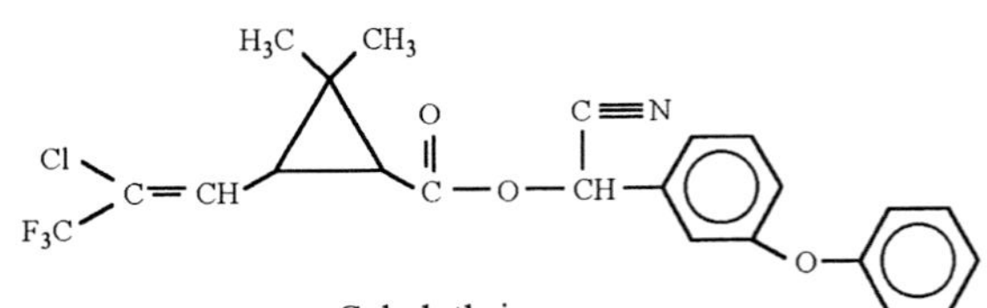

Cyhalothrin

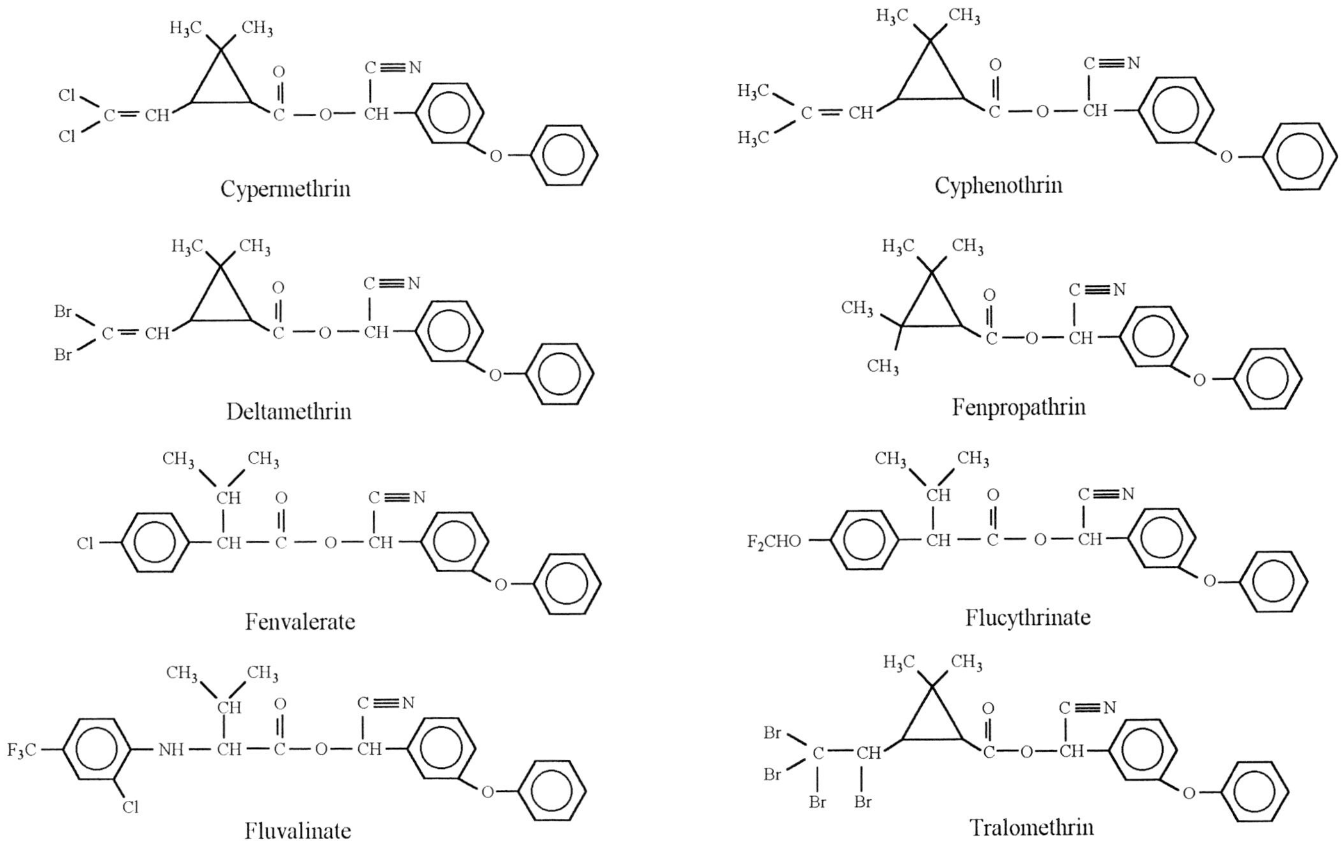

Figure 2.7 Chemical structures of some pyrethroid insecticides.

Deltamethrin did not result in Na^+ uptake into rainbow trout synaptoneurosomes, but in the presence of the Na^+ channel activators VTD and BTX it did increase the amount of Na^+ uptake into rainbow trout synaptosomes (Stuart *et al.*, 1987) and into rainbow trout synaptoneurosomes (Rubin and Soderlund, 1992). As described previously, rainbow trout brain Na^+ channels exhibit a much lower sensitivity to both activators (BTX, VTD, and ACN) and compounds which prolong opening (α-ScV and ATX) than do rat brain Na^+ channels. However, both permethrin (type I) and deltamethrin and cypermethrin (type II) increase the VTD-induced uptake of the lipophilic cation [^{3}H]-tetraphenylphosphonium through the Na^+ channel of rainbow trout brain synaptosomes to a much greater extent than in rat brain synaptosomes (Eells *et al.*, 1993). This suggests that the Na^+ channel of rainbow trout is much more sensitive to the effects of pyrethroids than that of mammals and that this differential may be the basis for their greater susceptibility to the neurotoxic actions of pyrethroids.

As stated earlier, the stimulation of Na^+ uptake by ACN is inhibited by deltamethrin in mouse (Rubin and Soderlund, 1992) and by permethrin, deltamethrin, and cypermethrin in rat (Eells *et al.*, 1993), but in rainbow trout the stimulation of Na^+ uptake by ACN is increased by these pyrethroids. Additionally, *cis*-permethrin was toxic to both mice and rainbow trout, but *trans*-permethrin is toxic only to rainbow trout (Glickman and Lech, 1982). This toxicity difference was present even when all metabolizing enzymes were inhibited in the mouse. When correlating the toxic signs of pyrethroid exposure with the level of pyrethroid in the brain of rainbow trout, mice and other species, the level required to induce toxic signs in fish was significantly lower than that required to induce signs in the other species (Glickman and Lech, 1982; Edwards *et al.*, 1986). These data further suggest that differences exist between mammalian Na^+ channels and fish Na^+ channels and that these differences may play an important role in the higher susceptibility of fish to pyrethroids.

Pyrethroids have been demonstrated to be good inhibitors of Ca^{2+},Mg^{2+}- and Ca^{2+}-ATPases in squid nerves (Clark and Matsumura, 1982). Type II pyrethroids are better inhibitors of Ca^{2+},Mg^{2+}-ATPase, whereas type I are better inhibitors of Ca^{2+}-ATPases. As described earlier, Ca^{2+},Mg^{2+}-ATPase is thought to be identical to the Ca^{2+} ion pump that maintains proper low internal Ca^{2+} concentration and the gradient across membranes (Shankland, 1983), and Ca^{2+}-ATPase is thought to represent the ATP-dependent phosphorylation and dephosphorylation system associated with Na^+/Ca^{2+} exchange (Matsumura and Clark, 1982). Pyrethroids can also decrease the Ca^{2+}-carrying ability of calmodulin in rat brain (Rashatwar and Matsumura, 1984; Enan and Matsumura, 1993). This effect on Ca^{2+} maintenance could have an impact on the nervous system. DDT has been shown to alter Ca^{2+} maintenance and to cause spontaneous release of neurotransmitters. Likewise, type II pyrethroids have been reported to stimulate the release of several neurotransmitters, including GABA (Matsumura, 1985) and norepinephrine (Clark and Brooks, 1989). This spontaneous release could be the result of a pyrethroid-induced increase in internal Ca^{2+} levels, thereby stimulating neurotransmitter release. However, as with DDT, it is not known whether pyrethroids will disrupt

Ca^{2+} maintenance in the nervous tissue of teleosts and, if it does, what role in toxicity this will play.

It has been suggested that some of the toxicity of the type II pyrethroids is mediated through binding to the GABA receptor in rat brain (Lawrence and Casida, 1983). However, it has been demonstrated that cyclodiene-resistant mosquitofish, which possessed an altered picrotoxin/TBPS site as the basis for their resistance, were not resistant to the toxic effects of the type II pyrethroid cypermethrin (Bonner and Yarbrough, 1989). However, Eshleman and Murray (1990) demonstrated that both type I (permethrin, kadethrin and allethrin) and type II (cypermethrin and deltamethrin) pyrethroids inhibit the binding of [^{3}H]-TBPS in rainbow trout brain. In addition, both type I (permethrin) and type II (deltamethrin and cypermethrin) pyrethroids have been shown to inhibit GABA-stimulated $^{36}Cl^-$ flux into rainbow trout brain synaptoneurosomes (Eshleman and Murray, 1991). However, these pyrethroids increased the basal flow of $^{36}Cl^-$ inward in the absence of GABA. Using deltamethrin as a model compound, it was found that the Na^+ channel blocker TTX blocked the increased basal $^{36}Cl^-$ influx caused by deltamethrin and eliminated the inhibition of GABA-stimulated $^{36}Cl^-$ influx induced by deltamethrin. Thus, it appeared that pyrethroid insecticides may interfere with the function of $GABA_A$ receptor indirectly, through action on the Na^+ channels. It was further found that the Na^+ channel agonist VTD produced similar inhibition of GABA-stimulated $^{36}Cl^-$ influx and enhanced basal flow of $^{36}Cl^-$, as was observed with deltamethrin. These actions could also be blocked by TTX, leading to the hypothesis that the pyrethroids are acting on the Na^+ channel to increase Na^+ flow inward, causing the opening of voltage-gated Cl^- channels. This opening would account for the increase in basal flow of $^{36}Cl^-$ into the synaptoneurosome which destroys the Cl^- gradient such that activation of the $GABA_A$ receptor by GABA would open the ionophore, but there would be no gradient present to force Cl^- to flow. Thus, the lack of an inhibitory response from the GABAergic system would potentiate the effects of the action of pyrethroids on the Na^+ channel.

It has also been proposed that an alternative mechanism of toxicity with pyrethroid insecticides may occur through destruction of the secondary lamella of the gill (Leahey, 1985). Fenvalerate has been shown to affect respiratory surfaces and renal ion regulation during acute exposure in rainbow trout (Bradbury *et al.*, 1987). However, the increased muscular activity and intense seizures associated with pyrethroid toxicity in fish suggest that these signs are due to disruption of the CNS. Although many physiologic signs in pyrethroid-intoxicated fish suggest a high oxygen demand that would suggest decreased oxygen uptake efficiency at the gills, these may be the result of the increased activity and convulsions which would increase oxygen demand (Bradbury *et al.*, 1991). Similarly, the pyrethroid-induced convulsions in mammals cause similar oxygen demand-related physiologic responses (Bradbury and Coats, 1989). It may be feasible to assume that the acute toxicity of pyrethroids to fish can be attributed mainly to the effects on the CNS, but it is not totally clear what role disruption of the respiratory membrane plays in acute toxicity. This disruption may be more important in chronic toxicity. During long-term sublethal exposure, disruption of the integrity of the gills has been reported in rainbow trout exposed to permethrin in the water and in the food (Kumaraguru *et al.*, 1982).

Another proposed effect of pyrethroid exposure is the inhibition of AChE. *In vitro* and *in vivo* inhibition of rat brain AChE has been reported by permethrin and cypermethrin (Rao and Rao, 1995). Cypermethrin has been reported to inhibit brain AChE in *Tilapia mossambica*, but the time of inhibition was relatively short (Reddy *et al.*, 1991). Fenvalerate (Reddy *et al.*, 1992) and cypermethrin (Reddy and Phillip, 1994) produced AChE inhibition in the carp, with inhibition increasing as exposure time increased. During two separate incidents of deltamethrin exposure in a large lake, the AChE of freshwater eel was inhibited compared with eels from other non-affected regions of the lake (Bálint *et al.*, 1997). However, sublethal exposure of deltamethrin to carp induced characteristic signs of toxicity, but did not result in significant AChE inhibition (Szegletes *et al.*, 1995). As not all pyrethroids inhibit AChE, it is unlikely that the inhibition of AChE is a primary mechanism of toxicity in fish. However, the inhibition of AChE by a pyrethroid in combination with its effect on the Na^+ channels could very well potentiate the toxicity of that pyrethroid. This could possibly be a determining factor in toxicity differences among different pyrethroids.

Anticholinesterase insecticides

Historical perspective

The organophosphorous (OP) insecticides were first developed in Germany in the 1930s by Gerhard Schrader at Farbenfabriken Bayer AG (Matsumura, 1985; Chambers, 1992; Ecobichon, 1996). Although OPs were developed as part of the nerve gas research ongoing at that time, Schrader was more interested in insecticides and was responsible for identification of the insecticidal properties of dimefox and the first systemic OP insecticide octamethylpyrophosphoramidate (OMPA) in 1941 and for the first commercial OP insecticide tetraethyl-pyrophosphate (TEPP) in 1942. However, TEPP was highly toxic and hydrolyzed easily in the presence of moisture; further development of the OPs as insecticides continued with the development of parathion in 1944. Although the OPs are much more toxic than the organochlorine insecticides, they are environmentally labile, and thus do not possess the problems with environmental stability that are seen with the organochlorines. Thus, in the 1950s and 1960s, they replaced many of the organochlorines, and by 1970 they became the dominant insecticide class. Although the introduction of the synthetic pyrethroids has led to an appreciable decrease in the use of OP insecticides, they are still widely used agriculturally and domestically throughout the world.

OPs generally contain four atoms attached to a phosphorus atom, three by single bonds and one by a coordinate covalent bond (which is usually expressed as a double bond). Based on their reactivity with the target once in an organism, OPs can be classified into two categories. The first category, such as the phosphates (Figure 2.8), have an oxygen atom attached to the phosphorus atom by a coordinate covalent bond. These are highly reactive and are capable of attacking a toxicologic target (i.e. AChE) in their original form. The second category, the phosphorothionates (Figure 2.9), have a sulfur atom attached to the phosphorus atom by a co-ordinate covalent bond. These must be metabolically activated by a cytochrome P450 desulfuration reaction to their P=O metabolite to be effective anticholinesterases.

The carbamate insecticides are synthetic derivatives of the anticholinesterase physostigmine, which is the principal alkaloid of the calabar bean plant (*Physostigma venenosum*) (Matsumura, 1985). Synthesized in the 1930s and used as fungicides, these analogues were too polar to penetrate the cuticle of an insect. Thus, there was not much interest in developing their possible insecticide potential. In the 1950s, the need for anticholinesterase insecticides which had lower mammalian toxicity stimulated the development of new analogues (Figure 2.10) which had polar moieties removed and allowed penetration of the insect cuticle. The carbamates are used agriculturally and domestically.

Signs of toxicity

Initially after exposure of fish to toxic levels of anticholinesterases, there are increased ventilation rates and amplitudes (McKim *et al.*, 1987). There is increased hyperactivity, overreaction to stimuli, spiral swimming, and loss of normal schooling behavior. Soon after, tremors are observed which progress to violent seizures, including head shaking and twisting. During this time, the opercula become flared. Additionally, piping may occur, even though dissolved oxygen is high. Finally, a point is reached where the fish becomes inactive and unable to maintain its position in the water column, which is soon followed by death.

Mechanism of action

Anticholinesterase insecticides, or their metabolites in the case of those requiring bioactivation, react covalently with the serine hydroxyl group located in the active site of AChE. *In vivo* exposure of teleosts to anticholinesterase pesticides results in inhibition of AChE in both the PNS and CNS of teleosts (Weiss, 1958; Murphy *et al.*, 1968; Guilbault *et al.*, 1972; Benke and Murphy, 1974; Benke *et al.*, 1974; Coppage and Matthews, 1974; Coppage *et al.*, 1975; Coppage and Braidech, 1976; Thirugnanam and Forgash, 1977; Goodman *et al.*, 1979; Jarvinen *et al.*, 1983; Ghosh, 1987; Subburaju and Selvarajan, 1988; Carr *et al.*, 1995, 1997; Straus and Chambers, 1995; Boone and Chambers, 1996). The inhibition of AChE results in the accumulation of ACh in the synapse, which overstimulates the cholinergic system. The signs of toxicity associated with anticholinergic poisoning in fish can be correlated with this overstimulation. In fish, as in mammals, there are differences in toxicity among different anticholinesterases within the same animal species (Ghosh, 1987; Boone and Chambers, 1996, 1997; Carr *et al.*, 1995). Conversely, there are differences in the toxicity of a single anticholinesterase among several species of fish (Linn, 1968; Murphy *et al.*, 1968; Washino *et al.*,

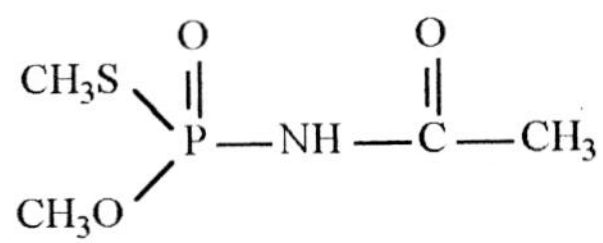

Acephate

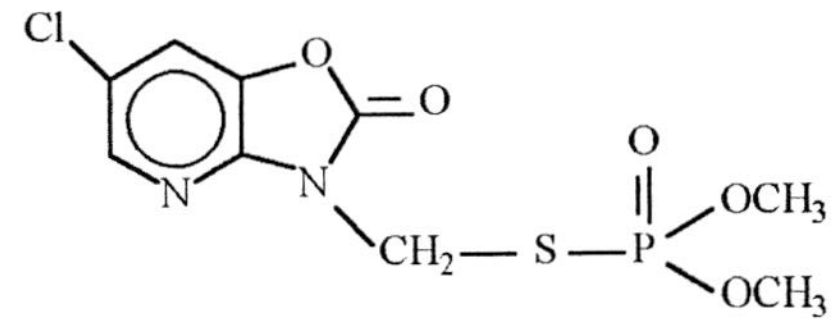

Azamethiphos

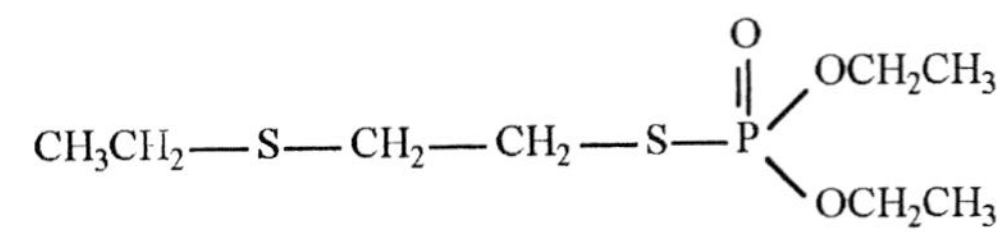

Demeton II (O)

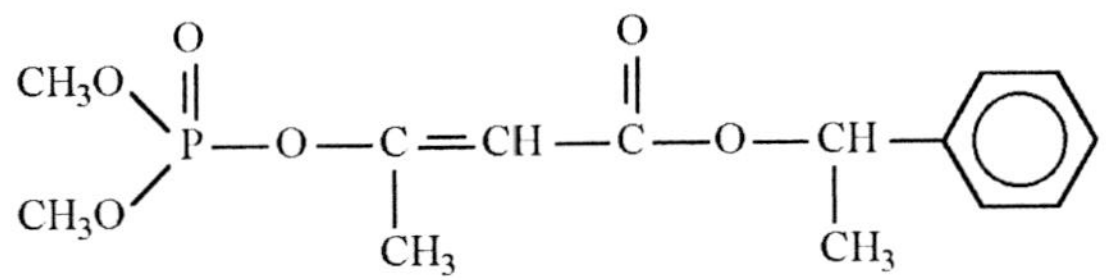

Crotoxyphos

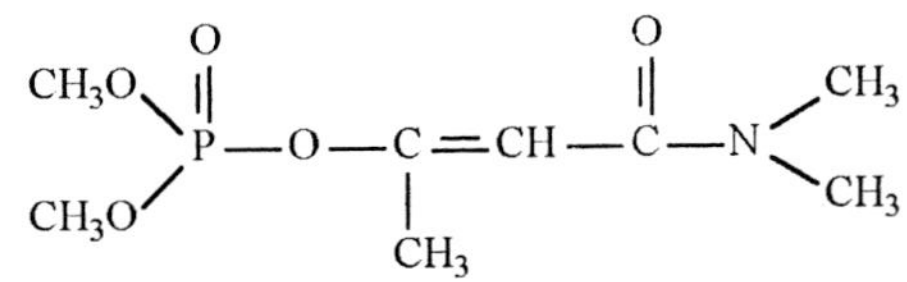

Dicrotophos

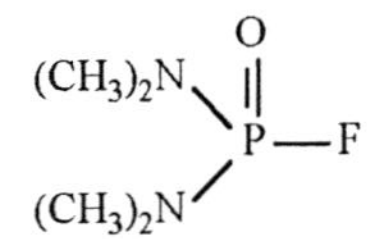

Dimefox

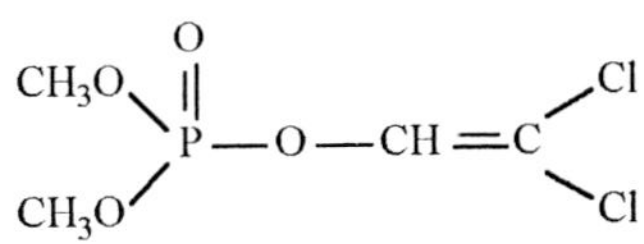

Dichlorvos

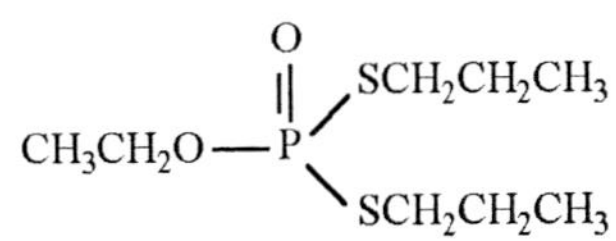

Ethoprophos

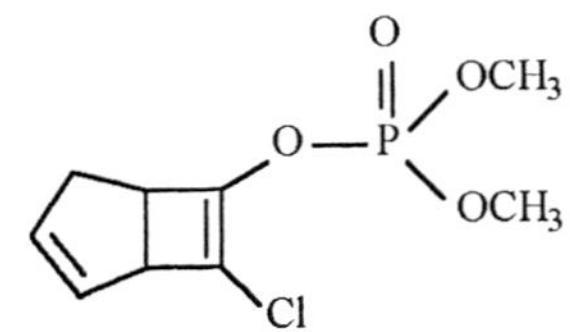

Heptenophos

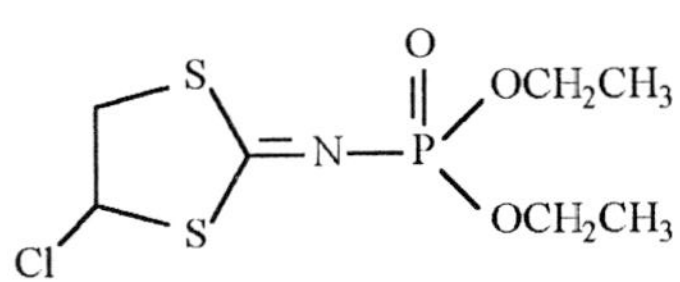

Mephosfolan

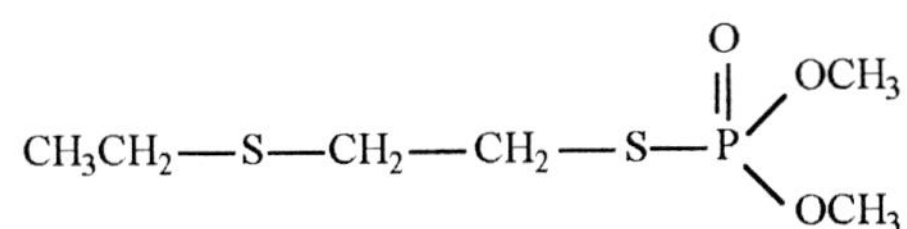

Methyl Demeton

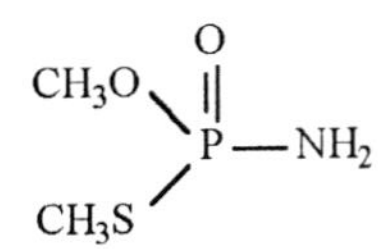

Methamidophos

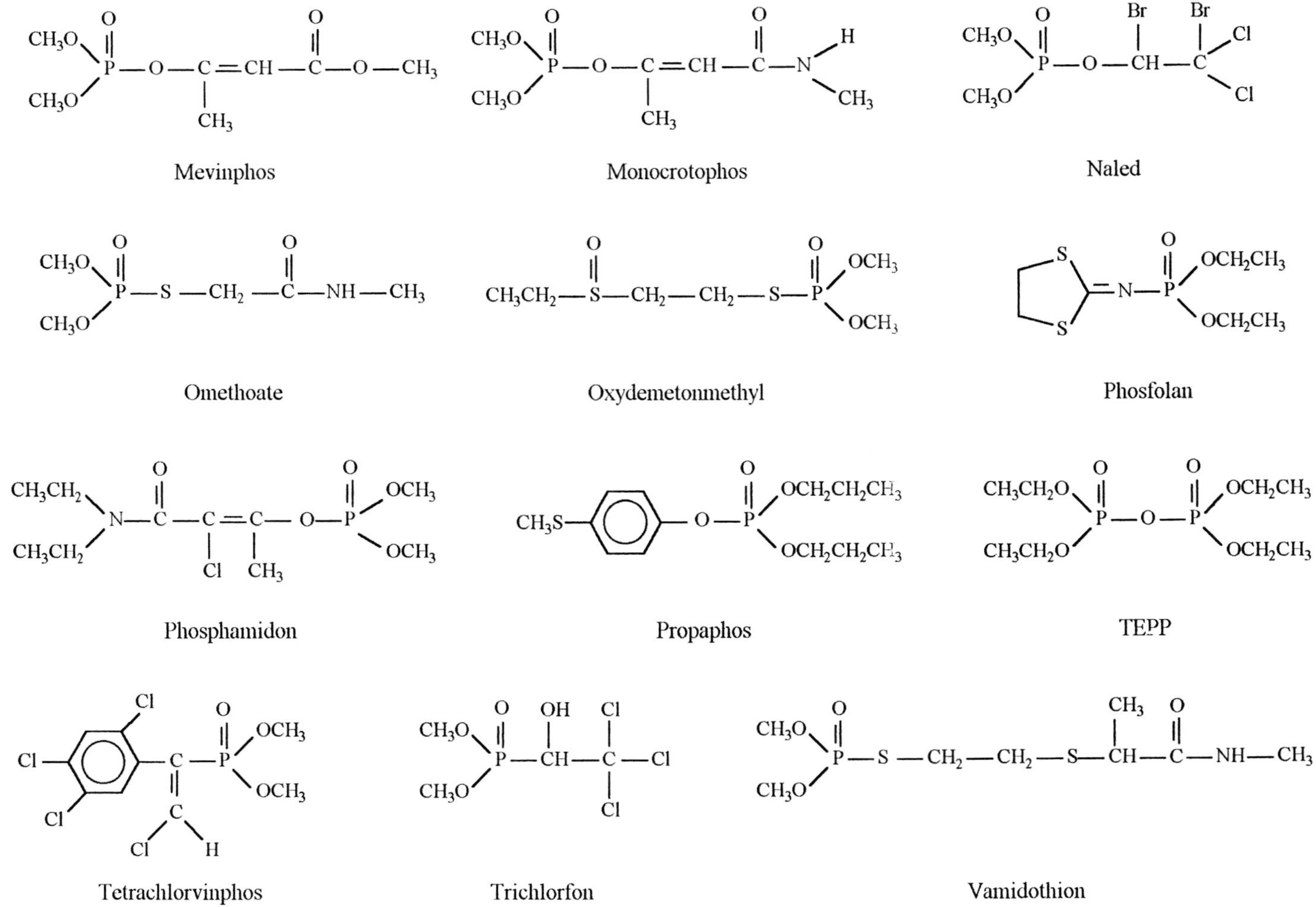

Figure 2.8 Chemical structures of some phosphate insecticides.

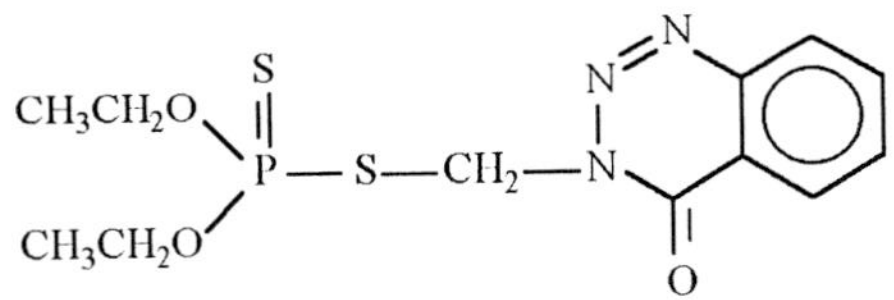

Azinphos-ethyl

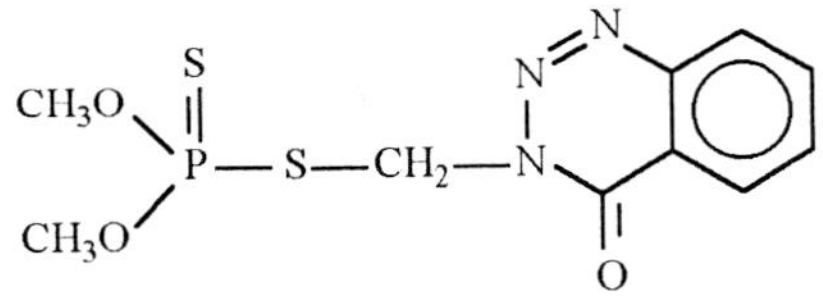

Azinphos-methyl

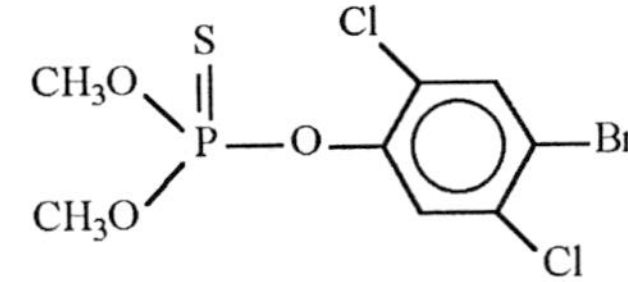

Bromophos

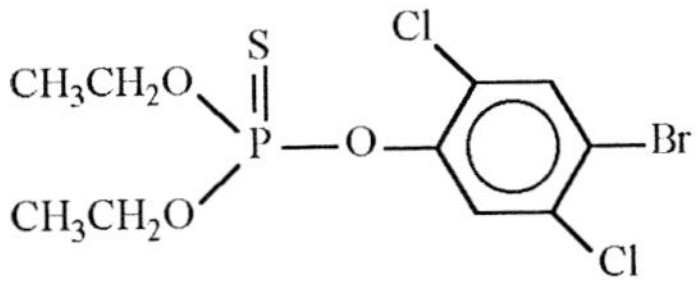

Bromophos-ethyl

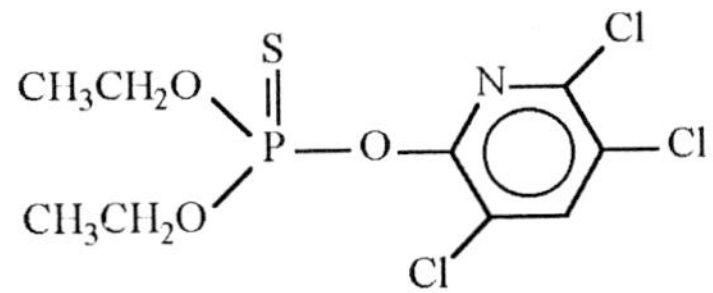

Chlorpyrifos

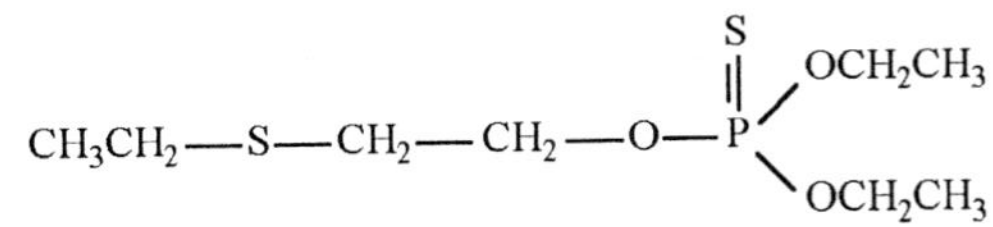

Demeton I (S)

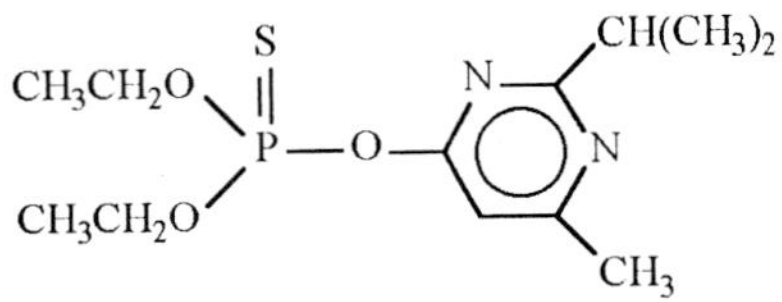

Diazinon

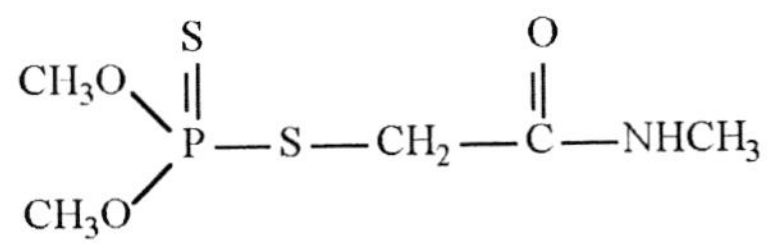

Dimethoate

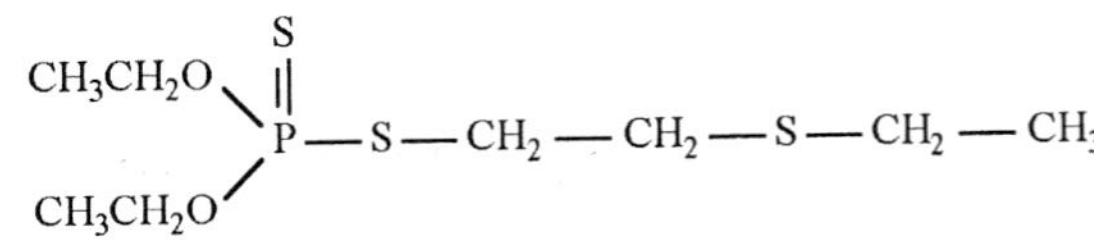

Disulfoton

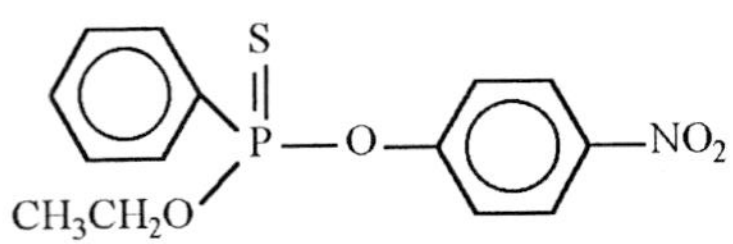

EPN

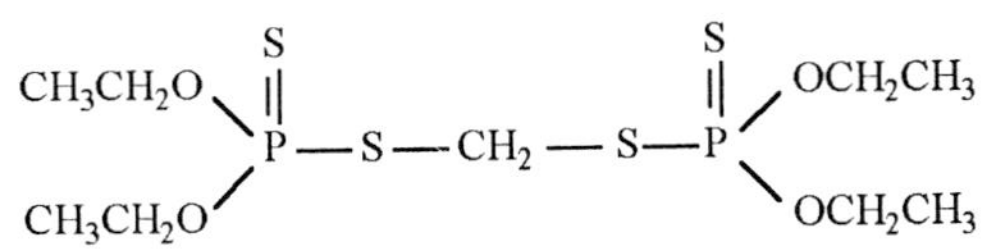

Ethion

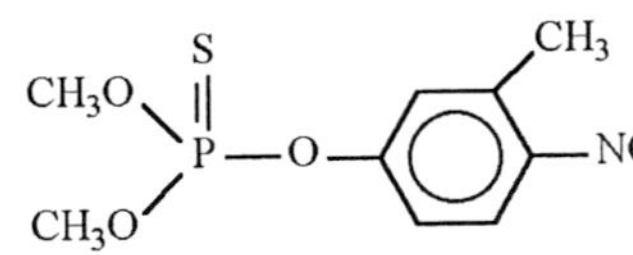

Fenitrothion

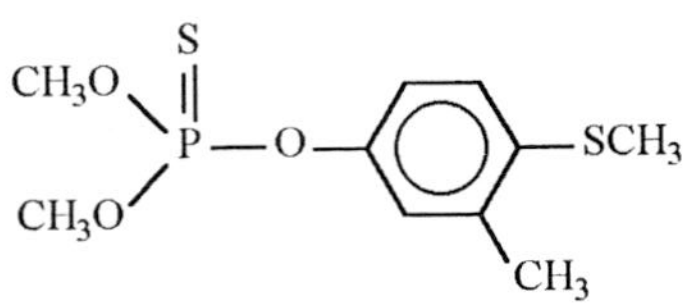

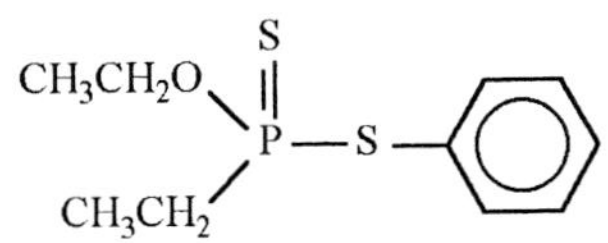

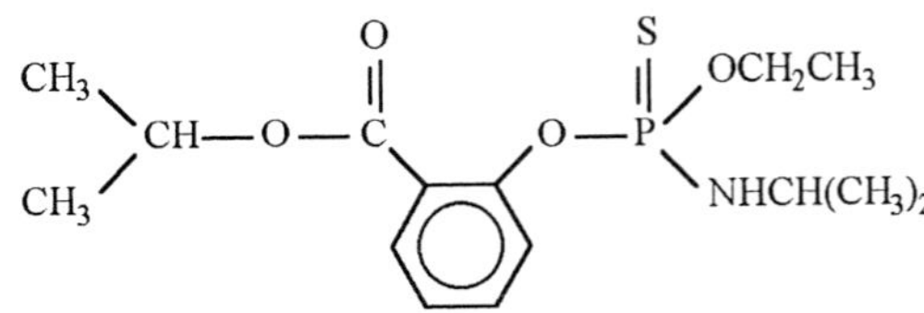

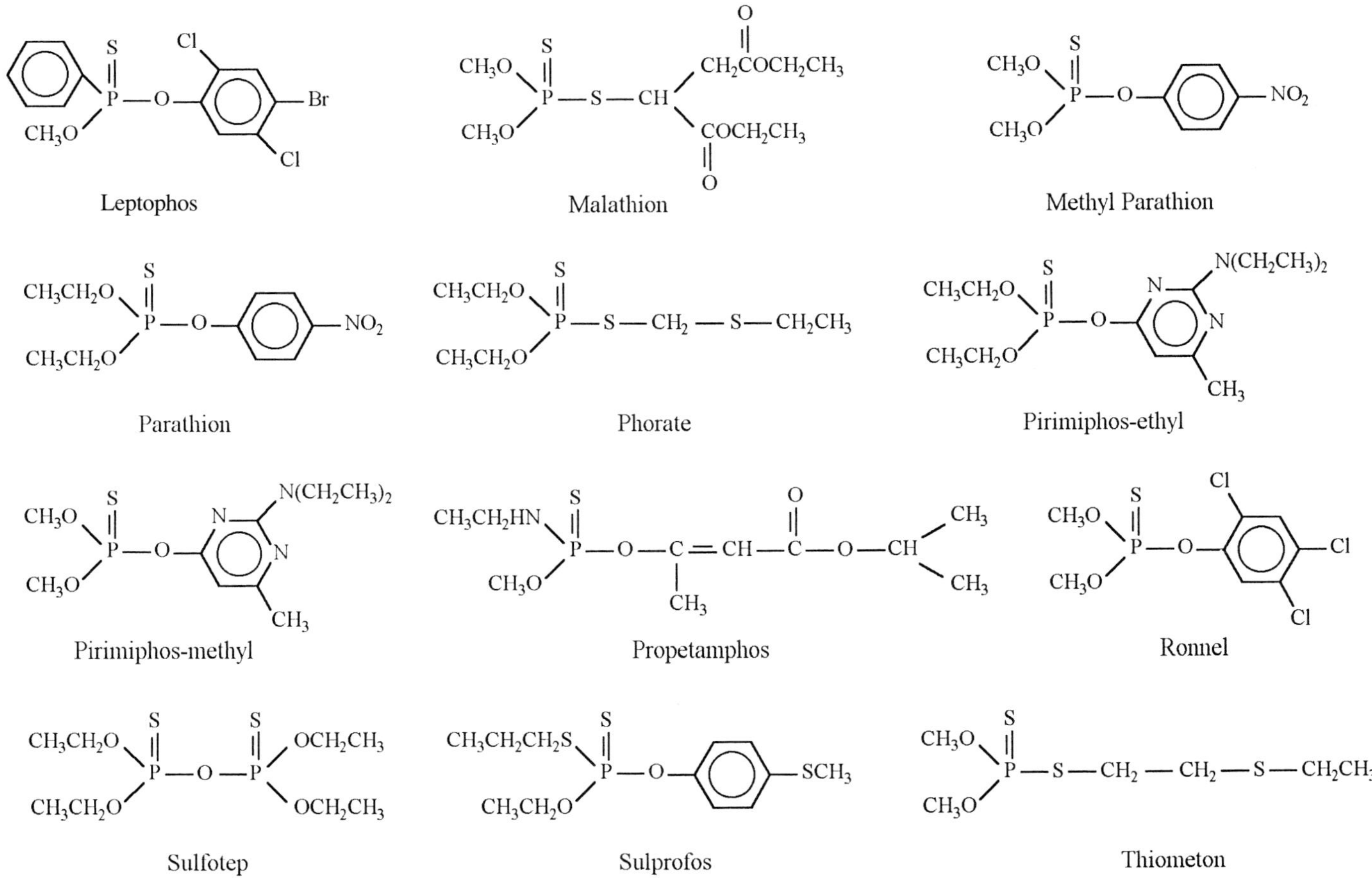

Figure 2.9 Chemical structures of some phosphorothionate insecticides.

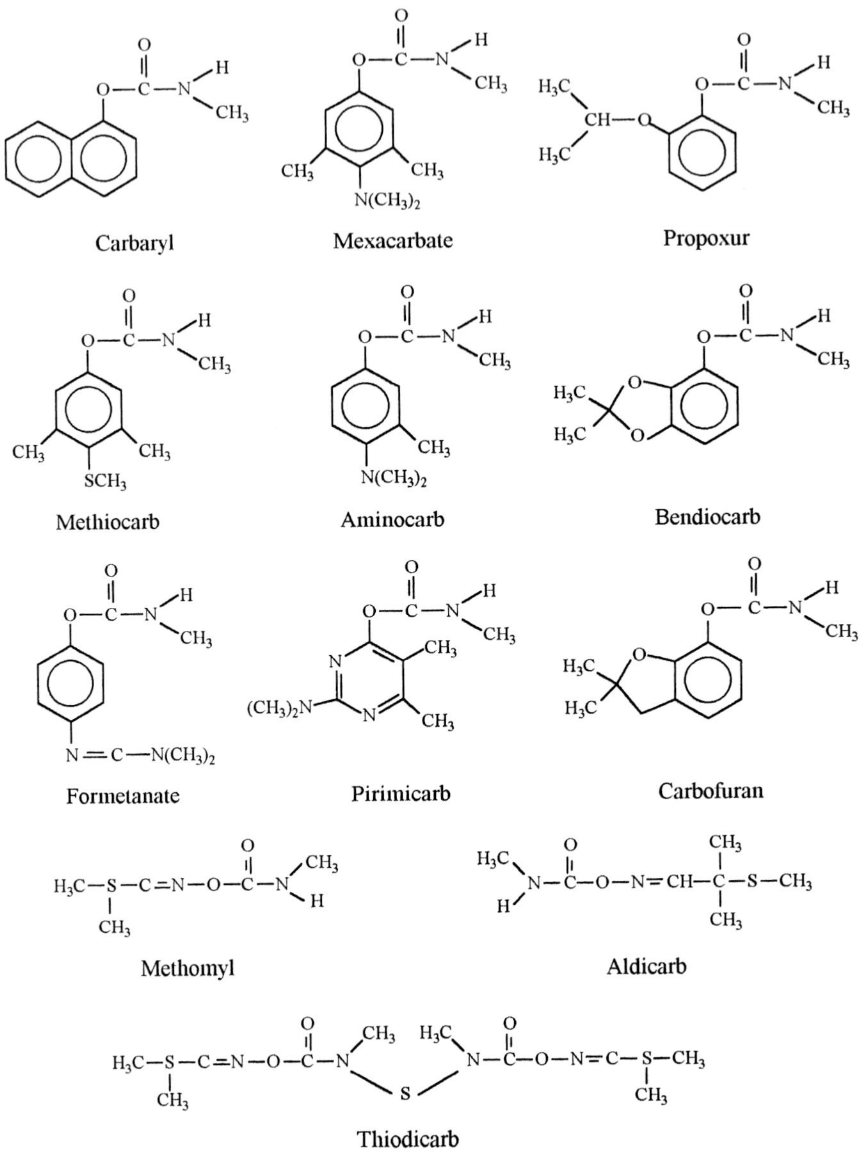

Figure 2.10 Chemical structures of some carbamate insecticides.

1972; Davy *et al.*, 1976; Shoa-Non and De-Fang, 1996; Carr *et al.*, 1997). Anticholinesterase biotransformation, bioconcentration, uptake, and AChE sensitivities may all be species specific, indicating dispositional processes may play a significant role in susceptibility to anticholinesterases (de Bruijn and Hermens, 1991; Keizer *et al.*, 1993, 1995; Shoa-Non and De-Fang, 1996).

The ultimate cause of lethality in fish exposed to an anticholinesterase insecticide is uncertain. In mammals, the ultimate cause of death is asphyxiation resulting from AChE inhibition. However, the respiratory exchange system in fish is obviously significantly different from mammals, such that the impact of an anticholinesterase in fish might not be similar to that observed in mammals. The adverse effects of anticholinesterases on the respiratory system of fish have been demonstrated in many species of fish (Bansal *et al.*, 1979; Rath and Misra, 1980; McKim *et al.*, 1987; Bradbury *et al.*, 1991). Zinkl *et al.* (1991) proposed that the probable cause of death in fish was the result of asphyxiation as a result of a combination of factors, including paralysis of muscles required for movement of water through the gills, bradycardia induced by hypercholinergic activity, and increased branchial resistance, shunting blood through filamentous sinuses. As regulation of all these factors is under the control of the cholinergic system, disruption of that system by inhibition of AChE could lead to impaired regulation of processes involved in respiration such as blood flow and ventilation patterns (McKim *et al.*, 1987; Pavlov, 1994). Some researchers have proposed that the effects on respiration result from irritation or other effects on the respiratory surfaces of the gill (Lunn *et al.*, 1976; Carlson, 1990).

Following exposure to anticholinesterases, AChE activity in fish tissues can be virtually eliminated without any resultant mortality (Carr *et al.*, 1995; Straus and Chambers, 1995; Boone and Chambers, 1996). This leads to questioning the importance of AChE inhibition in the toxicity of anticholinesterases in fish. However, Pavlov *et al.* (1992) reported that exposure of bream (*Abramis brama*) to sublethal concentrations of the OP insecticide dichlorvos resulted in inhibited brain AChE and decreased amount of food consumed. When the bream were injected with either the muscarinic receptor blocker atropine or the AChE oxime reactivator TMB-4 (the two drugs used as therapy for OP poisoning in mammals), feeding efficiency recovered. Pavlov (1994) also reported that sublethal dichlorvos exposure in perch decreased oxygen consumption and injection with atropine and the nicotinic receptor blocker pediphen resulted in the recovery of oxygen uptake by the perch. The ability of atropine and pediphen to reverse the signs of anticholinesterase toxicity suggests that, regardless of the ultimate cause of toxicity of anticholinesterases in fish, inhibition of AChE is an integral event in the toxicity of these compounds to fish. Alternative mechanisms of action include inhibition of other critical enzymes resulting in tissue necrosis (Nemcsók *et al.*, 1987), oxidative stress (Hai *et al.*, 1995, 1997), impairment of gas transfer (Hughes *et al.*, 1997), disruption of hormonal status (Guhathakurta and Bhattacharya, 1988), induced gluconeogenesis and cholesterol synthesis (Prasada Rao and Ramana Rao, 1984a), and inhibition of active transport and oxidative phosphorylation (Prasada Rao and Ramana Rao, 1984b). However, it is likely that the muscular and neural disturbances resulting from the AChE inhibition would be more critical in determining the ultimate toxicity of anticholinesterases than these non-cholinergic effects, which may play a role in chronic anticholinesterase toxicity (Hughes *et al.*, 1997).

It is not clear whether impacting teleost AChE in the PNS or the CNS would play a more important role in the mediation of the toxic effects. Survival can occur with greater than 90 percent AChE inhibition in the brain (Ansari and Kumar, 1984; Carr *et al.*, 1995, 1997, Straus and Chambers, 1995). However, some researchers have suggested that peripheral AChE inhibition is more important in anticholinesterase activity (Boone and Chambers, 1996). In contrast, Carr *et al.* (1997) suggested that the inhibition of brain AChE may be more important in the acute toxicity of chlorpyrifos and that skeletal muscle AChE may serve as an alternative binding site for the toxicant, preventing it from reaching the brain. In this field study, significant species differences in lethality were observed following exposure to chlorpyrifos in a residential pond. Bluegill sunfish, largemouth bass (*Micropterus salmoides*), and golden shiner (*Notemigonus crysoleucas*) were either dead or near death, whereas mosquitofish, although exhibiting signs of intoxication, survived. The brain AChE inhibition in the three susceptible species was very high (> 90 percent), while that in mosquitofish was much lower. Skeletal muscle AChE inhibition in mosquitofish was maximally inhibited. Brain AChE in the mosquitofish was less sensitive to inhibition by the active metabolite of chlorpyrifos, chlorpyrifos-oxon, than AChE of the three susceptible species. In contrast, mosquitofish skeletal muscle AChE was much more sensitive than AChE of the three susceptible species. The greater sensitivity of skeletal muscle AChE of the mosquitofish may have allowed it to be more efficient at providing an alternative binding site than the other species, thereby affording more protection.

Recovery

Although carbamates can significantly inhibit AChE, recovery of AChE activity is rapid once the fish is moved to uncontaminated water (Mukhopadhyay *et al.*, 1982; Szeto *et al.*, 1985). In contrast, recovery is much slower in inhibition of AChE by an OP (Koelle and Gilman, 1949; Benke and Murphy, 1974; Post and Leasure, 1974; Coppage *et al.*, 1975; Sanders *et al.*, 1981; Jash and Bhattacharya, 1983; Ansari and Kumar, 1984; van der Wel and Welling, 1989; Zinkl *et al.*, 1991; da Silva *et al.*, 1993; Carr *et al.*, 1995, 1997; Thangnipon *et al.*, 1995: Boone and Chambers, 1996; Sancho *et al.*, 1997). In some fish species, the basis for slow recovery following OP inhibition appears to be the inability of AChE in the brain to reactivate spontaneously. In others, AChE will spontaneously reactivate, but at a very slow rate compared with non-teleost species. Wallace and Herzberg (1988) and Carr *et al.* (1997) have demonstrated that brain AChE in rainbow trout and channel catfish, respectively, does not reactivate following *in vitro* inhibition by paraoxon and chlorpyrifos-oxon. However, fathead minnow brain AChE inhibited *in vitro* by paraoxon does reactivate at a 11- to 15-fold slower rate than that of mammals (Wallace and Herzberg, 1988). Thus, following inhibition of fish AChE during an OP exposure, recovery of AChE activity to control levels would have to occur through *de novo* synthesis of AChE.

Other factors

Organophosphates/carbamates have been shown to bind directly to nicotinic

(Eldefrawi *et al.*, 1971, 1988; Kuba *et al.*, 1973; Aracava *et al.*, 1987; Bakry *et al.*, 1988) and muscarinic (Volpe *et al.*, 1985; Bakry *et al.*, 1988; Jett *et al.*, 1991; Katz and Marquis, 1989) receptors in mammalian brain. Additionally, many of these studies have demonstrated alteration of second messenger functions in mammalian brain. However, there is currently no information on the ability of OPs or carbamates to bind to cholinergic receptors or disrupt second messenger systems in fish tissue.

Metals

Many heavy metals are normal components of the aquatic environment brought about by the natural weathering of rocks. However, the increased discharge of wastes from industrial, agricultural, and mining activities have pushed the levels of heavy metals well beyond normal trace amounts. As metals can interact with a variety of biochemical pathways and cellular macromolecules, the increased metal load in aquatic systems would most certainly impact the inhabitants at all trophic levels. Although is a large database detailing the metal residues present in fish caught in contaminated waters and the concentrations of various metals which result in mortality in fish, the effects of metals on fish CNS are uncharacterized.

Mercury

Mercury exists as several inorganic forms [i.e. metallic, mercurous (Hg_2Cl_2), and mercuric ($HgCl_2$)], and as alkylated organic forms (i.e. methylmercury). Each form of mercury has unique toxicokinetic and adverse effects. Generally, mercury and its derivatives are thought to bind effectively to sulfur and disulfide groups, inactivating critical proteins and replacing the trace element zinc. Consequently, structural alteration of proteins, inhibition of enzymes, binding to nucleic acids, and disruption of cellular membranes in often observed in mercury toxicity (Baatrup, 1991). Mercury is highly toxic to fish and impacts the fish nervous system in various areas. The majority of research on mercury toxicity in fish nervous tissue has focused on the effects of $HgCl_2$ and methylmercury.

$HgCl_2$ disrupts taste receptor binding (Zelson and Cagan, 1979), suppresses taste receptor activity (Hidaka, 1970), and causes morphologic degeneration of the system (Pevzner *et al.*, 1986). $HgCl_2$ has also been demonstrated to disrupt binding to olfactory receptors (Sutterlin and Sutterlin, 1971; Rehnberg and Schreck, 1986) and to generally disrupt olfaction (Hara *et al.*, 1976; Baatrup *et al.*, 1990). Likewise, methylmercury has been shown to disrupt olfactory function (Baatrup *et al.*, 1990), but accumulates within the receptor cells rather than at the borders of neighboring non-receptor cells as does $HgCl_2$ (Baatrup and Døving, 1990). Disruption of olfactory function caused by methylmercury was not reversible, whereas that caused by $HgCl_2$ was reversible (Baatrup *et al.*, 1990). It is thought that $HgCl_2$ acts on the Ca^{2+} gate of the olfactory receptor while methylmercury affects the active transport of Ca^{2+} (Baatrup *et al.*, 1990).

Various forms of mercury also cause behavioral alterations in fish which may be explained by the sensory damage (Smith, 1984; Heath, 1995). Although mercury affects the mammalian central nervous system at many levels, few of these effects have been documented in fish. $HgCl_2$ inhibits monoamine oxidase in the brain of the snakeheaded fish (murrel) (*Channa punctatus*), potentially altering catecholamine levels in fish brain (Ram and Sathyanesan, 1985). $HgCl_2$ inhibits Na^+,K^+-ATPases in the snakeheaded fish (Sastry and Sharma, 1980) and in the knifefish (*Notopterus notopterus*) (Verma *et al.*, 1983), possibly altering signal transmission in neurons. Ca^{2+}-ATPase in the brain of channel catfish is also inhibited by $HgCl_2$ (Reddy *et al.*, 1988). As Ca^{2+} is very important in regulating cellular homeostasis in neurons (i.e. stabilization of the axonal membrane, second messenger systems, and ion channels), inhibition of Ca^{2+}-ATPase could lead to significant alteration of the nervous system. $HgCl_2$ has also been reported to lower AChE activity in the brain, gill, kidney, intestine, liver, and muscle of freshwater carp (Gill *et al.*, 1990; Suresh *et al.*, 1992), indicating altered cholinergic effect as well. Apparently, multiple targets exist for mercury, but what contribution these targets make in mercury-induced CNS toxicity is unclear.

Lead

Lead is ubiquitous and is found in almost all components of the environment. No known physiologic role for lead has been demonstrated. In mammals, lead accumulates in the gray matter and in certain nuclei of the central nervous system (Goyer, 1996). Lead functions to disrupt neurotransmitter release at the synapse, mimicking Ca^{2+} in ion-dependent events in the brain. Lead can also activate the Ca^{2+}-dependent second messenger protein kinase C in the brain, resulting in disruption of the maintenance of normal cellular metabolism (Markovac and Goldstein, 1988). Replacement of Ca^{2+} in calmodulin-dependent reactions, inhibition of Na^+,K^+-ATPase, and disruption of Ca^{2+} release from mitochondria have also been observed (Simons, 1986). Although there are few data available on the effects of lead on the biochemical systems in the teleost brain, similarities between the processes in the mammalian nervous system and that of fish suggest that lead could easily target fish nervous systems as well.

It is presumed that lead is not very toxic to fish because of limited uptake resulting from low solubility across and limited access to cell membranes (Baatrup, 1991). However, lead exposure has resulted in behavioral alterations in fish (Heath, 1995), and long-term exposure of lead in fish has caused alterations in brain neurotransmitter functions (Katti and Sathyanesan, 1986).

Copper

Copper has frequently been used therapeutically for the control of parasites in fish (Cardeilhac and Whitaker, 1988; Noga, 1996) and is an essential cofactor for many enzymes such as tyrosinase, cytochrome oxidase, superoxide dismutase, amine oxidases, and uricase. However, at high levels, copper is toxic to fish (Brown and Dalton, 1970; Smith and Heath, 1979; Alam and Maughan, 1995). The toxicity of copper in mammals has been attributed to disruption of cellular membranes, leading to apoptosis via inactivation of sulfhydryl groups by free radicals formed during the H_2O_2-mediated conversion of Cu^+ to Cu^{2+} (Kumar *et al.*, 1978; Cataldi de Flombaum and Stoppani, 1986; Kanazawa *et al.*, 1994; Held *et al.*, 1996).

Although not a 'traditional neurotoxicant', copper affects nervous tissues in fish and causes behavioral alterations (Smith, 1984; Heath, 1995). In the brain of non-teleost fish, copper has been shown to cause mitochondrial degeneration (Toraro *et al.*, 1986) and decreased the amount of Golgi apparatus (Enesco *et al.*, 1989). This type of damage could disrupt the metabolic processes in the brain and could result in behavioral alterations and possibly mortality. Copper also damages fish sensory organs including those associated with sight, smell, and taste (Gardner and LaRoche, 1973; Baatrup, 1991).

Additionally, copper has been shown to inhibit AChE in the brain and muscle of fish (Verma *et al.*, 1981; Nemcsók and Hughes, 1988; Szabó *et al.*, 1992). While inhibition of AChE could cause some behavioral alterations and result in toxicity, it is unlikely that the ultimate target for copper toxicity is inhibition of AChE. Overall, the mechanistic basis for behavioral alterations and toxicity of copper in fish is open to conjecture.

Cadmium

Cadmium is traditionally not considered to be neurotoxic in fish but has been reported to disrupt behavior (Smith, 1984; Heath, 1995). Various CNS effects have been observed in mammalian species but very little has been reported in teleosts. Brown trout (*Salmo trutta*) living in cadmium- and zinc-contaminated waters have been reported to demonstrate evidence of chronic stimulation of the hypothalamo-pituitary-inter-renal axis of the brain, disrupting the neuroendocrine control mechanisms and resulting in both hypertrophy and hyperplasia of the inter-renal cells in the kidney (Norris *et al.*, 1997). Cadmium has also been reported to stimulate AChE activity in the brain and skeletal muscle of the rosy barb (*Barbus conchonius*) but depress AChE in the gills (Gill *et al.*, 1991).

Cadmium alters the appearance and composition of the population of sensory dendrites on the olfactory lamellae in the carp (*Alburnus alburnus*) (Hernádi, 1993). In pike (*Esox lucius*) exposed to cadmium, cadmium is transported along the olfactory neurons and accumulates in the receptor cell-containing olfactory rosettes of the olfactory bulbs (Gottofrey and Tjälve, 1991). However, only excessive concentrations of cadmium disrupted binding of L-alanine to its olfactory receptor in Atlantic salmon (*Salmo salar*) (Lo *et al.*, 1991). Limited data are available on the effects of cadmium on taste and optic systems, but cadmium causes disruption of mitochondrial respiration in rat optic nerves (Fern *et al.*, 1996). Thus, it may be feasible to assume that the altered behavior observed following cadmium exposure in teleosts could be the result of damage to the sensory tissues.

Zinc

Zinc is naturally present in the teleost CNS and is highly regulated (Pinuela, 1992a,b; Serra *et al.*, 1996). Exposure to excessive concentrations of zinc has caused behavioral alterations in fish (Smith, 1984; Heath, 1995).

In the brain of brown trout (*Salmo trutta*) living in cadmium- and zinc-contaminated waters, there is evidence of chronic stimulation of the hypothalamo-pituitary-inter-renal (HPI) axis, disrupting the neuroendocrine control mechanisms of the inter-renal cells (Norris *et al.*, 1997). Zinc has also been reported to inhibit AChE in the brain of carp (Suresh *et al.*, 1992). In teleosts, there is some indication that high levels of zinc will affect sensory tissue function. Short-term exposure of catfish olfactory mucosa to zinc induced extensive degeneration of olfactory receptor cells with subsequent regeneration (Cancalon, 1982). However, long-term exposure resulted in irreversible destruction with minimal regeneration. Exposure to excessive concentrations of zinc disrupted binding of L-alanine to its olfactory receptor in Atlantic salmon (*Salmo salar*) (Lo *et al.*, 1991). Thus, the basis for the neurotoxic effects of zinc in teleosts has not been defined.

Organics

Solvents

Solvents have been reported in the literature to cause 'narcosis syndrome'. Narcosis is a non-specific reversible state of arrested activity of protoplasmic structures caused by a wide variety of organic chemicals (Veith and Broderius, 1990). Many quantitative structure–activity relationship studies have been performed with these compounds using the 1-octanol/water partition coefficient as a descriptor. Many other studies have been performed to determine the distribution and elimination of these compounds. However, the actual neurologic target(s) for many solvents in teleosts has not been elucidated and little information is available in the literature.

The effect of alcohols in teleosts is perhaps the most studied. It has been proposed that alcohols disrupt membrane structure in teleosts (Ingram *et al.*, 1982). For example, ethanol, 1-propanol, and *t*-butanol, all of which increase membrane fluidity, have no effect on the activity of the soluble form of AChE (Lasner *et al.*, 1995). However, at low levels, the alcohols increase the activity of membrane-bound AChE, and at high levels they decrease the activity. The stimulatory effects at low levels could be a favorable conformational change induced by ethanol, whereas the inhibitory effects at high levels could be the perturbation of the structure of water around hydrophobic areas of AChE causing a conformational instability (Shin *et al.*, 1991) The replacement of the water molecules with ethanol on the membrane has been observed in the brain of carp exposed to ethanol (Isobe *et al.*, 1994). Additionally, there is some evidence in non-teleost fish (*Torpedo*) that alcohols bind to the nicotinic receptor and modulate its associated channel (Boyd and Cohen, 1984; Nagata *et al.*, 1996). However, there is current conflict between those who consider the action on receptors as more important toxicologically and those who consider alteration of the membrane adjacent to the receptors as more important.

Polychlorinated biphenyls

The polychlorinated biphenyls (PCBs) are common environmental contaminants but limited data are available on the neurologic effects in teleosts. In rat brain, exposure to PCBs causes decreased dopamine concentrations (Seegal *et al.*, 1986). Accumulation of PCBs in teleost brain does occur (Ingebrigtsen *et al.*, 1990) and similar effects on neurotransmitter levels have been demonstrated in teleosts. For example, following exposure to the PCB Aroclor 1254, decreased norepinephrine and dopamine levels were observed in the brain of gulf killifish (*Fundulus grandis*) (Fingerman and Russell, 1980) and channel catfish (Fingerman and Short, 1983). In Aroclor 1254-exposed Atlantic croaker (*Micropogonias undulatas*), there was decreased dopamine in the hypothalamus and increased levels of its metabolite, suggesting altered metabolism (Khan and Thomas, 1997). Similar alteration of monoamine metabolism has been identified in rats (Seegal *et al.*, 1986). However, the areas of the teleost brain which are most affected have not been identified.

Natural toxins

There are many natural neurotoxins which fish may encounter in their environment. As these are not from anthropogenic sources, there has not been a great concern in toxicology with their effects. These are part of the natural processes which occur in the environment. While some are not affected by man's activities (e.g. conotoxins), others can be (e.g. fertilizers or animal waste which increase populations of toxic algae). Nonetheless, these compounds are neurotoxic to fish.

Conotoxins

The slow-moving marine snails of the genus *Conus* utilize venom to capture fish. This venom, conotoxin, contains several classes of peptides which have different neurologic targets (Olivera *et al.*, 1985), including mainly the neuromuscular junction and to a lesser extent the brain. Classes of conotoxins include: the alpha-, alphaA-, and psi-conotoxins which block the nicotinic receptor by binding to its α subunit (Hopkins *et al.*, 1995; Shon *et al.*, 1997); the omega-conotoxins which presynaptically prevent the voltage-activated entry of calcium into the nerve terminal and inhibit the release of acetylcholine (Olivera *et al.*, 1984; Reily *et al.*, 1995); the mu-O-conotoxins block Na^+ channels which directly abolish muscle action potentials (McIntosh *et al.*, 1995); the kappa-conotoxins which bind to K^+ channels (Shon *et al.*, 1998); and the delta-conotoxins which bind to voltage-gated Na^+ channels and interfere with the opening and closing (Fainzilber *et al.*, 1994). There is diversity within these classes such that different subtypes are present which bind to different sites on their respective targets. Additionally, another class of toxins from the genus *Conus* venom are the conantokins. Two type have been isolated, conantokin-G and conantokin-T, and these are antagonists which selectively inhibit the NMDA glutamate receptor by binding to an allosteric site (Haack *et al.*, 1990; Hammerland *et al.*, 1992; Skolnick *et al.*, 1992).

Cyanotoxins

Cyanobacteria are very prevalent in most bodies of water and, in fact, become the dominant algal species in freshwater ponds and lakes during the summer months. Several cyanobacterial species produce neurotoxins which most certainly could affect planktivorous fish. The presence of these cyanobacterial neurotoxins has been reported mainly in North America with a few reports from Great Britain, Australia, and Scandinavia (Carmichael, 1994). The main genera of cyanobacteria which have been found to produce these neurotoxins are *Anabaena*, *Aphanizomenon*, and *Oscillatoria* (Carmichael, 1992). There are four types of cyanobacteria neurotoxins: anatoxin-a, anatoxin-a(s), saxitoxin, and neosaxitoxin (Figure 2.11).

Anatoxin-a was the first cyanotoxin from freshwater cyanobacteria to be identified (Huber, 1972) and exerts its toxicity by mimicking acetylcholine at nicotinic receptors in neuromuscular synapses (Carmichael *et al.*, 1979). Anatoxin-a has mainly been reported to be produced by strains of the cyanobacteria *Anabaena flos-aquae* (Gleason and Wood, 1987). However, it is produced by other *Anabaena* species including *Anabaena spiroides*, *Anabaena circinalis*, and even in *Oscillatoria* species (Carmichael, 1992).

Anatoxin-a(s) exerts its toxicity by inhibiting AChE (Mahmood and Carmichael, 1987). The chemical structure of anatoxin-a(s) is that of an organophosphate. Anatoxin-a(s) is the only known naturally occurring organophosphate and, at present, has only been identified in the cyanobacterium *Anabaena flos-aquae*.

While saxitoxin and neosaxitoxin are most commonly associated with the marine dinoflagellates and human paralytic shellfish poisoning, they have been found to be produced by certain species of cyanobacteria, mainly *Aphanizomenon flos-aquae* (Sawyer *et al.*, 1968) but also by *Anabaena flos-aquae* (Carmichael, 1994). The importance of finding the production of saxitoxin and neosaxitoxin in cyanobacteria is that it has allowed investigators to identify the pathways of synthesis (Shimizu *et al.*, 1984) because dinoflagellates are very difficult to culture in a laboratory setting.

More is known about the effects of neurotoxic cyanotoxins on wildlife and domestic mammals than on fish. Eel AChE has been used to identify and characterize anatoxin-a(s) (Mahmood and Carmichael, 1987; Hyde and Carmichael, 1991) and detect its presence in bodies of water where mammalian toxicity has occurred (Mahmood *et al.*, 1988). Although fish kills attributed to toxins produced by cyanobacteria are considered to be rare (Tucker and Robinson, 1990), there are several cases involving the effect of toxic algae in aquaculture ponds which are handled each year by the Fish Diagnostic Laboratories in south-east USA (L. Petrie-Hanson, personal communication). One such case of exposure reported in the literature involved fingerling channel catfish feeding on floating feed in a dense surface scum of cyanobacteria (Schwedler *et al.*, 1985). The fish exhibited neurologic signs of toxicity, including tetany and convulsions. However, the exposure did not result in high mortality. Many of the additional cases occurring have resulted in mortality. Overall, the direct effect of cyanotoxins on fish has not been determined.

Kainic acid

Kainic acid (Figure 2.11) is a neuroexcitatory amino acid that can be produced by the red algae *Digenea simplex* (Norton, 1996). The brain (Migani, 1990; Tong *et al.*, 1992) and retina (Hankins and Ruddock, 1984; Peng *et al.*, 1995) of fish are a rich source of binding sites for kainic acid and have been used as a source of tissue in elucidating the neurotoxicity associated with kainic acid. In the vertebrate CNS, kainic acid binds to ionotropic glutamate receptors (kainate receptors), inducing the opening of Ca^{2+} channels (Linn and Christensen, 1992) and possibly the release of excitatory amino acids (Brown and Nijjar, 1995). The binding can be displaced by glutamic acid. However, additional kainic acid sites which are insensitive to glutamic acid displacement have been found in the fish cerebellum and cerebellar crest (Davis *et al.*, 1992), which suggests that the receptors for kainic acid in fish brain may not always be glutamate receptors.

Domoic acid

Domoic acid (Figure 2.11) is a neuroexcitatory amino acid similar to kainic acid that can be produced by the green algae *Chondria armata* and the diatom *Pseudonitzshia pungens* (Norton, 1996). Domoic acid can be accumulated in filter feeding mollusks and has been associated with amnesic shellfish poisoning (ASP) in humans. It has been reported that domoic acid is an agonist for the glutamate receptors (kainate receptors) in the brain and causes the cell to fire uncontrollably and release excitatory amino acids (glutamic acid) (Brown and Nijjar, 1995). These excitatory amino acids affect neighboring cells, thus causing the neurotoxicity (Berman and Murray, 1997). Since domoic acid can accumulate in mollusks, it can be assumed that any fish which feeds on those mollusks could be susceptible to the toxic effects of domoic acid. Domoic acid has been identified in fin fish in the Pacific northwest. However, little work has been carried out to characterize the effects of domoic acid in fish nervous systems.

Saxitoxin and neosaxitoxin

The marine red tide dinoflagellates responsible for human paralytic shellfish poisoning frequently produce toxicity in fish (Mills and Klein-Macphee, 1979; White, 1981). The toxins produced and released by these dinoflagellates include saxitoxin, neosaxitoxin, and their analogues such as the gonyautoxins (Figure 2.11) and others. Only certain species of specific genera produce these toxins (i.e.

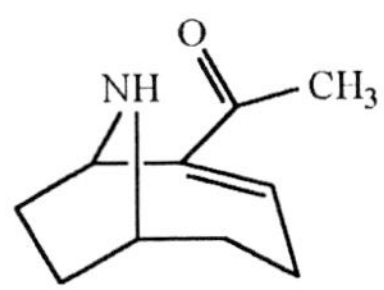

Anatoxin-a

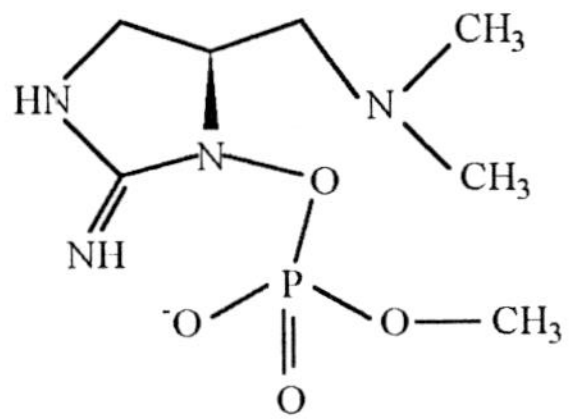

Anatoxin-a(s)

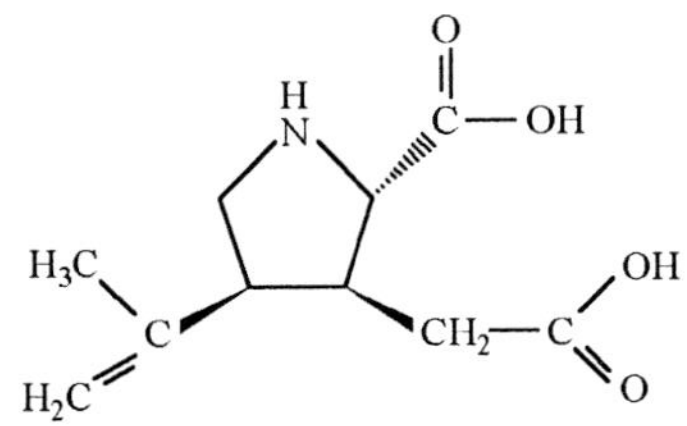

Kainic Acid

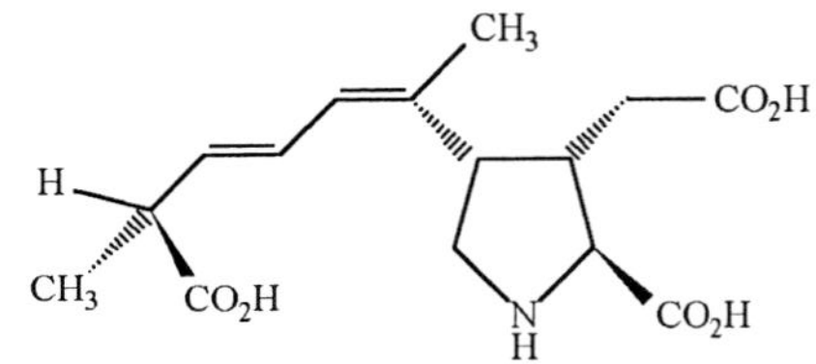

Domoic Acid

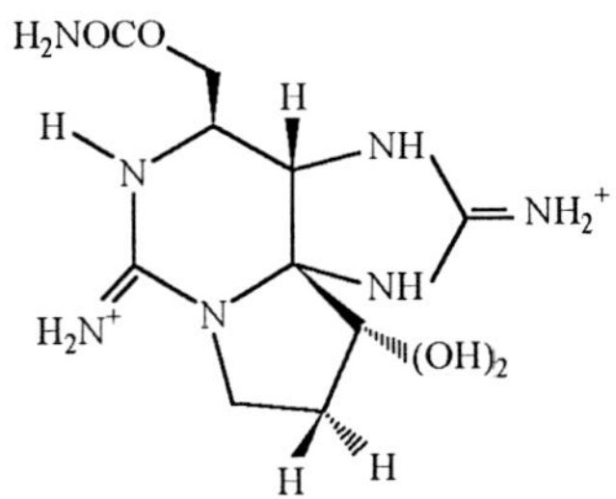

Saxitoxin

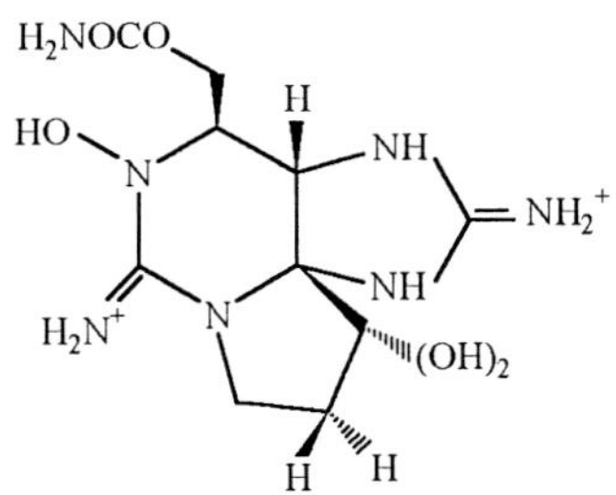

Neosaxitoxin

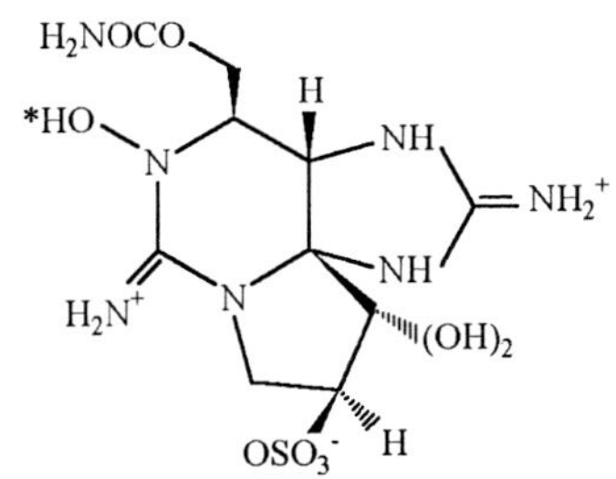

Gonyautoxin I

*Gonyautoxin II (substitute H)

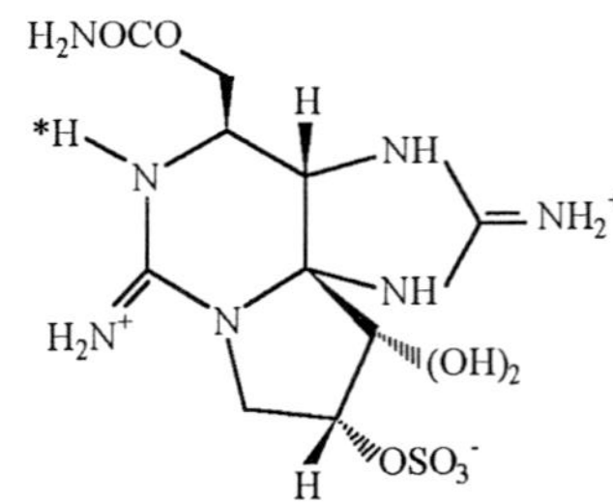

Gonyautoxin III

*Gonyautoxin IV (substitute OH)

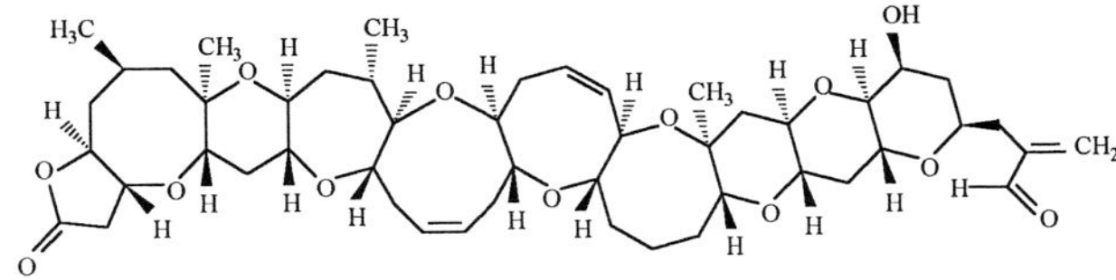

Brevetoxin 1
(PbTx 1)

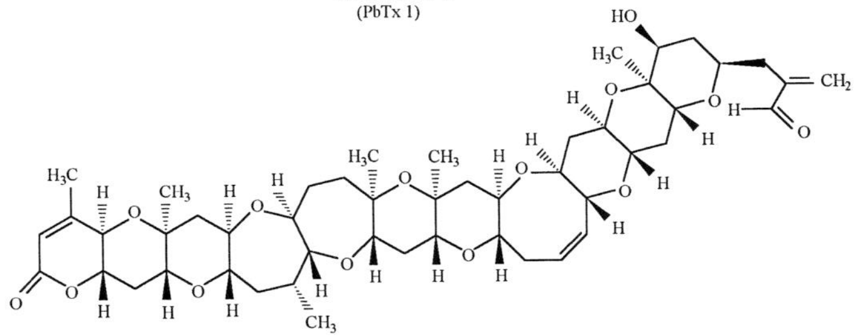

Brevetoxin 2
(PbTx 2)

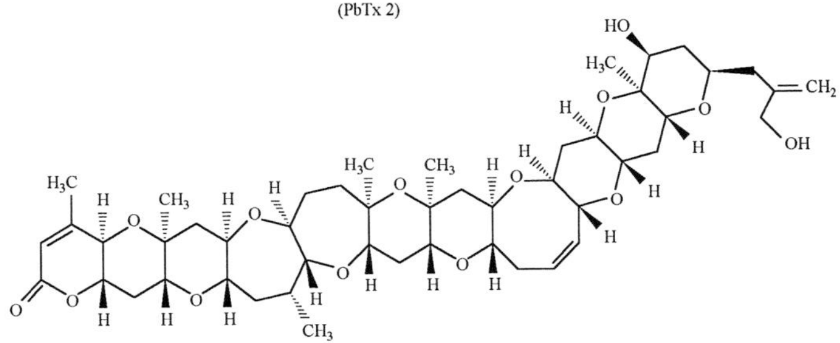

Brevetoxin 3
(PbTx 3)

Figure 2.11 Chemical structures of some natural toxins. Anatoxin-a and anatoxin-a(s) are from cyanobacteria, mainly the genera *Anabaena*, *Aphanizomenon*, and *Oscillatoria*. Kainic acid is from the red algae *Digenea simplex*. Domoic acid is from the green algae *Chondria armata* and the diatom *Pseudonitzshia pungens*. Saxitoxin, neosaxitoxin, and the gonyautoxins are from dinoflagellates, mainly the genera *Gonyaulax*, *Gymnodinium*, *Pyrodinium*, and *Protogonyaulax*. Brevetoxins are from the red tide dinoflagellate.

Gonyaulax, *Gymnodinium*, *Pyrodinium*, and *Protogonyaulax*), but not all members of these genera produce the toxins. Saxitoxin, neosaxitoxin, and their analogues are potent blockers of site 1 of the voltage-gated Na^+ channel (Frace *et al.*, 1986; Catterall, 1988). These toxins are confined as endotoxins in the cells of the dinoflagellate and are released by breakage of the cell wall in response to physical action or chemical processes (Shumway, 1990; Glasgow *et al.*, 1995).

While the occurrence of red tide blooms can result in fish kills, identification of the main toxins responsible for the mortality of the fish is difficult. The different toxins produced have different potencies with respect to blockage of the Na^+ channel (Frace *et al.*, 1986). However, a single species of red tide dinoflagellate can produce multiple toxins, with the ratio of various toxins being dependent upon environmental factors such as light, pH, nutrient availability, salinity, growth factors, and temperature (Iwasaki, 1984; Usup *et al.*, 1994). Also, other toxins with different modes of action, such as brevetoxin (discussed below), may be produced by other dinoflagellates present in a red tide bloom.

Brevetoxins

Brevetoxins are produced and released by the red tide dinoflagellate *Ptychodiscus brevis* and bind to site 5 of the Na^+ channel in rat brain (Sharkey *et al.*, 1987). Brevetoxins have been implicated in massive fish kills in Florida (Steidinger *et al.*, 1972). Intoxication signs in fish produced by *Ptychodiscus brevis* cells include violent twisting and corkscrew swimming, and pectoral and caudal fin paralysis progressing to loss of equilibrium, with subsequent respiratory paralysis (Baden and Mende, 1982). At least three brevetoxins (Figure 2.11) have been identified (Poli *et al.*, 1986) and their ability to bind to *Tilapia* brain synaptosomes has been demonstrated (Stuart and Baden, 1988). The potencies of these brevetoxins to bind to *Tilapia* brain are directly correlated to their toxicity to mosquitofish (Baden *et al.*, 1988).

Ciguatoxins

Ciguatoxins are produced by the benthic dinoflagellate *Gambierdiscus toxicus* (Murata *et al.*, 1990; Holmes *et al.*, 1991) and bind to site 5 of the Na^+ channel (Catterall, 1988). Ciguatoxins are most commonly associated with human ciguatera poisoning following consumption of ciguateric fish who have accumulated the toxin (Russell, 1996). At least three ciguatoxins (designated 1, 2, and 3) have been found using fish tissue as a source (Lewis *et al.*, 1991) (Figure 2.12). However, there have not been any fish kills in which ciguatoxins have been implicated. There have been reports of toxicity in sharks from ingesting ciguatoxin. A tissue fraction containing ciguatoxins has been reported to increase the influx of Na^+ into fish nerves *in vitro* (Capra *et al.*, 1987). A tissue fraction containing ciguatoxins injected intraperitoneally in fish was highly toxic (Capra *et al.*, 1988), whereas feeding ciguatoxin-contaminated flesh to fish did not result in any toxicity (Helfrich and Banner, 1963). Exposure of mosquitofish to purified ciguatoxin 1 and 2 resulted

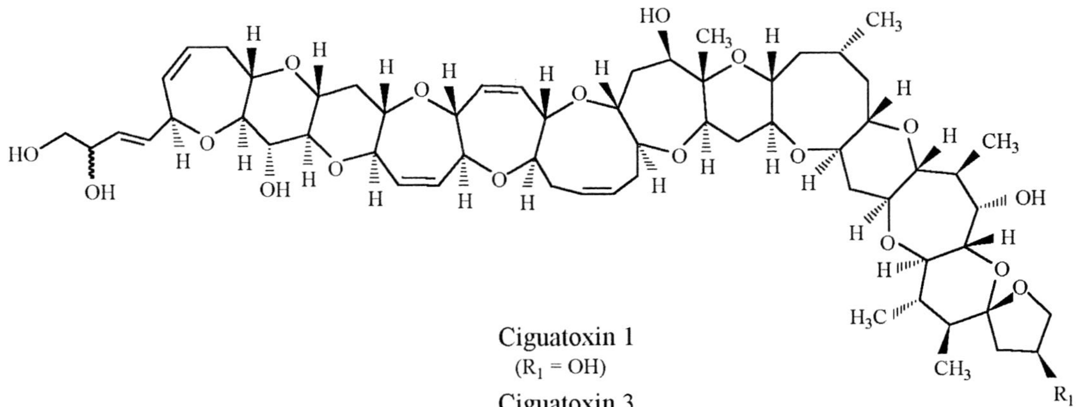

Ciguatoxin 1
(R_1 = OH)
Ciguatoxin 3
(R_1 = H)

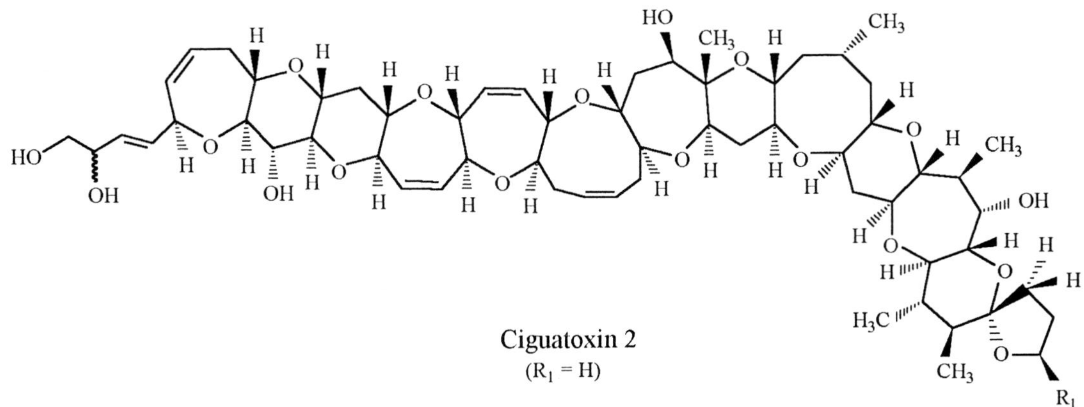

Ciguatoxin 2
(R_1 = H)

Figure 2.12 Chemical structure of ciguatoxin. Ciguatoxins are from the dinoflagellate *Gambierdiscus toxicus*. The structural differences between the toxins are located within the M-ring (far right).

in toxicity (Lewis, 1992). Intoxication signs included pronounced opercular movement, inactivity with bursts of uncoordinated swimming when disturbed, and a loss of righting reflex. The level of ciguatoxin 1 which was lethal to mosquitofish was higher than the levels found in the flesh of ciguateric fish. This suggests that, while the ciguateric fish can carry residues of the toxin, the levels which they can carry is limited by the toxicity of ciguatoxin to the fish. Surprisingly, even though no fish kills have been attributed to ciguatoxins, both ciguatoxin 1 and 2 were more toxic to mosquitofish than brevetoxin 2, to which fish kills have been attributed.

Pfiesteria piscicida

A large number of fish kills have occurred in North Carolina estuarine environments. The causative agent was discovered to be the dinoflagellate *Pfiesteria piscicida* (Burkholder *et al.*, 1992). This organism is extremely toxic to many species of fish and implicated as the causative agent behind over 50 percent of all estuarine fish kills in this area (Glasgow *et al.*, 1995). The increase in *Pfiesteria*-induced fish kills has been reported to be due to increased anthropogenic nutrient loading (Glasgow *et al.*, 1995). Unlike the other endotoxin-producing dinoflagellates, *Pfiesteria* produces an exotoxin which, by design, targets fish and other aquatic species. Production of this exotoxin is evidently required to complete the *Pfiesteria* life cycle. The exotoxin causes the fish to become lethargic, disrupts their osmoregulatory system, disrupts the nervous system, and produces bleeding ulcerations (Burkholder and Glasgow, 1997). The dinoflagellates then feed on the skin, blood, and tissues. The fish lose their ability to maintain position in the water column, undergo respiratory distress, and die. However, the neurologic target associated with this toxin has not been elucidated.

Conclusions

Many neurotoxic agents (both natural and anthropogenic) have resulted in fish kills. In searching for the mechanistic mode of action of these chemicals, the commonality in biochemistry and physiology in the different vertebrate classes suggests that the actions of many of these agents are similar in fish to what is already documented in other species (usually mammalian species). However, while the targets may be the same, the sensitivity of a target for a given toxic compound will differ among species. This difference in target site sensitivity may explain the increased sensitivity of fish to certain compounds (e.g. pyrethroids) compared with other species. Regardless, it can be assumed that sublethal levels of neurotoxic agents, while not resulting in mortality, can be expected to cause perturbations in physiology and behavior such that they will result in negative impacts on the natural aquatic populations.

References

Alam, M.K. and Maughan, O.E. 1995. Acute toxicity of heavy metals to common carp (*Cyprinus carpio*). *Journal of Environmental Science and Health* A30: 1807–1816.

Alford, S. and Grillner, S. 1991. The involvement of $GABA_B$ receptors and coupled G-proteins in spinal GABAergic presynaptic inhibition. *Journal of Neuroscience* 11: 3718–3726.

Anderson, J.M. and Peterson, M.R. 1969. DDT: Sublethal effects on brook trout nervous system. *Science* 164: 440–441.

Anderson, J.M. and Prins, H.B. 1970. Effects of sublethal DDT on a single reflex in brook trout. *Journal of the Fisheries Research Board of Canada* 27: 331–334.

Ansari, B.A. and Kumar, K. 1984. Malathion toxicity: *In vivo* inhibition of acetylcholinesterase in the fish *Brachidanyo rerio* (Cyprinidae). *Toxicology Letters* 20: 283–287.

Aracava, Y., Deshpande, S.S., Rickett, D.K., Brossi, A., Schonenberger, B. and Albuquerque, E.X. 1987. The molecular basis of anticholinesterase actions on nicotinic and glutamatergic synapses. *Annals of the New York Academy of Science* 505: 2261–255.

Ashkenazi, A. and Peralta, E.G. 1994. Muscarinic acetylcholine receptors. In *Handbook of Receptors and Channels*. Peroutka, S.J. (ed.), pp. 1–27. CRC Press, Boca Raton.

Baatrup, E. 1991. Structural and functional effects of heavy metals on the nervous system, including sense organs, of fish. *Comparative Biochemistry and Physiology C* 100: 253–257.

Baatrup, E. and Døving, K.B. 1990. Histochemical demonstration of mercury in the olfactory system of salmon (*Salmo salar* L.) following treatments with dietary methylmercuric chloride and dissolved mercuric chloride. *Ecotoxicology and Environmental Safety* 20: 277–289.

Baatrup, E., Døving, K.B. and Winberg, S. 1990. Differential effects of mercurial compounds on the electro-olfactogram (EOG) of salmon (*Salmo salar* L.). *Ecotoxicology and Environmental Safety* 20: 269–276.

Baden, D.G. and Mende, T.J. 1982. Toxicity of two toxins from the Florida red tide dinoflagellate *Ptychodiscus brevis*. *Toxicon* 20: 457–461.

Baden, D.G., Mende, T.J., Szmant, A.M., Trainer, V.L., Edwards, R.A. and Roszell, L.E. 1988. Brevetoxin binding: Molecular pharmacology versus immunoassay. *Toxicon* 26: 97–103.

Bahn, S., Harvey, R.J., Darlinson, M.G. and Wisden, W. 1996. Conservation of γ-aminobutyric acid type A receptor α6 subunit gene expression in cerebellar granule cells. *Journal of Neurochemistry* 66: 1810–1818.

Bahr, B.A., Hoffman, K.B., Kessler, M., Hennegriff, M., Park, G.Y., Yamamoto, R.S., Kawasaki, B.T., Vanderklish, P.W., Hall, R.A. and Lynch, G. 1996. Distinct distributions of alpha-amino-3-hydroxy-5-methyl-4-isoxazolepropionate (AMPA) receptor subunits and a related 53,000 M(R) antigen (GR53) in brain tissue. *Neuroscience* 74: 707–721.

Bakry, N.M.S., El-Rashidy, A.H., Eldefrawi, A.T. and Eldefrawi, M.E. 1988. Direct actions of organophosphate anticholinesterases on nicotinic and muscarinic acetylcholine receptors. *Journal of Biochemical Toxicology* 2: 235–259.

Bálint, T., Ferenczy, J., Kátai, F., Kiss, I., Kráczer, L., Kufcsák, O., Láng, G., Polyhos, C., Szabó, I., Szegletes, T. and Nemcsók, J. 1997. Similarities and differences between the massive eel (*Anguilla anguilla* L.) devastations that occurred in Lake Balaton in 1991 and 1995. *Ecotoxicology and Environmental Safety* 37: 17–23.

Bansal, S.K., Verma, S.R., Gupta, A.K., Rani, S. and Dalela, R.C. 1979. Pesticide-induced alterations in the oxygen uptake rate of a freshwater major carp *Ladeo rohita*. *Ecotoxicology and Environmental Safety* 3: 374–382.

Barhanin, J., Meiri, H., Romey, G., Pauron, D. and Lazdunski, M. 1985. A monoclonal immunotoxin acting on the Na^+ channel, with properties similar to those of a scorpion toxin. *Proceedings of the National Academy of Sciences of the United States of America* 82: 1842–1846.

Barnes, J.M. and Henley, J.M. 1994. Quantitative analysis of the distributions of glutamatergic ligand binding sites in goldfish brain. *Brain Research* 637: 323–327.

Barnes, J.M., Murphy, P.A., Kirkham, D. and Henley, J.M. 1993. Interaction of guanine nucleotides with [^{3}H]kainate and 6-[^{3}H]cyano-7-nitroquinoxaline-2,3-dione binding in goldfish brain. *Journal of Neurochemistry* 61: 1685–1691.

Behrens, U.D. and Wagner, H.J. 1995. Localization of dopamine D_1-receptors in vertebrate retinae. *Neurochemistry International* 27: 497–507.

Bengtsson, B.E. and Larsson, Å. 1981. Hyperactivity and changed diurnal activity in flounders, Platichthys flesus, exposed to DDT. *Marine Pollution Bulletin* 12: 100–102.

Benke, G.M. and Murphy, S.D. 1974. Anticholinesterase action of methyl parathion, parathion and azinophosmethyl in mice and fish: onset and recovery of inhibition. *Bulletin of Environmental Contamination and Toxicology* 12: 117–122.

Benke, G.M., Cheever, K.L., Mirer, F.E. and Murphy, S.D. 1974. Comparative toxicity, acetylcholinesterase action and metabolism of methyl parathion and parathion in sunfish and mice. *Toxicology and Applied Pharmacology* 28: 97–109.

Bergstrom, A. and Wikberg, J.E. 1986. Structural and pharmacological differences between codfish and rat brain alpha 1-adrenergic receptors revealed by photoaffinity labeling with ^{125}I-APDQ. *Acta Pharmacologica et Toxicologica (Copenhagen)* 58: 148–155.

Berman, F.W. and Murray, T.F. 1997. Domoic acid neurotoxicity in cultured cerebellar granule neurons is mediated predominantly by NMDA receptors that are activated as a consequence of excitatory amino acid release. *Journal of Neurochemistry* 69: 693–703.

Bindokas, V.P. and Ishida, A.T. 1991. (–)-Baclofen and gamma-aminobutyric acid inhibit calcium currents in isolated retinal ganglion cells. *Proceedings of the National Academy of Sciences of the United States of America* 88: 10759–10763.

Bloomquist, J.M. and Soderlund, D.M. 1988. Pyrethroid insecticides and DDT modify alkaloid-dependent sodium channel activation and its enhancement by sea anemone toxin. *Molecular Pharmacology* 33: 543–550.

Bonner, J.C. and Yarbrough, J.D. 1987. Alteration of the t-butylbicyclophosphorothionate binding site as a mechanism of vertebrate cyclodiene insecticide resistance. *Pesticide Biochemistry and Physiology* 29: 260–265.

Bonner, J.C. and Yarbrough, J.D. 1988. Vertebrate cyclodiene insecticide resistance: Role of γ-aminobutyric acid and diazepam binding sites. *Archives of Toxicology* 62: 311–315.

Bonner, J.C. and Yarbrough, J.D. 1989. Role of the brain *t*-butylbicyclophosphorothionate receptor in vertebrate resistance to endrin, 1,1,1-trichloro-2,2-Bis(*p*-chlorophenyl)ethane and cypermethrin. *Journal of Pharmacology and Experimental Therapeutics* 249: 149–154.

Boone, J.S. and Chambers, J.E. 1996. Time course inhibition of cholinesterase and aliesterase activities, and nonprotein sulfhydryl levels following exposure to organophosphorus insecticides in mosquitofish (*Gambusia affinis*). *Fundamental and Applied Toxicology* 29: 202–207.

Boone, J.S. and Chambers, J.E. 1997. Biochemical factors contributing to toxicity differences among chlorpyrifos, parathion, and methyl parathion in mosquitofish (*Gambusia affinis*). *Aquatic Toxicology* 39: 333–343.

Bottai, D., Dunn, R.J., Ellis, W. and Maler, L. 1997. *N*-methyl-D-aspartate receptor 1 mRNA distribution in the central nervous system of the weakly electric fish *Apteronotus leptorhynchus*. *Journal of Comparative Neurology* 389: 65–80.

Bowery, N. 1989. $GABA_A$ receptors and their significance in mammalian pharmacology. *Trends in Pharmacological Science* 10: 401–407.

Boyd, N.D. and Cohen, J.B. 1984. Desensitization of membrane-bound Torpedo acetylcholine receptor by amine noncompetitive antagonists and aliphatic alcohols: studies of [^{3}H]acetylcholine binding and $^{22}Na^+$ ion fluxes. *Biochemistry* 23: 4023–4033.

Bradbury, S.P. and Coats, J.R. 1989. Toxicokinetics and toxicodynamics of pyrethroid insecticides in fish. *Environmental Toxicology and Chemistry* 8: 373–380.

Bradbury, S.P., McKim, J.M. and Coats, J.R. 1987. Physiological response of rainbow trout (*Salmo gairdneri*) to acute fenvalerate intoxication. *Pesticide Biochemistry and Physiology* 27: 275–288.

Bradbury, S.P., Carlson, R.W., Niemi, G.J. and Henry, T.R. 1991. Use of respiratory-cardiovascular responses of rainbow trout (*Oncorhynchus mykiss*) in identifying acute toxicity syndromes in fish. 4. Central nervous system seizure agents. *Environmental Toxicology and Chemistry* 10: 115–131.

Brown, J.A. and Nijjar, M.S. 1995. The release of glutamate and aspartate from the rat brain synaptosomes in response to domoic acid (amnesic shellfish poisoning) and kainic acid. *Molecular and Cellular Biochemistry* 151: 49–54.

Brown, V.M. and Dalton, R.A. 1970. The acute lethal toxicity to rainbow trout of mixtures of copper, phenol, zinc, and nickel. *Journal of Fish Biology* 2: 211–216.

de Bruijn, J. and Hermens, J. 1991. Uptake and elimination kinetics of organophosphorus pesticides in the guppy (*Poecilia reticulata*): Correlations with the octanol/water partition coefficient. *Environmental Toxicology and Chemistry* 10: 791–804.

Burka, J.F., Blair, R.M. and Hogan, J.E. 1989. Characterization of the muscarinic and serotonergic receptors of the intestine of the rainbow trout (*Salmo gairdneri*). *Canadian Journal of Physiology and Pharmacology* 67: 477–482.

Burkholder, J.M. and Glasgow, Jr, H.B. 1997. Trophic controls on stage transformation of a toxic ambush-predator dinoflagellate. *Journal of Eukaryotic Microbiology* 44: 200–205.

Burkholder, J.M., Noga, E.J., Hobbs, C.W., Glasgow, Jr, H.B. and Smith, S.A. 1992. New 'phantom' dinoflagellate is the causative agent of major estuarine fish kills. *Nature* 358: 407–410.

Burt, D.R. and Kamatchi, G.L. 1991. $GABA_A$ receptor subtypes: from pharmacology to molecular biology. *Federation of the American Societies of Experimental Biologist Journal* 5: 2916–2923.

Cancalon, P. 1982. Degeneration and regeneration of olfactory cells induced by ZnSO4 and other chemicals. *Tissue Cell* 14: 717–733.

Cao, Y., Oh, B.C. and Stryer, L. 1998. Cloning and localization of two multigene receptor families in goldfish olfactory epithelium. *Proceedings of the National Academy of Sciences of the United States of America* 95: 11987–11992.

Capra, M.F., Flowers, A.E. and Cameron, J. 1987. The effects of ciguatoxin on the rate of Na^+ efflux in unmyelinated olfactory nerves of teleosts. In *Progress in Venom and Toxin Research.* Gopalakrishnakone, P. and Tan, C.K. (eds), pp. 418–422. National University of Singapore, The Venom and Toxin Research Group, Singapore.

Capra, M.F., Cameron, J., Flowers, A.E., Coombe, C.G. and Hahn, S.T. 1988. The effects of ciguatoxin on teleosts. *Proceedings of the Sixth International Coral Reef Symposium* 3: 37–41.

Cardeilhac, P.T. and Whitaker, B.R. 1988. Tropical fish medicine. Copper treatments. Uses and precautions. *Veterinary Clinical North American Small Animal Practices* 18: 435–448.

Cardinaud, B., Sugamori, K.S., Coudouel, S., Vincent, J.D., Niznik, H.B. and Vernier, P. 1997. Early emergence of three dopamine D_1 receptor subtypes in vertebrates. Molecular phylogenetic, pharmacological, and functional criteria defining D_{1A}, D_{1B}, and D_{1C} receptors in European eel *Anguilla anguilla*. *Journal of Biological Chemistry* 272: 2778–2787.

Carlson, R.W. 1990. Ventilatory patterns of bluegill (*Lepomis macrochirus*) exposed to organic chemicals with different mechanisms of toxic action. *Comparative Biochemistry and Physiology C* 95: 181–196.

Carmichael, W.W. 1992. A review: Cyanobacteria secondary metabolites – The cyanotoxins. *Journal of Applied Bacteriology* 72: 445–459.

Carmichael, W.W. 1994. The toxins of cyanobacteria. *Scientific American* 270: 78–86.

Carmichael, W.W., Biggs, D.F. and Peterson, M.A. 1979. Pharmacology of anatoxin-a, produced by the freshwater cyanophyte, *Anabaena flos-aquae* NRC-44–1. *Toxicon* 17: 229–236.

Carr, R.L. and Chambers, J.E. 1996. Kinetic analysis of the *in vitro* inhibition, aging and reactivation of brain acetylcholinesterase from rat and channel catfish by paraoxon and chlorpyrifos-oxon. *Toxicology and Applied Pharmacology* 139: 365–373.

Carr, R.L., Straus, D.L. and Chambers, J.E. 1995. Inhibition and aging of channel catfish brain acetylcholinesterase following exposure to two phosphorothionate insecticides and their active metabolites. *Journal of Toxicology and Environmental Health* 45: 325–336.

Carr, R.L., Ho, L.L. and Chambers, J.E. 1997. Selective toxicity of chlorpyrifos to several species of fish during an environmental exposure: An investigation into the biochemical mechanisms behind this toxicity. *Environmental Toxicology and Chemistry* 16: 2369–2374.

Carr, R.L., Atterberry, T.T., Liu, J., Coats, J.R. and Chambers, J.E. 1998. The interaction of chlorinated alicyclic insecticides with brain $GABA_A$ receptors in channel catfish (*Ictalurus punctatus*). Submitted.

Casida, J.E., Gammon, D.W., Glockman, A.H. and Lawrence, L.J. 1983. Mechanisms of selective action of pyrethroid insecticides. *Annual Reviews in Pharmacology and Toxicology* 23: 413–438.

Cataldi de Flombaum, M.A. and Stoppani, A.O. 1986. Inactivation of mitochondrial adenosine triphosphatase from *Trypanosoma cruzi* by oxygen radicals. *Biochemical International* 12: 785–793.

Catterall, W.A. 1988. Structure and function of voltage-sensitive ion channels. *Science* 242: 50–61.

Catterall, W.A., Morrow, C.S. and Hartshorne, R.P. 1979. Neurotoxin binding to receptor sites associated with voltage-sensitive sodium channels in intact, lysed, and detergent solubilized membranes. *Journal of Biological Chemistry* 254: 11379–11387.

Chambers, H.W. 1992. Organophosphorus compounds: An overview. In *Organophosphates: Chemistry, Fate, and Effects*. Chambers, J.E. and Levi, P.E. (eds), pp. 3–17. Academic Press, San Diego.

Chang, H.M., Wu, Y.M., Chang, Y.C., Hsu, Y.C., Hsu, H.Y., Chen, Y.C. and Chow, W.Y. 1998. Molecular and electrophysiological characterizations of fGluR3 alpha, an ionotropic glutamate receptor subunit of a teleost fish. *Brain Research and Molecular Brain Research* 57: 211–220.

Christenson, J. and Grillner, S. 1991. Primary afferents evoke excitatory amino acid receptor-mediated EPSPs that are modulated by presynaptic $GABA_B$ receptors in lamprey. *Journal of Neurophysiology* 66: 2141–2149.

Clark, J.M. and Brooks, M.W. 1989. Neurotoxicology of pyrethroids: Single or multiple mechanisms of action. *Environmental Toxicology and Chemistry* 8: 361–372.

Clark, J.M. and Matsumura, F. 1982. Two different types of inhibitory effects of pyrethroids in nerve Ca– and Ca+ Mg ATPase activity in the squid, *Loligo poali*. *Pesticide Biochemistry and Physiology* 18: 180–190.

Cole, L.M., Lawrence, L.J. and Casida, J.E. 1984. Similar properties of [^{35}S]*t*-butylbicyclophosphorothionate receptor and coupled components of the GABA receptor–ionophore complex in brains of human, cow, rat, chicken, and fish. *Life Science* 35: 1755–1762.

Conti-Tronconi, B.M., Hunkapiller, M.W., Lindstrom, J.M. and Raftery, M.A. 1984. Multisubunit structure and amino-terminal sequences of piscine muscle acetylcholinesterase receptor. *Journal of Receptor Research* 4: 801–816.

Coppage, D.L. and Braidech, T.E. 1976. River pollution by anticholinesterase agents. *Water Research* 10: 19–24.

Coppage, D.L. and Matthews, E. 1974. Short-term effects of organophosphate pesticides on cholinesterases of estuarine fishes and pink shrimp. *Bulletin of Environmental Contamination and Toxicology* 11: 483–488.

Coppage, D.L., Matthews, E., Cook, G.H. and Knight, J. 1975. Brain acetylcholinesterase inhibition in fish as a diagnosis of environmental poisoning by malathion, *O,O*-dimethyl *S*-(1,2-dicarbethoxyethyl) phosphorodithioate. *Pesticide Biochemistry and Physiology* 5: 536–542.

Corda, M.G., Longoni, B., Cau, A., Paci, S., Salvadori, S., Laudani, U. and Biggio, G. 1989. Distribution and pharmacological properties of the $GABA_A$/benzodiazepine/chloride ionophore receptor complex in the brain of the fish *Anguilla anguilla*. *Journal of Neurochemistry* 52: 1025–1034.

Corda, M.G., Orlandi, M., Deiana, A.M., Salvadori, S. and Giorgi, O. 1991. Age-related decrease in the density of benzodiazepine recognition sites in the brain of the fresh water eel (*Anguilla anguilla*). *Neuroscience Letters* 133: 168–170.

Cutkomp, L.K., Yap, H.H., Cheng, E.Y. and Koch, R.B. 1971. ATPase activity in fish tissue homogenates and inhibitory effects of DDT and related compounds. *Chemical-Biological. Interactions* 3: 439–447.

Cutkomp, L.K., Desaiah, D. and Koch, R.B. 1972. The *in vitro* sensitivity of fish brain ATPases to organochlorine acaricides. *Life Science [II]* 11: 1123–1133.

Dalela, R.C., Bhatnagar, M.C., Tyagi, A.K. and Verma, S.R. 1978. Adenosine triphosphatase activity in few tissues of a fresh water teleost, *Channa gachua* following *in vivo* exposure to endosulfan. *Toxicology* 11: 361–368.

Davis, P.W. and Wedemeyer, G.A. 1971. Na^+, K^+-activated-ATPase inhibition in rainbow trout: a site for organochlorine pesticide toxicity? *Comparative Biochemistry and Physiology B* 40: 823–827.

Davis, P.W., Friedhoff, J.M. and Wedemeyer, G.A. 1972. Organochlorine insecticide, herbicide, and polychlorinated biphenyl (PCB) inhibition of NaK-ATPase in rainbow trout. *Bulletin of Environmental Contamination and Toxicology* 8: 69–72.

Davis, R.E., Wilmont, G.R. and Cha, J.H. 1992. Glutamic acid-insensitive [^{3}H]kainic acid binding in goldfish brain. *Brain Research* 571: 73–78.

Davy, F.B., Kleerekoper, H. and Gensler, P. 1972. Effects of exposure to sublethal DDT on the locomoter behavior of goldfish (*Carassius auratus*). *Journal of the Fisheries Research Board of Canada* 29: 1333–1336.

Davy, R.B., Meisch, M.V. and Carter, F.L. 1976. Toxicity of five ricefield pesticides to the mosquitofish, *Gambusia affinis*, and green sunfish, *Lepomis cyanellus*, under laboratory and field conditions in Arkansas. *Environmental Entomology* 5: 1053–1056.

Dearry, A. and Burnside, B. 1988. Stimulation of distinct D_2 dopaminergic and alpha 2-adrenergic receptors induces light-adaptive pigment dispersion in teleost retinal pigment epithelium. *Journal of Neurochemistry* 51: 1516–1523.

DeLory, T.M. and Olsen, R.W. 1994. GABA and glycine. In *Basic Neurochemistry*. Siegel, G.J., Agranoff, B.W., Albers, R.W. and Molinoff, P.B. (eds), pp. 389–400. Raven Press, New York.

Deng, L., Nielsen, M. and Olsen, R.W. 1991. Pharmacological and biochemical properties of the gamma-aminobutyric acid-benzodiazepine receptor protein from codfish brain. *Journal of Neurochemistry* 56: 968–977.

Deplano, S. 1988. Binding pattern of alpha-bungarotoxin on horizontal cells of a marine teleost retina. *Journal of Comparative Neurology* 277: 541–548.

Desaiah, D. and Koch, R.B. 1975a. Inhibition of ATPases activity in channel catfish brain by Kepone® and its reduction product. *Bulletin of Environmental Contamination and Toxicology* 13: 153–158.

Desaiah, D. and Koch, R.B. 1975b. Toxaphene inhibition of ATPase [adenosine triphosphatase] activity in catfish, Ictalurus punctatus, tissues. *Bulletin of Environmental Contamination and Toxicology* 13: 238–244.

Desaiah, D. and Koch, R.B. 1975c. Inhibition of fish brain ATPases by aldrin-transdiol, aldrin, dieldrin and photodieldrin. *Biochemical and Biophysical Research Communications* 64: 13–19.

Desaiah, D. and Koch, R.B. 1977. Influence of solvents on the pesticide inhibition of ATPase activities in fish and insect tissue homogenates. *Bulletin of Environmental Contamination and Toxicology* 17: 74–81.

Desaiah, D., Cutkomp, L.K., Koch, R.B. and Jarvinen, A. 1975. DDT: effect of continuous exposure on ATPase activity in fish, Pimephales promelas. *Archives of Environmental Contamination and Toxicology* 3: 132–141.

Dingledine, R. and McBain, C.J. 1994. Excitatory amino acid transmitters. In *Basic Neurochemistry*. Siegel, G.J., Agranoff, B.W., Albers, R.W. and Molinoff, P.B. (eds), pp. 367–387. Raven Press, New York.

Dixon, D.B. and Copenhagen, D.R. 1997. Metabotropic glutamate receptor-mediated suppression of an inward rectifier current is linked via a cGMP cascade. *Journal of Neuroscience* 17: 8945–8954.

Dong, C.J. and Werblin, F.S. 1996. Use-dependent and use-independent blocking actions of picrotoxin and zinc at the $GABA_C$ receptor in retinal horizontal cells. *Vision Research* 36: 3997–4005.

Dong, C.J., Picaud, S.A. and Werblin, F.S. 1994. GABA transporters and $GABA_C$-like receptors on catfish cone – but not rod – driven horizontal cells. *Journal of Neuroscience* 14: 2648–2658.

Dubuc, R., Bongianni, F., Ohta, Y. and Grillner, S. 1993. Dorsal root and dorsal column mediated synaptic inputs to reticulospinal neurons in lampreys: involvement of glutamatergic, glycinergic, and GABAergic transmission. *Journal of Comparative Neurology* 327: 251–259.

Ecobichon, D.J. 1996. Toxic effects of pesticides. In *Casarett and Doull's Toxicology: The Basic Science of Poisons*. Klaassen, C.D. (ed.), pp. 643–689. McGraw-Hill, New York.

Edwards, R., Millburn, P. and Hutson, D.H. 1986. Comparative toxicity of cis-cypermethrin in rainbow trout, frog, mouse, and quail. *Toxicology and Applied Pharmacology* 84: 512–522.

Eells, J.T., Rasmussen, J.L., Bandettini, P.A. and Propp, J.M. 1993. Differences in the neuroexcitatory actions of pyrethroid insecticides and sodium channel-specific neurotoxins in rat and trout brain synaptosomes. *Toxicology and Applied Pharmacology* 123: 107–119.

Eldefrawi, M.D., Britten, A.G. and O'Brien, R.D. 1971. Action of organophosphates on binding of cholinergic ligands. *Pesticide Biochemistry and Physiology* 1: 101–108.

Eldefrawi, M.E., Schweizer, G., Bakry, N.M. and Valdes, J.J. 1988. Desensitization of the nicotinic acetylcholine receptor by diisopropylfluorophosphate. *Journal of Biochemical Toxicology* 3: 21–32.

Ellgaard, E.F., Ochsner, J.C. and Cox, J.K. 1977. Locomoter hyperactivity induced in the bluegill sunfish, *Lepomis macrochirus*, by sublethal concentrations of DDT. *Canadian Journal of Zoology* 55: 1077–1081.

Enan, E. and Matsumura, F. 1993. Action of deltamethrin on the calcium/calmodulin-dependent protein kinase from rat brain. *Pesticide Science* 37: 21–30.

Enesco, H.E., Pisanti, F.A. and Toraro, E.A. 1989. The effect of copper on the ultrastructure of *Torpedo marmorata* neurons. *Marine Pollution Bulletin* 20: 232–235.

Eshleman, A.J. and Murray, T.F. 1989. Differential binding properties of the peripheral-type benzodiazepine ligands [^{3}H]PK 11195 and [^{3}H]Ro 5–4864 in trout and mouse brain membranes. *Journal of Neurochemistry* 53: 494–502.

Eshleman, A.J. and Murray, T.F. 1990. Dependence on γ-aminobutyric acid of pyrethroid and 4′-chlorodiazepam modulation of the binding of *t*-[^{35}S]butylbicyclophosphorothionate in piscine brain. *Neuropharmacology* 29: 641–648.

Eshleman, A.J. and Murray, T.F. 1991. Pyrethroid insecticides indirectly inhibit GABA-dependent $^{36}Cl^-$ influx in synaptoneurosomes from the trout brain. *Neuropharmacology* 12: 1333–1341.

Fainzilber, M., Kofman, O., Zlotkin, E. and Gordon, D. 1994. A new neurotoxin receptor site on sodium channels is identified by a conotoxin that affects sodium channel inactivation in molluscs and acts as an antagonist in rat brain. *Journal of Biological Chemistry* 269: 2574–2580.

Fern, R., Black, J.A., Ransom, B.R. and Waxman, S.G. 1996. Cd^{2+}-induced injury in CNS white matter. *Journal of Neurophysiology* 76: 3264–3273.

Fernholm, B., Nielsen, M. and Braestrup, C. 1979. Absence of brain specific benzodiazepine receptors in cylcostomes and elasmobranchs. *Comparative Biochemistry and Physiology C* 62: 209–211.

Fingerman, S.W. and Russell, L.C. 1980. Effects of the polychlorinated biphenyl Aroclor 1242 on locomotor activity and on the neurotransmitters dopamine and norepinephrine in the brain of the gulf killifish, *Fundulus grandis. Bulletin Environmental Contamination Toxicology* 25: 682–687.

Fingerman, S.W. and Short, Jr, E.C. 1983. Changes in neurotransmitter levels in channel catfish after exposure to benzo(a)pyrene, naphthalene, and Aroclor 1254. *Bulletin Environmental Contamination Toxicology* 30: 147–151.

Frace, A.M., Hall, S., Brodwick, M.S. and Eaton, D.C. 1986. Effects of saxitoxin analogues and ligand competition on sodium currents of squid axons. *American Journal of Physiology* 251: 159–166.

Frail, D.E., Manelli, A.M., Witte, D.G., Lin, C.W., Steffey, M.E. and Mackenzie, R. 1993. Cloning and characterization of a truncated dopamine D_1 receptor from goldfish retina: stimulation of cyclic AMP production and calcium mobilization. *Molecular Pharmacology* 44: 1113–1118.

Francis, A. and Schechter, N. 1980. Regional and subcellular distribution of cholinergic enzyme and receptor activity in goldfish brain. *Neuroscience* 5: 293–301.

Friedl, W., Hebebrand, J., Rabe, S. and Propping, P. 1988. Phylogenetic conservation of the benzodiazepine binding sites: Pharmacological evidence. *Neuropharmacology* 27: 163–170.

Gähwiler, B.H. and Brown, D.A. 1985. $GABA_B$-receptor-activated K^+ current in voltage-clamped CA3 pyramidal cells in hippocampal cultures. *Proceedings of the National Academy of Sciences of the United States of America* 82: 1558–1562.

Gammon, D.W., Brown, M.A. and Casida, J.E. 1981. Two classes of pyrethroid action in the cockroach. *Pesticide Biochemistry and Physiology* 15: 181–191.

Garcia-Sainz, J.A., Olivares-Reyes, J.A., Macias-Silva, M. and Villalobos-Molina, R. 1995. Characterization of the alpha 1B-adrenoceptors of catfish hepatocytes: functional and binding studies. *General Comparative Endocrinology* 97: 111–120.

Gardner, G.R. and LaRoche, G. 1973. Copper induced lesions in estuarine teleosts. *Journal of Fisheries Research Board of Canada*. 30: 363–368.

Ghiasuddin, S.M. and Matsumura, F. 1982. DDT inhibition of Ca^{++}-Mg^+ ATPase from peripheral nerves and muscles of lobster, *Homarus americanus. Biochemical and Biophysical Research Communications* 103: 31–37.

Ghosh, T.K. 1987. Comparative toxicogenic evaluation of the three organophosphate pesticides on some selected enzyme in the fresh water teleost, *Sarotherodon mossambicus. Pollution Research* 6: 133–138.

Giannaccini, G., Betti, L., Correani, M., Dini, M., Giusti, L. and Lucacchini, A. 1997. Binding of the benzodiazepine ligand [^{3}H]-Ro 15–1788 to brain membrane of the saltwater fish *Mullus surmuletus*. *Neurochemical International* 31: 781–787.

Gill, T.S., Tewari, H. and Pande, J. 1990. Use of fish enzyme system in monitoring water quality: Effects of mercury on tissue enzymes. *Comparative Biochemistry and Physiology C* 97: 287–292.

Gill, T.S., Tewari, H. and Pande, J. 1991 In vivo and in vitro effects of cadmium on selected enzymes in different organs of the fish *Barbus conchonius* Ham. (rosy barb). *Comparative Biochemistry and Physiology C* 100: 501–505.

Glasgow, Jr, H.B., Burkholder, J.M., Schmechel, D.E., Tester, P.A. and Rublee, P.A. 1995. Insidious effects of a toxic estuarine dinoflagellate on fish survival and human health. *Journal of Toxicology and Environmental Health* 46: 501–522.

Gleason, F.K. and Wood, J.M. 1987. Secondary metabolism in the cyanobacteria. In *The Cyanobacteria*. Fay, P. and Van Baalen, C. (eds). Elsevier Publishers B.V. (Biomedical Division), Amsterdam.

Glickman, A.H. and Lech, J.J. 1982. Differential toxicity of *trans*-permethrin in rainbow trout and mice. II. Role of target organ sensitivity. *Toxicology and Applied Pharmacology* 66: 162–171.

Glickman, A.H., Weitman, S.D. and Lech, J.J. 1982. Differential toxicity of *trans*-permethrin in rainbow trout and mice. II. Role of biotransformation. *Toxicology and Applied Pharmacology* 66: 153–161.

Goodman, L.R., Hansen, D.J., Coppage, D.L., Moore, J.C. and Matthews, E. 1979. Diazinon: Chronic toxicity to, and brain acetylcholinesterase inhibition in, the sheepshead minnow, *Cyprinodon variegatus*. *Transactions of the American Fisheries Society* 108: 479–488.

Gottofrey, J. and Tjälve, H. 1991. Axonal transport of cadmium in the olfactory nerve of the pike. *Pharmacology and Toxicology* 69: 242–252.

Goyer, R.A. 1996. Toxic effects of metals. In *Casarett and Doull's Toxicology: The Basic Science of Poisons*. Klaassen, C.D. (ed.), pp. 691–736. McGraw-Hill, New York.

Guhathakurta, S. and Bhattacharya, S. 1988. Target and nontarget actions of phenthoate and carbofuran: Brain acetylcholinesterase, kidney iodide peroxidase, and blood thyroxine profiles in *Channa punctatus*. *Biomedical and Environmental Science* 1: 59–63.

Guilbault, G.G., Lozes, R.L., Moore, W. and Kuan, S.S. 1972. Effect of pesticides on cholinesterase from aquatic species: Crayfish, trout, and fiddler crab. *Environmental Letters* 3: 235–245.

Haack, J.A., Rivier, J., Parks, T.N., Mena, E.E., Cruz, L.J. and Olivera, B.M. 1990. Conantokin-T. A gamma-carboxyglutamate containing peptide with N-methyl-D-aspartate antagonist activity. *Journal of Biological Chemistry* 265: 6025–6029.

Hai, D.Q., Varga, S.I. and Matkovics, B. 1995. Effects of organophosphate on the antioxidant systems of fish tissues. *Acta Biologica Hungary* 46: 39–50.

Hai, D.Q., Varga, S.I. and Matkovics, B. 1997. Organophosphate effects on antioxidant system of carp (*Cyprinus carpio*) and catfish (*Ictalurus nebulosus*). *Comparative Biochemistry and Physiology C* 117: 83–88.

Hammerland, L.G., Olivera, B.M. and Yoshikami, D. 1992. Conantokin-G selectively inhibits N-methyl-D-aspartate-induced currents in Xenopus oocytes injected with mouse brain mRNA. *European Journal of Pharmacology* 226: 239–244.

Han, M.H., Li, Y. and Yang, X.L. 1997. Desensitizing $GABA_C$ receptors on carp retinal bipolar cells. *Neuroreport* 8: 1331–1335.

Hankins, M.W. and Ruddock, K.H. 1984. Hyperpolarization of fish retinal cells by kainate and quisqualate. *Nature* 308: 360–362.

Hara, T.J., Law, Y.M.C. and MacDonald, S. 1976. Effects of mercury and copper on the olfactory response in rainbow trout, *Salmo gairdneri*. *Journal of the Fisheries Research Board of Canada* 33: 1568–1573.

Haya, K. 1989. Toxicity of pyrethroid insecticides to fish. *Environmental Toxicology and Chemistry* 8: 381–391.

Hayashi, H. and Fujii, R. 1994. Pharmacological profiles of the subtypes of muscarinic cholinoceptors that mediate aggregation of pigment in the melanophores of two species of catfish. *Pigment and Cellular Research* 7: 175–183.

Heath, A.G. 1995. *Water Pollution and Fish Physiology*. CRC Press, Boca Raton.

Hebebrand, J., Friedl, W., Breidenbach, B. and Propping, P. 1987. Phylogenetic comparison of the photoaffinity-labeled benzodiazepine receptor subunits. *Journal of Neurochemistry* 48: 1103–1108.

Heidelberger, R. and Matthews, G. 1991. Inhibition of calcium influx and calcium current by γ-aminobutyric acid in single synaptic terminals. *Proceedings of the National Academy of Sciences of the United States of America* 88: 7135–7139.

Held, K.D., Sylvester, F.C., Hopcia, K.L. and Biaglow, J.E. 1996. Role of Fenton chemistry in thiol-induced toxicity and apoptosis. *Radiation Research* 145: 542–553.

Helfrich, P. and Banner, A.H. 1963. Experimental induction of ciguatera toxicity in fish through diet. *Nature* 197: 1025–1026.

Henderson, R., Ritchie, J.M. and Strichartz, G.R. 1973. The binding of labelled saxitoxin to the sodium channels in nerve membranes. *Journal of Physiology* 235: 783–804.

Henley, J.M. and Oswald, R.E. 1988. Characterization and regional distribution of glutamatergic and cholinergic ligand binding sites in goldfish brain. *Journal of Neuroscience* 8: 2101–2107.

Hernádi, L. 1993. Fine structural characterization of the olfactory epithelium and its response to divalent cations Cd2+ in the fish *Alburnus alburnus* (*Teleostei, Cyprinidae*): A scanning and transmission electron microscopic study. *Neurobiology* 1: 11–31.

Hidaka, J. 1970. The effects of transition metals on the palatal chemoreceptors of the carp. *Japanese Journal of Physiology* 20: 599–609.

Hillman, D.W., Lin, D. and Burnside, B. 1995. Evidence for D_4 receptor regulation of retinomotor movement in isolated teleost cone inner–outer segments. *Journal of Neurochemistry* 64: 1326–1335.

Hirano, J., Archer, S.N. and Djamgoz, M.B. 1998. Dopamine receptor subtypes expressed in vertebrate (carp and eel) retinae: cloning, sequencing and comparison of five D_1-like and three D_2-like receptors. *Receptors and Channels* 5: 387–404.

Holcombe, G.W., Phipps, G.L. and Tanner, D.K. 1982. The acute toxicity of kelthane, dursban, disulfoton, pydrin, and permethrin to fathead minnows (*Pimephales promelas*) and rainbow trout (*Salmo gairdneri*). *Environmental Pollution* 29A: 167–178.

Holmes, M.J., Lewis, R.J., Poli, M.A. and Gillespie, N.C. 1991. Strain dependent production of ciguatoxin precursors (gambiertoxins) by *Gambierdiscus toxicus* (Dinophyceae) in culture. *Toxicon* 29: 761–775.

Hopkins, C., Grilley, M., Miller, C., Shon, K.J., Cruz, L.J., Gray, W.R., Dykert, J., Rivier, J., Yoshikami, D. and Olivera, B.M. 1995. A new family of *Conus* peptides targeted to the nicotinic acetylcholine receptor. *Journal of Biological Chemistry* 270: 22361–22367.

Hubbard, P.C., Balment, R.J. and McCrohan, C.R. 1996. Adrenergic receptor activation hyperpolarizes the caudal neurosecretory cells of the flounder, *Platichthys flesus*. *Journal of Neuroendocrinology* 8: 153–159.

Huber, C.S. 1972. The crystal structure and absolute configuration of 2,9-diacetyl-9-azabicyclo(4,2,1)non-2,3-ene. *Acta Crystallographica* B28: 2577–2582.

Hughes, G.M., Szegletes, T. and Nemcsók, J. 1997. Study of the effects of brief exposure to an organophosphorus insecticide (methidathion) on blood characteristics of carp (*Cyprinus carpio*). *Acta Biologica Hungary* 48: 157–166.

Hyde, E.G. and Carmichael, W.W. 1991. Anatoxin-a(s), a naturally occurring organophosphate, is an irreversible active site-directed inhibitor of acetylcholinesterase (EC 3.1.1.7). *Journal of Biochemical Toxicology* 6: 195–201.

Iga, T. 1983. Basic properties of beta adrenergic receptors mediating melanosome dispersion of melanophores in a cyprinid fish, *Zacco temmincki*. *Comparative Biochemistry and Physiology C* 76: 297–303.

Ingebrigtsen, K., Hektoen, H., Andersson, T., Bergman, A. and Brandt, I. 1990. Species-specific accumulation of the polychlorinated biphenyl (PCB) 2,3,3′,4,4′-pentachlorobiphenyl in fish brain: a comparison between cod (*Gadus morhua*) and rainbow trout (*Oncorhynchus mykiss*). *Pharmacology and Toxicology* 67: 344–345.

Ingram, L.O., Carey, V.C. and Dombek, K.M. 1982. On the relationship between alcohol narcosis and membrane fluidity. *Substance and Alcohol Actions Misuse* 3: 213–224.

Ishida, A.T. and Cohen, B.N. 1988. GABA-activated whole cell currents in isolated retinal ganglion cells. *Journal of Neurophysiology* 60: 381–396.

Ishida, A.T. and Neyton, J. 1985. Quisqualate and L-glutamate inhibit retinal horizontal-cell responses to kainate. *Proceedings of the National Academy of Sciences of the United States of America* 82: 1837–1841.

Isobe, S., Hazlewood, C.F., Misra, L.K. and Klemm, W.R. 1994. Acute ethanol decreases NMR relaxation times of water hydrogen protons in fish brain. *Alcohol* 11: 571–576.

Iwasaki, H. 1984. Growth physiology of red-tide microorganisms. *Microbiological Science* 1: 179–182.

Janssen-Bienhold, U., Wenzel, A., Hannken, T. and Weiler, R. 1994. Involvement of metabotropic and ionotropic glutamate receptors in inositol polyphosphate formation in carp retinal slices. *European Journal of Neuroscience* 6: 1897–1902.

Jarvinen, A.W., Nordling, B.R. and Henry, M.E. 1983. Chronic toxicity of dursban (chlorpyrifos) to the fathead minnow (*Pimephales promelas*) and the resultant acetylcholinesterase inhibition. *Ecotoxicology and Environmental Safety* 7: 423–434.

Jash, N.B. and Bhattacharya, S. 1983. Phenthoate-induced changes in the profiles of acetylcholinesterase and acetylcholine in the brain of *Anabas testudineus* (Bloch): Acute and delayed effect. *Toxicology Letters* 15: 349–356.

Jett, D.A., Abdallah, E.A.M., El-Fakahany, E.E., Eldefrawi, M.E. and Eldefrawi, A.T. 1991. High-affinity activation by paraoxon of a muscarinic receptor subtype in rat brain striatum. *Pesticide Biochemistry and Physiology* 39: 149–157.

Jolly, Jr, A.L., Avault, Jr, J.W., Koonce, K.L. and Graves, J.B. 1978. Acute toxicity of permethrin to several aquatic animals. *Transactions of the American Fisheries Society* 107: 825–827.

Jones, S.B. and King, L.B. 1995. Muscarinic cholinergic receptors in brain and atrial membranes of adult brook trout (*Salvelinus fontinalis*) measured by radioligand binding techniques. *Comparative Biochemistry and Physiology C* 112: 43–50.

Jozefowski, S.J. and Plytycz, B. 1998. Characterization of beta-adrenergic receptors in fish and amphibian lymphoid organs. *Developmental and Comparative Immunology* 22: 587–603.

Kah, O., Trudeau, V.L., Sloley, B.D., Chang, J.P., Dubourg, P., Yu, K.L. and Peter, R.E. 1992. Influence of GABA on gonadotrophin release in the goldfish. *Neuroendocrinology* 55: 396–404.

Kanazawa, H., Fujimoto, S. and Ohara, A. 1994. On the mechanism of inactivation of active papain by ascorbic acid in the presence of cupric ions. *Biological Pharmacology Bulletin* 17: 789–793.

Karlsson, J.O., Andersson, R.G., Elwing, H. and Grundstrom, N. 1987. Comparative studies on nerve- and noradrenaline-induced melanosome aggregation within different species of fish. *Comparative Biochemistry and Physiology C* 88: 287–291.

Katti, S.R. and Sathyanesan, A.G. 1986. Lead nitrite induced changes in the brain constituents of the freshwater fish, *Clarias batrachus* (L). *Neurotoxicology* 7: 47–52.

Katz, L. and Marquis, J.K. 1989. Modulation of central muscarinic receptor binding in vitro by ultralow levels of the organophosphate paraoxon. *Toxicology and Applied Pharmacology* 101: 114–123.

Kebabian, J.W., DeNinno, M.P., Schoenleber, R., MacKenzie, R., Britton, D.R. and Asin, K.E. 1992. A68930: A potent agonist specific for the dopamine D_1 receptor. *Neurochemistry International* 20 (Suppl.): 157S–160S.

Keizer, J., D'Agostino, G., Nagel, R., Gramenzi, F. and Vittozzi, L. 1993. Comparative diainon toxicity in guppy and zebra fish: Different role of oxidative metabolism. *Environmental Toxicology and Chemistry* 12: 1243–1250.

Keizer, J., D'Agostino, G., Nagel, R., Volpe, T., Gnemi, P. and Vittozzi, L. 1995. Enzymological differences of AChE and diazinon hepatic metabolism: Correlation of *in vitro* data with the selective toxicity of diazinon to fish species. *Science of the Total Environment* 171: 213–220.

Kemp, J.R. and Wallace, K.B. 1990. Molecular determinants of the species-selective inhibition of brain acetylcholinesterase. *Toxicology and Applied Pharmacology* 104: 1–12.

Khan, I.A. and Thomas, P. 1997. Aroclor 1254-induced alterations in hypothalamic monoamine metabolism in the Atlantic croaker (*Micropogonias undulatas*): Correlation with pituitary gonadotropin release. *Neurotoxicology* 18: 553–560.

Khokhryakova, O.N. 1981. Effect of DDT and kepone on Ca2+-dependent calcium ion ATPase from fragments of sarcoplasmic reticulum of cod and wolffish skeletal muscles *Gadus morhua morhua, Anarhikas lupus. Hydrobiology Journal* 17: 72–76.

Koch, R.B. 1969. Inhibition of animal tissue ATPase activities by chlorinated hydrocarbon pesticides. *Chemical-Biological Interactions* 1: 199–209.

Koch, R.B., Cutkomp, L.K. and Yap, H.H. 1971. Inhibition of oligomycin sensitive and insensitive fish adenosine triphosphatase activity by chlorinated hydrocarbon insecticides. *Biochemical Pharmacology* 20: 3243–3245.

Koelle, G.B. and Gilman, A. 1949. Anticholinesterase drugs. *Journal of Pharmacology and Experimental Therapeutics* 95: 166–216.

Kozlovskaya, V.I., Mayer, F.L., Menzikova, O.V. and Chuyko, G.M. 1993. Cholinesterases of aquatic animals. *Reviews in Environmental Contamination and Toxicology* 132: 117–142.

Kuba, K., Albuquerque, E.X. and Barnard, E.A. 1973. Diisopropylfluorophosphate: Suppression of ionic conductance of the cholinergic receptor. *Science* 181: 853–856.

Kubokawa, K., Miyashita, T., Nagasawa, H. and Kubo, Y. 1996. Cloning and characterization of a bifunctional metabotropic receptor activated by both extracellular calcium and glutamate. *Federation of European Biochemical Societies Letters* 392: 71–76.

Kumar, K.S., Rowse, C. and Hochstein, P. 1978. Copper-induced generation of superoxide in human red cell membranes. *Biochemical and Biophysical Research Communications* 83: 587–592.

Kumaraguru, A.K., Beamish, F.W.H. and Ferguson, H.W. 1982. Direct and circulatory paths of permethrin (NRDC-143) causing histopathological changes in the gills of rainbow trout, *Salmo gairdneri* Richardson. *Journal of Fish Biology* 20: 87–91.

Kung, S.S., Wu, Y.M. and Chow, W.Y. 1996. Characterization of two fish glutamate receptor cDNA molecules: absence of RNA editing at the Q/R site. *Brain Research and Molecular Brain Research* 35: 119–130.

Kuzhikandathil, E.V. and Oxford, G.S. 1999. Activation of human D3 dopamine receptor inhibits P/Q-type calcium channels and secretory activity in AtT-20 cells. *Journal of Neuroscience* 19: 1698–1707.

Lamers, A.E., Groneveld, D., de Kleijn, D.P., Geeraedts, F.C., Leunissen, J.A., Flik, G., Wendelaar Bonga, S.E. and Martens, G.J. 1996. Cloning and sequence analysis of a hypothalamic cDNA encoding a D_{1c} dopamine receptor in tilapia. *Biochimica et Biophysica Acta* 1308: 17–22.

Lamers, A.E., Ter Brugge, P.J., Flik, G. and Wendelaar Bonga, S.E. 1997. Acid stress induces a D_1-like dopamine receptor in pituitary MSH cells of *Oreochromis mossambicus*. *American Journal of Physiology* 273: 387–392.

Lasner, M., Roth, L.G. and Chen, C.H. 1995. Structure–functional effects of a series of alcohols on acetylcholinesterase-associated membrane vesicles: Elucidation of factors contributing to the alcohol action. *Archives of Biochemistry and Biophysics* 317: 391–396.

Lawrence, L.J. and Casida, J.E. 1983. Stereospecific action of pyrethroid insecticides on the γ-aminobutyric acid receptor-ionophore complex. *Science* 221: 1399–1401.

Leahey, J.P. 1985. *The Pyrethroid Insecticides*, pp. 201–211. Taylor and Francis, Philadelphia, PA.

Leonard, J.P. and Wickelgren, W.O. 1986. Prolongation of calcium action potentials by gamma-aminobutyric acid in primary sensory neurones of lamprey. *Journal of Physiology* 375: 481–497.

Lewis, R.J. 1992. Ciguatoxins are potent ichthyotoxins. *Toxicon* 2: 207–211.

Lewis, R.J., Sellin, M., Poli, M.A., Norton, R.S., MacLeod, J.K. and Sheil, M.M. 1991. Purification and characterization of ciguatoxins from moray eel (*Lycodontis javanicus*, Muraenidae). *Toxicon* 29: 1115–1127.

Linn, C.P. and Christensen, B.N. 1992. Excitatory amino acid regulation of intracellular Ca^{2+} in isolated catfish cone horizontal cells measured under voltage- and concentration-clamp conditions. *Journal of Neuroscience* 12: 2156–2164.

Linn, J.D. 1968. Effects of low volume aerial spraying of dursban and fenthion on fish. *Down to Earth* 24: 28–30.

Lo, Y.H., Bradley, T.M. and Rhoads, D.E. 1991. L-alanine binding sites and Na+, K(+)-ATPase in cilia and other membrane fractions from olfactory rosettes of Atlantic salmon. *Comparative Biochemistry and Physiology B* 98: 121–126.

Lukasiewicz, P.D. 1996. $GABA_C$ receptors in the vertebrate retina. *Molecular Neurobiology* 12: 181–194.

Lund, A.E. and Narahashi, T. 1981. Interaction of DDT with sodium channels in squid giant axon membranes. *Neuroscience* 6: 2253–2258.

Lund, A.E. and Narahashi, T. 1982. Dose-dependent interaction of the pyrethroid isomers with sodium channels of squid axon membranes. *Neurotoxicology* 3: 11–24.

Lunn, C.R., Toews, D.P. and Pree, D.J. 1976. Effects of three pesticides on respiration, coughing, and heart rates of rainbow trout (*Salmo gairdneri*). *Canadian Journal of Zoology* 54: 214–219.

McIntosh, J.M., Hasson, A., Spira, M.E., Gray, W.R., Li, W., Marsh, M., Hillyard, D.R. and Olivera, B.M. 1995. A new family of conotoxins that blocks voltage-gated sodium channels. *Journal of Biological Chemistry* 270: 16796–16802.

McKim, J.M., Schmieder, P.K., Niemi, G.J., Carlson, R.W. and Henry, T.R. 1987. Use of respiratory-cardiovascular responses of rainbow trout (*Salmo gairdneri*) in identifying acute toxicity syndromes in fish, 2. Malathion, carbaryl, acrolein, and benzaldehyde. *Environmental Toxicology and Chemistry* 6: 295–312.

McNicholl, P.G. and Mackay, W.C. 1975a. Effect of DDT and M.S. 222 on learning a sample conditioned response in rainbow trout (*Salmo gairdneri*). *Journal of the Fisheries Research Board of Canada* 32: 661–665.

McNicholl, P.G. and Mackay, W.C. 1975b. Effect of DDT on discriminating ability of rainbow trout (*Salmo gairdneri*). *Journal of the Fisheries Research Board of Canada* 32: 785–788.

Magour, S., Maser, H. and Steffen, I. 1984. Effect of lindane on synaptosomal Na^+/K^+-ATPase in relation to its subcellular distribution in the brain. *Acta Pharmacologica et Toxicologica* 54: 299–303.

Mahmood, N.A. and Carmichael, W.W. 1987. Anatoxin-a(s), an anticholinesterase from the cyanobacterium *Anabaena flos-aquae* NRC-525–17. *Toxicon* 25: 1221–1227.

Mahmood, N.A., Carmichael, W.W. and Pfahler, D. 1988. Anticholinesterase poisonings in dogs from a cyanobacterial (blue–green algae) bloom dominated by *Anabaena flos-aquae*. *American Journal of Veterinary Research* 49: 500–503.

Maler, L. and Monaghan, D. 1991. The distribution of excitatory amino acid binding sites in the brain of an electric fish, *Apteronotus leptorhynchus*. *Journal of Chemical Neuroanatomy* 4: 39–61.

Markovac, J. and Goldstein, G.W. 1988. Lead activates protein kinase C in immature rat brain microvessels. *Toxicology and Applied Pharmacology* 96: 14–23.

Mathis, C.A. and Tunnicliff, G. 1984. The γ-aminobutyric acid receptor in catfish brain. *Comparative Biochemistry and Physiology C* 78: 479–481.

Matsumura, F. 1985. *Toxicology of Insecticides*. Plenum Press. New York.

Matsumura, F. and Clark, J.M. 1982. ATP-utilizing systems in the squid axons: a review on the biochemical aspects of ion-transport. *Progress in Neurobiology* 18: 231–255.

Matsumura, F. and Ghiasuddin, S.M. 1979. Characteristics of DDT-sensitive Ca-ATPase. In *Neurotoxicity of Insecticides and Pheromones*. Narahashi, T. (ed.), pp. 259–275. Plenum Press, New York.

Matsumura, F. and Ghiasuddin, S.M. 1983. Evidence for similarities between cyclodiene type insecticides and picrotoxinin in their action mechanisms. *Journal of Environmental Science and Health B* 18: 1–14.

Matsushima, T., Tegnér, J., Hill, R.H. and Grillner, S. 1993. $GABA_B$ receptor activation causes a depression of low- and high-voltage-activated Ca^{2+} currents, postinhibitory rebound, and postspike afterhyperpolarization in lamprey neurons. *Journal of Neurophysiology* 70: 2606–2619.

Matthews, G., Ayoub, G.S. and Heidelberger, R. 1994. Presynaptic inhibition by GABA is mediated via two distinct GABA receptors with novel pharmacology. *Journal of Neuroscience* 14: 1079–1090.

Medler, K.F., Tran, H.N., Parker, J.M., Caprio, J. and Bruch, R.C. 1998. Metabotropic glutamate receptor expression in olfactory receptor neurons from the channel catfish, *Ictalurus punctatus*. *Journal of Neurobiology* 35: 94–104.

Meissl, H. and Ekstrom, P. 1991. Action of γ-aminobutyric acid (GABA) in the isolated pineal organ. *Brain Research* 562: 71–78.

Migani, P. 1990. A competitive inhibitor of kainic acid binding from the goldfish nervous tissue. *Brain Research* 518: 179–185.

Mills, L.J. and Klein-Macphee, G. 1979. Toxicity of the New England red tide dinoflagellate to winter flounder larvae. *Developments in Marine Biology* 1: 389–394.

Miyamoto, J. 1976. Degradation, metabolism, and toxicity of synthetic pyrethroids. *Environmental Health Perspective* 14: 15–28.

Moon, T.W., Capuzzo, A., Puviani, A.C., Ottolenghi, C. and Fabbri, E. 1993. Alpha-mediated changes in hepatocyte intracellular calcium in the catfish, *Ictalurus melas*. *American Journal of Physiology* 264: 735–740.

Morishita, F., Katayama, H. and Yamada, K. 1985. Subtypes of beta adrenergic receptors mediating pigment dispersion in chromatophores of the medaka, *Oryzias latipes*. *Comparative Biochemistry and Physiology C* 81: 279–285.

Mukhopadhyay, P.K., Mukherji, A.P. and Dehadrai, P.V. 1982. Certain biochemical responses in air-breathing catfish *Clarias batrachus* exposed to sublethal carbofuran. *Toxicology* 23: 337–345.

Murata, M., Legrand, A.M., Ishibashi, Y., Fukui, M. and Yasumoto, T. 1990. Structures of ciguatoxin and its congener. *Journal of the American Chemical Society* 112: 43880–4386.

Murphy, P., Barnes, J.M., Nielsen, M. and Henley, J.M. 1993. Molecular size of the kainate binding protein in goldfish brain determined by radiation inactivation. *Biochemical Pharmacology* 46: 539–541.

Murphy, S.D., Laurwerys, R.R. and Cheever, K.L. 1968. Comparative anticholinesterase action of organophosphorus insecticides in vertebrates. *Toxicology and Applied Pharmacology* 12: 22–35.

Nagata, K., Aistrup, G.L., Huang, C.S., Marszalec, W., Song, J.H., Yeh, J.Z. and Narahashi, T. 1996. Potent modulation of neuronal nicotinic acetylcholine receptor-channel by ethanol. *Neuroscience Letters* 217: 189–193.

Narahashi, T. 1971. Mode of action of pyrethroids. *Bulletin of the World Health Organization* 44: 337–345.

Narahashi, T. 1979. Nerve membrane ionic channels as the target site of insecticides. In *Neurotoxicity of Insecticides and Pheromones*. Narahashi, T. (ed.), pp. 211–243. Plenum Press, New York.

Narahashi, T., Nishimura, K., Parmentier, J.L., Takeno, K. and Elliott, M. 1977. Neurophysiological study of the structure–activity relation of pyrethroids (insecticides). *American Chemical Society Symposium Series* 42: 85–97.

Nemcsók, J., Asztalos, B., Víg, E. and Orbán, L. 1987. The effect of an organophosphorus pesticide on the enzymes of carp (*Cyprinus carpio* L.). *Acta Biologica Hungary* 38: 77–85.

Nemscok, J.G. and Hughes, G.M. 1988. The effect of copper sulphate on some biochemical parameters of rainbow trout. *Environmental Pollution* 49: 77–85.

Nielsen, M., Braestrup, C. and Squires, R.F. 1978. Evidence for a late evolutionary appearance of brain specific benzodiazepine receptors: An investigation of 18 vertebrate and 5 invertebrate species. *Brain Research* 141: 342–346.

Noga, E.J. 1996. *Fish Disease: Diagnosis and Treatment*. Mosby-Year Book, St. Louis, MO.

Norris, D.O., Felt, S.B., Woodling, J.D. and Dores, R.M. 1997. Immunocytochemical and histological differences in the interrenal axis of feral brown trout, *Salmo trutta*, in metal-contaminated waters. *General Comparative Endocrinology* 108: 343–351.

Norton, S. 1996. Toxic effects of plants In *Casarett and Doull's Toxicology: The Basic Science of Poisons*. Klaassen, C.D. (ed.), pp. 841–853. McGraw-Hill, New York.

Numa, S. 1986. Molecular basis for the function of ionic channels. *Biochemical Society Symposium* 52: 119–143.

O'Connor, P., Kropf, R.B. and Dowling, J.E. 1989. Catecholamine-sensitive adenylate cyclase in the white perch (*Roccus americanus*) retina: evidence for beta-adrenergic and dopamine receptors linked to adenylate cyclase. *Journal of Neurochemistry* 53: 969–975.

Olivera, B.M., McIntosh, J.M., Cruz, L.J., Luque, F.A. and Gray, W.R. 1984. Purification and sequence of a presynaptic peptide toxin from *Conus geographus* venom. *Biochemistry* 23: 5087–5090.

Olivera, B.M., Gray, W.R., Zeikus, R., McIntosh, J.M., Varga, J., Rivier, J., de Santos, V. and Cruz, L.J. 1985. Peptide neurotoxins from fish-hunting cone snails. *Science* 230: 1338–1343.

Olsen, R.W., Bergman, M.O., Van Ness, P.C., Lummis, S.C., Watkins, A.E., Napias, C. and Greenlee, D.V. 1981. γ-aminobutyric acid receptor binding in mammalian brain: Heterogeneity of binding sites. *Molecular Pharmacology* 19: 217–227.

Omeljaniuk, R.J. 1995. Specific [^{3}H]spiperone binding sites in the pituitary of rainbow trout (*Oncorhynchus mykiss*) and goldfish (*Carassius auratus*). *Canadian Journal of Physiology and Pharmacology* 73: 585–593.

Omeljaniuk, R.J. and Peter, R.E. 1989. In vitro binding characteristics of [^{3}H]spiperone to the pituitary of the goldfish (*Carassius auratus*). *General Comparative Endocrinology* 74: 57–67.

Pang, J., Lo, Y.H., Chandlee, J.M. and Rhoads, D.E. 1994. A subtype of the metabotropic glutamate receptor family in the olfactory system of Atlantic salmon. *Federation of European Biochemical Societies Letters* 354: 301–304.

Pavlov, D.F. 1994. A simple pharmacological analysis of the cholinergic control of respiration in perch (*Perca fluviatilis* L.) exposed to DDVP, an organophosphorus insecticide. *Comparative Biochemistry and Physiology C* 108: 107–111.

Pavlov, D.F., Chuiko, G.M., Gerassimov, Y.V. and Tonkopiy, V.D. 1992. Feeding behavior and brain acetylcholinesterase activity in bream (*Abramis brama* L.) as affected by DDVP, an organophosphorus insecticide. *Comparative Biochemistry and Physiology C* 103: 563–568.

Peng, Y.W., Blackstone, C.D., Huganir, R.L. and Yau, K.W. 1995. Distribution of glutamate receptor subtypes in the vertebrate retina. *Neuroscience* 66: 483–497.

Pevzner, R.A., Hernadi, L. and Salanki, J. 1986. Effect of mercury on the fish (*Alburnus alburnus*) chemoreceptor taste buds: A scanning electron microscopic study. *Acta Biologica Hungary* 37: 159–167.

Pinuela, C., Baatrup, E. and Geneser, F.A. 1992a. Histochemical distribution of zinc in the brain of the rainbow trout, *Oncorhynchos myciss*. I. The telencephalon. *Anatomical Embryology* 185: 379–388.

Pinuela, C., Baatrup, E. and Geneser, F.A. 1992b. Histochemical distribution of zinc in the brain of the rainbow trout, *Oncorhynchos myciss*. II. The diencephalon. *Anatomical Embryology* 186: 275–284.

Poli, M.A., Mende, T.J. and Baden, D.G. 1986. Brevetoxins, unique activators of voltage-sensitive sodium channels, bind to specific sites in rat brain synaptosomes. *Molecular Pharmacology* 30: 129–135.

Post, G. and Leasure, R.A. 1974. Sublethal effect of malathion to three salmonid species. *Bulletin of Environmental Contamination and Toxicology* 12: 312–319.

Prasada Rao, K.S. and Ramana Rao, K.V. 1984a. Changes in the tissue lipid profiles of fish (*Oreochromis mossambicus*) during methyl parathion toxicity: A time course study. *Toxicology Letters* 21: 147–153.

Prasada Rao, K.S. and Ramana Rao, K.V. 1984b. Tissue alteration of aminotransferases and total ATPases in the fish (*Tilapia mossambica*) under methyl parathion impact. *Toxicology Letters* 20: 53–57.

Qian, H. and Dowling, J.E. 1995. $GABA_A$ and $GABA_C$ receptors on hybrid bass retinal bipolar cells. *Journal of Neurophysiology* 74: 1920–1928.

Qian, H., Hyatt, G., Schanzer, A., Hazra, R., Hackam, A.S., Cutting, G.R. and Dowling, J.E. 1997. A comparison of $GABA_C$ and rho subunit receptors from the white perch retina. *Vision Neuroscience* 14: 843–851.

Ram, R.N. and Sathyanesan, A.G. 1985. Mercurial induced brain monoamine oxidase inhibition in the teleost *Channa punctatus* (Bloch). *Bulletin of Environmental Contamination and Toxicology* 35: 620–626.

Rao, G.V. and Rao, K.S. 1995. Modulation in acetylcholinesterase of rat brain by pyrethroids in vivo and an in vitro kinetic study. *Journal of Neurochemistry* 65: 2259–2266.

Rashatwar, S.S. and Matsumura, F. 1984. Interaction of DDT and pyrethroids with calmodulin and its significance in the expression of enzyme activities of phosphodiesterase. *Biochemical Pharmacology* 15: 1689–1694.

Rashid, K., Baldridge, W.H. and Ball, A.K. 1993. Evidence for D_2 receptor regulation of dopamine release in the goldfish retina. *Journal of Neurochemistry* 61: 2025–2033.

Rath, S. and Misra, B. 1980. Age-related changes in oxygen consumption by gill, brain and muscle tissues of *Tilapia mossambica* (Peters) exposed to dichlorvos. *Environmental Pollution* 23: 95–101.

Reddy, P.M. and Philip, G.H. 1994. *In vivo* inhibition of AChE and ATPase activities in the tissues of freshwater fish, *Cyprinus carpio* exposed to technical grade cypermethrin. *Bulletin of Environmental Contamination and Toxicology* 52: 619–626.

Reddy, R.S., Jinna, R.R., Uzodinma, J.E. and Desaiah, D. 1988. *In vitro* effect of mercury and cadmium on brain Ca^{++}-ATPase of the catfish *Ictalurus punctatus*. *Bulletin of Environmental Contamination and Toxicology* 41: 324–328.

Reddy, A.T., Ayyanna, K. and Yellamma, K. 1991. Sensitivity of brain cholinesterase to cypermethrin toxicity in freshwater teleost *Tilapia mossambica*. *Biochemical International* 23: 959–962.

Reddy, P.M., Philip, G.H. and Bashamohideen, M. 1992. Regulation of AChE system of freshwater fish, *Cyprinus carpio*, under fenvalerate toxicity. *Bulletin of Environmental Contamination and Toxicology* 48: 18–22.

Rehnberg, B.C. and Schreck, C.B. 1986. Acute metal toxicology of olfaction in coho salmon: Behavior, receptors and odor-metal complexation. *Bulletin of Environmental Contamination and Toxicology* 36: 579–586.

Rehnberg, B.G., Bates, E.H., Smith, R.J.F., Sloley, B.D. and Richardson, J.S. 1989. Brain benzodiazepine receptors in fathead minnows and the behavioral response to alarm pheromone. *Pharmacology Biochemistry and Behavior* 33: 435–442.

Reily, M.D., Holub, K.E., Gray, W.R., Norris, T.M. and Adams, M.E. 1995. Structure–activity relationships for P-type calcium channel-selective omega-agatoxins. *Biochemistry* 34: 4913–4918.

Robinson, S.W. and Caron, M.G. 1997. Selective inhibition of adenylyl cyclase type V by the dopamine D_3 receptor. *Molecular Pharmacology* 52: 508–514.

Rubin, J.G. and Soderlund, D.M. 1992. Interaction of naturally occurring neurotoxins and the pyrethroid insecticide deltamethrin with rainbow trout (*Oncorhynchus mykiss*) brain sodium channels. *Environmental Toxicology and Chemistry* 11: 677–685.

Rubin, J.G. and Soderlund, D.M. 1993. Binding of [^{3}H]batrachotoxinin A-20-α-benzoate and [^{3}H]saxitoxin to receptor sites associated with sodium channels in trout brain synaptoneurosomes. *Comparative Biochemistry and Physiology C* 105: 231–238.

Russell, F.E. 1996. Toxic effects of animal toxins. In *Casarett and Doull's Toxicology: The Basic Science of Poisons*. Klaassen, C.D. (ed.), pp. 801–840. McGraw-Hill, New York.

Ruzo, L.O., Engel, J.L. and Casida, J.E. 1979. Decamethrin metabolites from oxidative, hydrolytic, and conjugative reactions in mice. *Journal of Agriculture and Food Chemistry* 27: 725–731.

Sammak, P.J., Adams, S.R., Harootunian, A.T., Schliwa, M. and Tsien, R.Y. 1992. Intracellular cyclic AMP not calcium, determines the direction of vesicle movement in melanophores: direct measurement by fluorescence ratio imaging. *Journal of Cell Biology* 117: 57–72.

Sancho, E., Ferrando, M.D. and Andreu, E. 1997. Response and recovery of brain acetylcholinesterase activity in the European eel, *Anguilla anguilla*, exposed to fenitrothion. *Ecotoxicology and Environmental Safety* 38: 205–209.

Sanders, H.O., Walsh, D.F. and Campbell, R.S. 1981. *Effects of the Organophosphate Insecticides on Bluegills and Invertebrates in Ponds*. US Department of the Interior/Fish and Wildlife Service, Technical Paper 104. US Department of the Interior/Fish and Wildlife Service, Washington, DC.

Sastry, K.V. and Sharma, K. 1980. Effect of mercuric chloride on the activities of brain enzymes in a freshwater teleost, *Ophiocephalus (Channa) punctatus*. *Archives of Environmental Contamination and Toxicology* 9: 425–430.

Sawyer, P.J., Gentile, J.H. and Sasner Jr, J.J. 1968. Demonstration of a toxin from *Aphanizomenon flos-aquae*. *Canadian Journal of Microbiology* 14: 1199–1204.

Schultz, K., Goldman, D.J., Ohtsuka, T., Hirano, J., Barton, L. and Stell, W.K. 1997. Identification and localization of an immunoreactive AMPA-type glutamate receptor subunit (GluR4) with respect to identified photoreceptor synapses in the outer plexiform layer of goldfish retina. *Journal of Neurocytology* 26: 651–666.

Schwedler, T.E., Tucker, C.S. and Beleau, M.H. 1985. Non-infectious diseases. In *Channel Catfish Culture*. Tucker, C.S. (ed.), pp. 497–541. Elsevier Press, Amsterdam.

Scott, J.G. and Matsumura, F. 1983. Evidence for two types of toxic actions of pyrethroids on susceptible and DDT resistant German cockroaches. *Pesticide Biochemistry and Physiology* 19: 141–150.

Seegal, R.F., Brosch, K.O. and Bush, B. 1986. Polychlorinated biphenyls produce regional alterations of dopamine metabolism in rat brain. *Toxicology Letters* 30: 197–202.

Serra, R., Isani, G., Cattani, O. and Carpené, E. 1996. Effects of different levels of dietary zinc on the gilthead, *Sparus aurata*, during the growing season. *Biology of Trace Elemental Research* 51: 107–116.

Shankland, D.L. 1979. Action of dieldrin and related compounds on synaptic transmission. In *Neurotoxicity of Insecticides and Pheromones*. Narahashi, T. (ed.), pp. 139–153. Plenum Press, New York.

Shankland, D.L. 1983. Neurotoxic action of chlorinated hydrocarbon insecticides. *Neurobehavioral Toxicology and Teratology* 4: 805–811.

Shankland, D.L. and Schroeder, M.E. 1973. Pharmacological evidence for a discrete neurotoxic action of dieldrin (HEOD) in the American cockroach, *Periplaneta americana* (L.). *Pesticide Biochemistry and Physiology* 3: 77–86.

Sharkey, R.G., Jover, E., Couraud, F., Baden, D.G. and Catterall, W.A. 1987. Allosteric modulation of neurotoxin binding to voltage-sensitive sodium channels by *Ptychodiscus brevis* toxin 2. *Molecular Pharmacology* 31: 273–278.

Shimizu, Y., Norte, M. Hori, A., Genenah, A. and Kobayushi, M. 1984. Biosynthesis of saxitoxin analogues: the unexpected pathway. *Journal of the American Chemical Society* 106: 6433–6434.

Shin, S., Wu, P. and Chen, C.H. 1991. Biochemical studies of the actions of ethanol on acetylcholinesterase activity: ethanol–enzyme–solvent interaction. *International Journal of Biochemistry* 23: 169–174.

Shoa-Non, L. and De-Fang, F. 1996. Correlation between biochemical parameters and susceptibility of freshwater fish to malathion. *Journal of Toxicology and Environmental Health* 48: 413–418.

Shon, K.J., Grilley, M., Jacobsen, R., Cartier, G.E., Hopkins, C., Gray, W.R., Watkins, M., Hillyard, D.R., Rivier, J., Torres, J., Yoshikami, D. and Olivera, B.M. 1997. A noncompetitive peptide inhibitor of the nicotinic acetylcholine receptor from *Conus purpurascens* venom. *Biochemistry* 36: 9581–9587.

Shon, K.J., Stocker, M., Terlau, H., Stuhmer, W., Jacobsen, R., Walker, C., Grilley, M., Watkins, M., Hillyard, D.R., Gray, W.R. and Olivera, B.M. 1998. kappa-Conotoxin PVIIA is a peptide inhibiting the shaker K^+ channel. *Journal of Biological Chemistry* 273: 33–38.

Shumway, S.E. 1990. A review of the effects of algal blooms on shellfish and aquaculture. *Journal of the World Aquaculture Society* 21: 65–104.

da Silva, H.C., Segundo, H., Medina, G., Fanta, E. and Bacila, M. 1993. Sub-lethal effects of the organophosphate Folidol 600 (methyl parathion) on *Callichthys callichthys* (Pisces: Teleostei). *Comparative Biochemistry and Physiology C* 105: 197–201.

Simons, T.J.B. 1986. Cellular interactions between lead and calcium. *British Medical Bulletin* 42: 431–434.

Skolnick, P., Boje, K., Miller, R., Pennington, M. and Maccecchini, M.L. 1992. Noncompetitive inhibition of N-methyl-D-aspartate by conantokin-G: Evidence for an allosteric interaction at polyamine sites. *Journal of Neurochemistry* 59: 1516–1521.

Smeraski, C.A., Dunwiddie, T.V., Diao, L. and Finger, T.E. 1998. Excitatory amino acid neurotransmission in the primary gustatory nucleus of the goldfish *Carassius auratus*. *Annals of the New York Academy of Science* 855: 442–449.

Smith, J.R. 1984. Fish Neurotoxicology. In *Aquatic Toxicology*. Weber, L.J. (ed.), pp. 107–151. Raven Press, New York.

Smith, M.J. and Heath, A.G. 1979. Acute toxicity of copper, chromate, zinc, and cyanide pollutants to freshwater fish: Effects of different temperatures. *Bulletin of Environmental Contamination and Toxicology* 22: 113–119.

Spiro, J.E. 1997. Differential activation of glutamate receptor subtypes on a single class of cells enables a neural oscillator to produce distinct behaviors. *Journal of Neurophysiology* 78: 835–847.

Staatz, C.G., Bloom, A.S. and Lech, J.J. 1980. A pharmacological analysis of mechanisms of permethrin neurotoxicity in mice. *Federal Proceedings* 39: 624.

Steffey, M.E., Snyder, G.L., Barrett, R.W., Fink, J.S., Ackerman, M., Adams, P., Bhatt, R., Gomez, E. and MacKenzie, R.G. 1991. Dopamine D_1 receptor stimulation of cyclic AMP accumulation in COS-1 cells. *European Journal of Pharmacology* 207: 311–317.

Steidinger, K.A., Burklew, M.A. and Ingle, R.M. 1972. The effects of *Gymnodinium breve* toxin on estuarine animals. In *Marine Pharmacognosy*. Martin, D.F., and Padillla, G.M. (eds), pp. 179–202. Academic Press, New York.

Straus, D.L. and Chambers, J.E. 1995. Inhibition of acetylcholinesterase and aliesterase of fingerling channel catfish by chlorpyrifos, parathion, and *S,S,S*-tributyl phosphorotrithionate (DEF). *Aquatic Toxicology* 33: 311–324.

Stuart, A.M. and Baden, D.G. 1988. Florida red tide brevetoxins and binding in fish brain synaptosomes. *Aquatic Toxicology* 13: 271–280.

Stuart, A.M., Bloomquist, J.R. and Soderlund, D.M. 1987. Pharmacological characterization of the voltage-dependent sodium channels of rainbow trout brain synaptosomes. *Brain Research* 437: 77–82.

Subburaju, S. and Selevarajan, V.R. 1988. Effect of acute toxic concentrations of chlorpyrifos on the acetylcholinesterase activity in different regions of the brain of edible fish, *Tilapia mossambica* (Peters). *Indian Journal of Comparative Animal Physiology* 6: 86–89.

Suresh, A., Sivaramkrisha, B., Victoriamma, P.C. and Radhakrishnaiah, K. 1992. Comparative study on the inhibition of acetylcholinesterase activity in the freshwater fish *Cyprinus carpio* by mercury and zinc. *Biochemical International* 26: 367–375.

Sutterlin, A.M. and Sutterlin, N. 1971. Electrical responses of the olfactory epithelium of Atlantic salmon (*Salmo salar*). *Journal of the Fisheries Research Board of Canada* 28: 565–572.

Svensson, S.P., Bailey, T.J., Pepperl, D.J., Grundstrom, N., Ala-Uotila, S., Scheinin, M., Karlsson, J.O. and Regan, J.W. 1993. Cloning and expression of a fish alpha 2-adrenoceptor. *British Journal of Pharmacology* 110: 54–60.

Svensson, S.P., Adolfsson, P.I., Grundstrom, N. and Karlsson, J.O. 1997. Multiple alpha 2-adrenoceptor signalling pathways mediate pigment aggregation within melanophores. *Pigment Cell Research* 10: 395–400.

Szabó, A., Nemcsók, J., Kása, P. and Gulya, K. 1989. Muscarinic cholinergic components in the carp brain. *Neurochemical International* 15: 511–516.

Szabó, A., Nemcsók, J., Asztalos, B., Rakonczay, Z., Kása, P. and Hieu, L.H. 1992. The effect of pesticides on carp (*Cyprinus carpio* L.). Acetylcholinesterase and its biochemical characterization. *Ecotoxicology and Environmental Safety* 23: 39–45.

Szegletes, T., Bálint, T., Szegletes, Z. and Nemcsók, J. 1995. *In vivo* effects of deltamethrin exposure on activity and distribution of molecular forms of carp AChE. *Ecotoxicology and Environmental Safety* 31: 258–263.

Szeto, S.Y., Sundaram, K.M. and Feng, J. 1985. Inhibition of brain AChE in brook trout by aminocarb and its toxic metabolites. *Journal of Environmental Science and Health* 20: 559–575.

Tachibana, M. 1985. Permeability changes induced by L-glutamate in solitary retinal horizontal cells isolated from *Carassius auratus*. *Journal of Physiology (London)* 358: 153–167.

Takahashi, K., Miyoshi, S. and Kaneko, A. 1994. Two components of GABA-induced currents in catfish retinal horizontal cells. *Japanese Journal of Physiology* 2: S141–S144.

Takahashi, K., Miyoshi, S. and Kaneko, A. 1995. GABA-induced chloride current in catfish horizontal cells mediated by non-$GABA_A$ receptor channels. *Japanese Journal of Physiology* 45: 437–456.

Tang, L., Todd, R.D., Heller, A. and O'Malley, K.L. 1994. Pharmacological and functional characterization of D_2, D_3 and D_4 dopamine receptors in fibroblast and dopaminergic cell lines. *Journal of Pharmacology and Experimental Therapeutics* 268: 495–502.

Tegnér, J., Matsushima, T., el Manira, A. and Grillner, S. 1993. The spinal GABA system modulates burst frequency and intersegmental coordination in the lamprey: differential effects of $GABA_A$ and $GABA_B$ receptors. *Journal of Neurophysiology* 69: 647–657.

Thangnipon, W., Thangnipon, W., Luangpaiboon, P. and Chinabut, S. 1995. Effects of the organophosphate insecticide, monocrotophos, on acetylcholinesterase activity in the nile tilapia fish (*Oreochromis niloticus*) brain. *Neurochemical Research* 20: 587–591.

Thirugnanam, M. and Forgash, A.J. 1977. Environmental impact of mosquito pesticides: Toxicity and anticholinesterase activity of chlorpyrifos to fish in a salt marsh habitat. *Archives of Environmental Contamination and Toxicology* 5: 415–425.

Tobias, J.M., Krollros, J.J. and Savit, J. 1946. Acetylcholine and related substrates in the cockroach, fly, and crayfish and the effect of DDT. *Journal of Cellular Comparative Physiology* 28: 159–182.

Tong, C.K., Pan, M.P. and Chang, Y.C. 1992. Characterization of L-glutamate and kainate binding sites in the brain of a freshwater fish, *Tilapilia monsambica*. *Neuroscience* 49: 237–246.

Toraro, E.A., Pisanti, F.A., Glees, P. and Continillo, A. 1986. The effect of copper pollution on mitochondrial degeneration. *Marine Environmental Research* 18: 245–253.

Trudeau, V.L., Sloley, B.D. and Peter, R.E. 1993. GABA stimulation of gonadotrophin-II release in goldfish: Involvement of $GABA_A$ receptors, dopamine, and sex steroids. *American Journal of Physiology* 265: 348–355.

Tucker, C.S. and Robinson, E.H. 1990. *Channel Catfish Farming Handbook*. Van Nostrand Reinhold, New York.

Tunnicliff, G. and Smith, J.A. 1981. Action of analogues on γ-aminobutyric acid recognition sites in rat brain. *Neurochemical International* 3: 371–376.

Usup, G., Kulis, D.M. and Anderson, D.M. 1994. Growth and toxin production of the toxic dinoflagellate *Pyrodinium bahamense var. compressum* in laboratory cultures. *Natural Toxins* 2: 254–262.

Veith, G.D. and Broderius, S.J. 1990. Rules for distinguishing toxicants that cause type I and type II narcosis syndromes. *Environmental Health Perspectives* 87: 207–211.

Verma, S.R., Gupta, A.K., Bansal, S.K. and Dalela, R.C. 1978a. *In vitro* disruption of ATP dependent active transport following treatment with aldrin and its epoxy analog dieldrin in a fresh water teleost, *Labeo rohita*. *Toxicology* 11: 193–201.

Verma, S.R., Bansal, S.K., Gupta, A.K. and Dalela, R.C. 1978b. *In vivo* effect on ATPase in certain tissues of *Labeo rohita* and *Saccobranchus fossilis*, following chronic chlordane intoxication. *Bulletin of Environmental Contamination and Toxicology* 20: 769–777.

Verma, S.R., Bansal, S.K., Gupta, A.K. and Dalela, R.C. 1979. Sensitivity of ATPase system in certain tissues of fish *Labeo rohita* and *Saccobranchus fossilis* following in vitro treatment with chlordane. *Environmental Research* 19: 14–22.

Verma, S.R., Tonk, I.P., Gupta, A.K. and Dalela, R.C. 1981. *In vivo* enzymatic alterations in certain tissues of *Saccobranchus fossilis* following exposure to four toxic substances. *Environmental Pollution* 26: 121–127.

Verma, S.R., Madhur, J. and Tonk, I.P. 1983. In vivo effect of mercuric chloride on tissue ATPases of *Notopterus notopterus*. *Toxicology Letters* 16: 305–309.

Villmann, C., Bull, L. and Hollmann, M. 1997. Kainate binding proteins possess functional ion channel domains. *Journal of Neuroscience* 17: 7634–7643.

Volpe, L.S., Biagioni, T.M. and Marquis, J.K. 1985. In vitro modulation of bovine caudate muscarinic receptor number by organophosphates and carbamates. *Toxicology and Applied Pharmacology* 78: 226–234.

Wallace, K.B. and Herzberg, U. 1988. Reactivation and aging of phosphorylated brain acetylcholinesterase from fish and rodents. *Toxicology and Applied Pharmacology* 92: 307–314.

Wang, Y., Harsanyi, K. and Mangel, S.C. 1997. Endogenous activation of dopamine D_2 receptors regulates dopamine release in the fish retina. *Journal of Neurophysiology* 78: 439–449.

Warngard, L., Fransson, R., Drakenberg, T.B., Flodstrom, S. and Ahlborg, U.G. 1988. Calmodulin involvement in TPA and DDT induced inhibition of intercellular communication. *Chemical-Biological Interactions* 65: 41–49.

Washino, R.K., Ahmed, W., Linn., J.D. and Whitesell, K.G. 1972. Rice field mosquito control studies with low volume Dursban® sprays in Colusa County, California. IV. Effects upon aquatic nontarget organisms. *Mosquito News* 32: 531–537.

Weiner, N. and Molinoff, P.B. 1994. Catecholamines. In *Basic Neurochemistry*. Siegel, G.J., Agranoff, B.W., Albers, R.W. and Molinoff, P.B. (eds), pp. 261–281. Raven Press, New York.

Weiss, C.M. 1958. The determination of cholinesterase in the brain tissue of three species of fresh water fish and its inactivation *in vivo*. *Ecology* 39: 192–199.

van der Wel, H. and Welling, W. 1989. Inhibition of acetylcholinesterase in guppies (*Poecilia reticulata*) by chlorpyrifos at sublethal concentrations: methodological aspects. *Ecotoxicology and Environmental Safety* 17: 205–215.

Werner, P., Hussy, N., Buell, G., Jones, K.A. and North, R.A. 1996. D_2, D_3, and D_4 dopamine receptors couple to G protein-regulated potassium channels in Xenopus oocytes. *Molecular Pharmacology* 49: 656–661.

White, A.W. 1981. Sensitivity of marine fishes to toxins from the red tide dinoflagellate *Gonyaulax excavata* and implications for fish kills: Herring, pollock, flounder, salmon, and cod. *Marine Biology* 65: 255–260.

Wilkinson, M., Wilkinson, D.A., Khan, I. and Crim, L.W. 1983. Benzodiazepine receptors in fish brain: [^{3}H]-Flunitrazepam binding and modulatory effects of GABA in rainbow trout. *Brain Research Bulletin* 10: 301–303.

Willard, J.M., Ziegra, C.J. and Oswald, R.E. 1991. The interaction of a kainate receptor from goldfish brain with a pertussis toxin-sensitive GTP-binding protein. *Journal of Biological Chemistry* 266: 10196–10200.

Witt, M.-R. and Nielsen, M. 1994. Characterization of the influence of unsaturated free fatty acids on brain GABA/benzodiazepine receptor binding *in vitro*. *Journal of Neurochemistry* 62: 1432–1439.

Wo, Z.G. and Oswald, R.E. 1996. Ligand-binding characteristics and related structural features of the expressed goldfish kainate receptors: Identification of a conserved disulfide bond and three residues important for ligand binding. *Molecular Pharmacology* 50: 770–780.

Wong, A.O., Chang, J.P. and Peter, R.E. 1993. Characterization of D_1 receptors mediating dopamine-stimulated growth hormone release from pituitary cells of the goldfish, *Carassius auratus*. *Endocrinology* 133: 577–584.

Wong, A.O., Le Drean, Y., Liu, D., Hu, Z.Z., Du, S.J. and Hew, C.L. 1996. Induction of chinook salmon growth hormone promoter activity by the adenosine 3′,5′-monophosphate (cAMP)-dependent pathway involves two cAMP-response elements with the CGTCA motif and the pituitary-specific transcription factor Pit-1. *Endocrinology* 137: 1775–1784.

Yamaguchi, I., Matsumura, F. and Kadous, A.A. 1979. Inhibition of synaptic ATPases by heptachlorepoxide in rat brain. *Pesticide Biochemistry and Physiology* 11: 285–293.

Yamaguchi, I., Matsumura, F. and Kadous, A.A. 1980. Heptachlor epoxide: effects on calcium-mediated transmitter release from brain synaptosomes in rat. *Biochemical Pharmacology* 29: 1815–1823.

Yan, Z., Song, W.J. and Surmeier, J. 1997. D2 dopamine receptors reduce N-type Ca^{2+} currents in rat neostriatal cholinergic interneurons through a membrane-delimited, protein-kinase-C-insensitive pathway. *Journal of Neurophysiology* 77: 1003–1015.

Yap, H.H., Desaiah, D., Cutkomp, L.K. and Koch, R.B. 1975. In vitro inhibition of fish brain ATPase activity by cyclodiene insecticides and related compounds. *Bulletin of Environmental Contamination and Toxicology* 14: 163–167.

Yasuoka, A., Abe, K., Arai, S. and Emori, Y. 1996. Molecular cloning and functional expression of the alpha1A-adrenoceptor of Medaka fish, *Oryzias latipes*. *European Journal of Biochemistry* 235: 501–507.

Yazejian, B. and Fain, G.L. 1992. Excitatory amino acid receptors on isolated retinal ganglion cells from the goldfish. *Journal of Neurophysiology* 67: 94–107.

Zelson, P.R. and Cagan, R.H. 1979. Biochemical studies of taste sensation. VIII. Partial characterization of alanine-binding taste receptor sites of catfish *Ictalurus punctatus* using mercurials: Sulfhydryl reagents, trypsin and phospholipase C. *Comparative Biochemistry and Physiology B* 64: 141–147.

Zhang, J., Desilets, M. and Moon, T.W. 1992. Adrenergic modulation of Ca^{2+} homeostasis in isolated fish hepatocytes. *General Comparative Endocrinology* 88: 267–276.
Ziegra, C.J., Willard, J.M. and Oswald, R.E. 1992. Biochemical characterization of kainate receptors from goldfish brain. *Molecular Pharmacology* 42: 203–209.
Zinkl, J.G., Lockhart, W.L., Kenny, S.A. and Ward, F.J. 1991. The effects of cholinesterase inhibiting insecticides on fish. In *Cholinesterase-Inhibiting Insecticides*. Mineau, P. (ed.), pp. 234–254. Elsevier, New York.

3 Fish immunotoxicology

Understanding mechanisms of action

Charles D. Rice

The purpose of this chapter is to provide a general overview of our current understanding of fish immunobiology, the possible mechanisms of toxicity associated with xenobiotics that target the immune system of fish, and explore new frontiers in the discipline. A summary of the literature devoted to descriptive and historical studies in fish immunotoxicology has been minimized. Instead, the reader is referred to a partial listing of key reviews from the past 25 years (Snieszko, 1974; Sinderman, 1979; Zeeman and Brindley, 1981; Sinderman *et al.*, 1982; Porter *et al.*, 1984; Zeeman, 1986; Anderson, D.P., 1990; Weeks *et al.*, 1992; Dunier and Siwicki, 1993; Anderson, R.S., 1994; Zeeman, 1994; Zelikoff, 1994; Anderson and Zeeman, 1995).

Introduction

The ability to distinguish between self, non-self, and altered self is critical for survival. To that end, a sensory system that recognizes non-self has been selected for in all animals. This system requires a division of labor among certain populations of cells and a mechanism to communicate this information throughout the host in a reasonably short period of time. As with other physiologic systems, part of this division of labor is dedicated to the continued renewal of specialized cell types that comprise the organs of the immune system. In terms of a sensory system, it is this collection of cells, tissues, and organs and their interactions that we call the immune system and it includes the various mucosal tissues, the integument, and the lymphoid and myeloid tissues, as well as circulating or trafficking blood leukocytes.

With regards to immunobiology, individuality is based on more than mere separateness. As discussed later, self is genetically determined at the highly polymorphic major and minor histocompatability gene complex loci. Moreover, the term 'immune' implies that once the information of non-self is encountered, processed, and presented to other cells of the immune system, self is capable of responding more rapidly, with greater efficiency and with less energy expenditure upon subsequent exposures. Not all cellular functions of the immune system co-ordinate enhanced immune responses to subsequent exposures to non-self. The innate, non-specific components of the immune system are equally critical. Such non-specific aspects of the immune system are the first-line defense and are not dependent on previous exposure to non-self.

The immune system(s) of the vertebrates is far better characterized than those of the invertebrates. Even so, compared with mammals, we still know very little about immune function in lower vertebrates. In fact, most of what we do know about the immune systems of lower vertebrates, and especially our approach to comparative immunobiology, is derived from mammalian biomedical research. As a consequence, comparative immunologists often impose rodent and human immunobiology on lower vertebrates. This may be appropriate in some aspects of comparative immunology, but not all.

The focus of this chapter is on the immunobiology of fish. There are over 20 000 different species ranging from the primitive agnathan cyclostomes (e.g. hagfish) to the gnathan fish including the chondrosteans (sturgeons, paddlefish), the elasmobranchs (sharks, skates, and rays), and the teleosts, or modern bony fish (cod, trout, catfish, tuna, killifish, zebrafish, blue gill, grouper, etc.). It is impossible to draw simple conclusions about the immunobiology of fish because the differences between immune systems of the major groups and between species within each group may be substantial.

Although primitive in comparison with mammals, the immune systems of fish are highly evolved to meet the demands of the environment in which they live. Surrounded by water of optimal temperature, fish exist in a relatively stable environment. Water, however, is teaming with microorganisms, some of which are pathogenic while others may share a commensal or mutualistic relationship with fish. The immune system must distinguish between these relationships. In addition, the water column contains dissolved inorganic and organic substances, some of which may affect immunocompetence. As a consequence, the mucosa of the alimentary canal, skin mucosa in contact with water, and gill lamellae collectively form the most important immune organ, the integument. The importance of the integument as an immune organ is maintained throughout the evolution of vertebrates, culminating in an extremely complex component of the immune system in homeothermic vertebrates.

As stated above, the immune system must be able to replenish immunocompetent cells and it must be in a sentinel location to process, present, and communicate information about the form and nature of foreign (non-self) material to other cells of the immune system. The cells of the immune system of all gnathan vertebrates are either myeloid or lymphoid in origin and function (Figure 3.1). It is generally accepted that myeloid and lymphoid cells of fish are derived from a single pleuripotent stem cell, as is the case with higher vertebrate systems.

In higher vertebrates, myeloid and lymphoid cell types are generated in two separate compartments, the lymphoid organs (lymph nodes, thymus, and spleen) and the myeloid organ (the bone marrow). Fish do not have bone marrow or lymph nodes. Instead, during development, the cellular organization of kidney differentiates into a hematopoietic anterior portion and a renal organ in the caudal region. In some species of fish, such as the channel catfish, *Ictalurus punctatus*, these two areas are anatomically distinct. In others, there is a gradation of cell

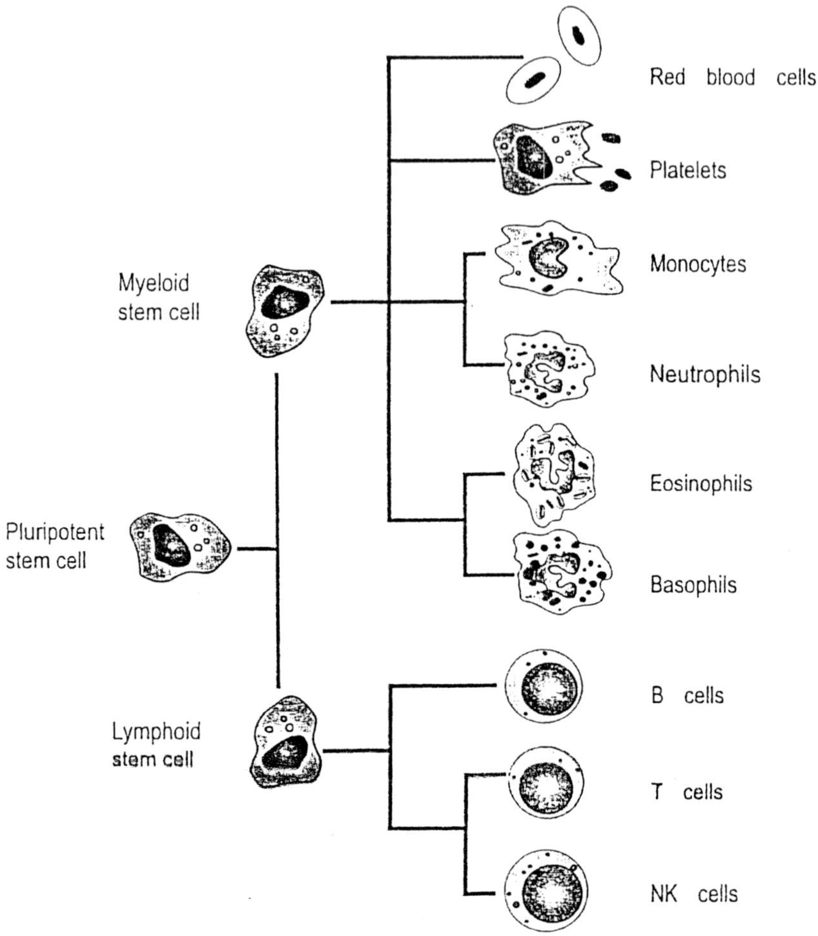

Figure 3.1 Origin of cells from the pleuripotent stem cell. Note that basophils are listed; however, this is a rare cell type in fish. Note also that in addition to natural killer (NK)-like cells fish have non-specific cytotoxic cells (NCC) (adapted from Tizard, 1996).

types within the organ, with hematopoiesis taking place in the most anterior portion and renal function taking place in the caudal portion. Unlike higher vertebrates, the primary lymphoid tissues in fish contain both lymphoid and myeloid cells in different stages of development. This is especially true of the anterior kidney. Moreover, as discussed later, the primary lymphomyeloid tissues of teleosts also contain neuroendocrine tissues.

The lymphoid structure of elasmobranchs (sharks, skates, and rays) is not only quite different from that of humans and rodents but also from that of modern bony fish (McKinney, 1992; Luer *et al.*, 1995) (Figure 3.2). Although typical sharks, rays, and skates have a true spleen, most have epigonal tissues and Leydig organs as the major lymphomyeloid organs. Hagfish are very primitive

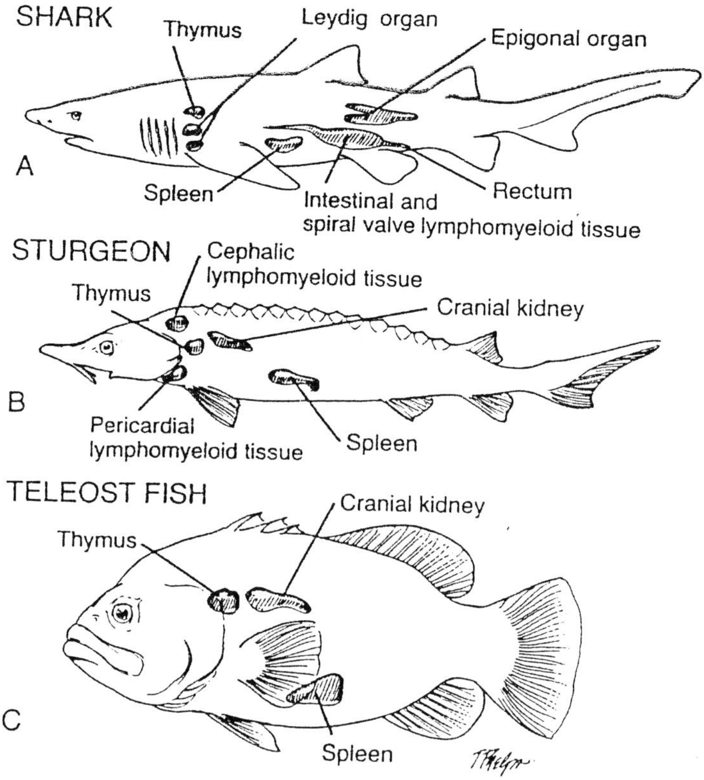

Figure 3.2 Locations of lymphopoietic tissues in elasmobranchs (A), sturgeons (B), and teleosts (C) (after Kenedy-Stoskopf, 1993).

representative vertebrates in that leukocyte stem differentiation seems to occur in the peripheral blood (Fange, 1994). Basic lymphoid anatomy is fairly uniform among teleosts. The primary lymphoid organs are the thymus and anterior kidney (the bone marrow equivalent of higher vertebrates), whereas the spleen functions as a secondary organ. Lymphomyeloid tissues of sturgeons and paddlefish are located in the cranial bones, in the cardiac muscle, as well as in the anterior kidney, thymus, and spleen. In addition, mucosal immunity seems to be present in all fish. As one can see, the differences in lymphomyeloid structures between the major groups of fish are in contrast to that of mammals, whose anatomy is comparatively uniform.

The variety of lymphoid or myeloid structures of fish serve to establish a network of phagocytes, endothelial cells, and stroma that process information regarding non-self. Body cavities also have wandering phagocytes. This entire network is referred to as the reticuloendothelial system (RES), or, from the older literature, the mononuclear phagocyte system. The reader is referred to a relatively recent review of the RES in fish by Dalmo *et al.* (1997), including the references therein.

The sentinel phagocytes of the RES are named according to their anatomic location. For example, microglia are located in the cranial cavity, and there are splenic macrophages, wandering serosal macrophages in the peritoneal cavity, and endothelium-fixed Kupffer cells in the liver. Blood monocytes are a developmental component of the RES as they are actively phagocytic and, upon maturation in various tissues, become fixed tissue macrophages. Although neutrophils and other granulocytes are not considered to be RES phagocytes in mammals, the neutrophil is the predominant phagocyte found in the anterior kidney and spleen of some fish (Ellis, 1977; Ainsworth, 1992).

The liver of mature fish is not considered to be as important to immune function as it is in mammals. Macrophage aggregations are common in the liver as well as in the spleen and kidney of most fish (Agius, 1985; Kenedy-Stoskopf, 1993). These reticulated regions of cells have a debated function. It is hypothesized that these regions are the phylogenetic precursors of germinal centers; however, they tend to increase in number with age and under oxidative stress, as well as with disease even in younger fish (Camp, 1997). These aggregations are highly pigmented with lipofuscin, melanin, hemosiderin, and, in some cases, ceroid. Lipofuscin and ceroid are byproducts of saturated fat oxidation, whereas hemosiderin is a byproduct of hemoglobin degradation. Melanin has antioxidant properties and is a predominant feature of macrophage aggregations in channel catfish during disease stress (Figure 3.3). Taken together, macrophage aggregations seem to be sites of inflammation, oxygen radical formation, and lipid oxidation. Furthermore, genetic resistance to certain diseases in channel catfish correlate with increased numbers of splenic macrophage aggregations (Camp, 1997; Camp *et al.*, 2000) (Figure 3.3). Kupffer cells (liver macrophages) have been described in a few species of fish based on morphologic criteria only. Faisal *et al.* (1995) isolated a continuous cell line from the liver of spot, *Leiostomus xanthurus*, that has enzymatic staining characteristics of macrophages and seems to proliferate in response to mitogens. As with mammals, the liver of fish is a source of acute phase proteins (Demers and Bayne, 1994) which augment the RES in times of inflammation (see next section). Secretory products (cytokines) from the RES signal the liver to secrete these factors which almost exclusively enhance innate, non-specific immune functions, including bacterial lysis, enhanced chemotaxis and phagocytosis, and inflammatory responses. Efforts to understand the impact of environmental contaminants on the intact RES of fish have given way to an emphasis on phagocyte biology, as evidenced by a proliferation of studies over the last 15 years that investigated neutrophils and macrophages (Weeks and Warinner, 1986; Weeks *et al.*, 1988, Payne and Fancey, 1989; Rice and Weeks, 1989; Seeley and Weeks-Perkins, 1991; Secombes, 1992; Bowser *et al.*, 1994; Wester *et al.*, 1994; Muhvich *et al.*, 1995; Rice *et al.*, 1996; Secombes *et al.*, 1997; Fournier *et al.*, 1998; Salo *et al.*, 1998).

With relevance to the RES, it is important to recognize that neutrophils and macrophages require the full milieu of contact with stromal, endothelial, and other cells (lymphocytes) via adhesion molecules in order to affect their functions fully, including activation (Ainsworth and Boyd, 1998). Although modern cellular and

(A)

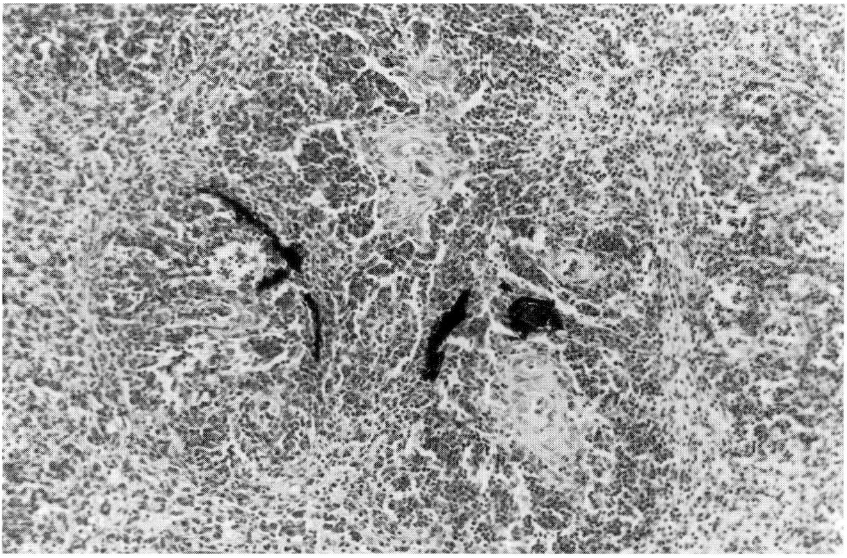

(B)

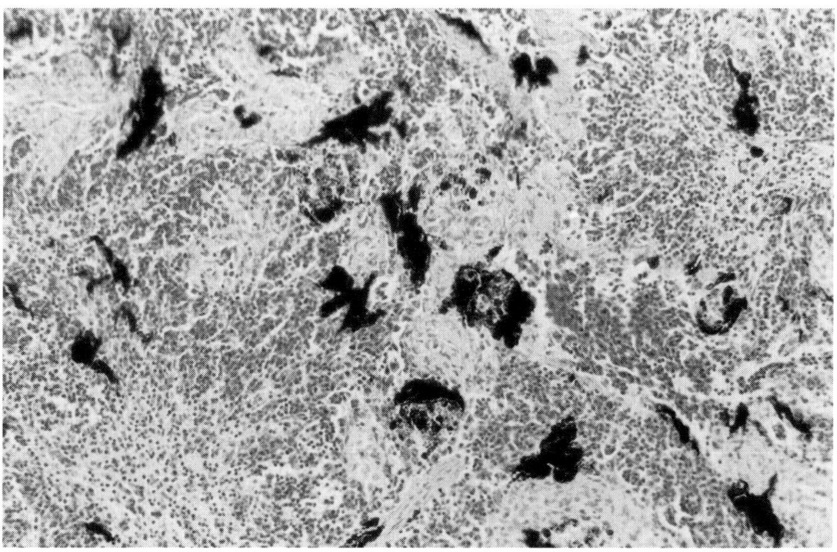

Figure 3.3 Macrophage aggregations in the spleen of channel catfish, *Ictalurus punctatus*. (A) Spleen of an *Edwardsiella ictalura*-sensitive fingerling (400×; hematoxylin and eosin staining). (B) Spleen of an *Edwardsiella ictaluri*-resistant fingerling. Note the larger and more numerous macrophage aggregations in the resistant fish (courtesy of Dr Karen Camp).

molecular approaches to understanding the biology of macrophages *in vitro* are indeed interesting and relatively reproducible, their true functions cannot be fully appreciated outside of the context of RES anatomy and physiology. As detailed by Dalmo *et al.* (1997), the architecture of the RES varies between groups of fish, therefore the immunophysiology of the RES between groups would be expected to vary as well.

Innate, non-specific, and first-line defenses

The simplest way to deal with pathogenic non-self or potential proinflammatory agents is to retard their entry in a non-specific fashion. The following overview to the subject is an adaptation of several investigators, but principally Dalmo *et al.* (1997). As mentioned earlier, the integument is important as a barrier. To this end, fish secrete a layer of glycoproteins and proteoglycans as a layer of mucus. Other components such as lysozyme, complement components, and unique antimicrobial peptides can be found in the mucus as well. Each of these components may be increased during exposure to a pathogen (Fletcher and White, 1973; Yang and Albright, 1994). The other surface exposed to the outside is the intestinal lining, a tissue that, in many fish, is the first to be exposed to environmental contaminants. Mucus secretion, the natural low pH of the stomach, and peptide-digesting enzymes in the intestine provide for a hostile environment for microbial invasion. Moreover, at least in teleosts, intraepithelial lymphocytes, macrophages, and granulocytes are found in the region of the gut called the lamina propria. Elasmobranchs also have gut-associated lymphoid tissues (Fange, 1994).

As with higher vertebrates, connective tissue helps to provide a barrier to the exterior and provides stroma for the organs and tissues of the immune system. To that end, various leukocytes are found in the heart wall and the gills of some fish (Ainsworth, 1992). Although expulsion of pathogens from mucosal-rich regions by mast cell products is common in mammals, especially humans, fish do not have mast cells. Instead, the eosinophilic granular cell may be the fish equivalent of mast cells and is more common in marine than in freshwater fish (Vallejo and Ellis, 1989; Roszell and Rice, 1998). These cells are found in the intestine and swim bladder of many fish as well as in the anterior kidney of estuarine killifish (Roszell and Rice, 1998).

As is the case with higher vertebrates, fish rely on circulating (humoral) non-specific defense mechanisms. Lysozyme is secreted into the circulation by leukocytes, mostly phagocytes, located either in the circulating pool of cells or in the various lymphomyeloid tissues. Lysozyme is active against the peptidoglycan layer of Gram-positive bacteria and is also active against certain Gram negatives. Both stress and pollution levels affect lysozyme levels in fish (Brousseau *et al.*, 1994; Marc *et al.*, 1995).

Not all humoral non-specific mediators of defense are secreted by the immune system. During inflammation, the liver secretes various acute phase proteins in response to leukocyte secretions (Figure 3.4) that subsequently modify leukocyte functions and enable the non-specific arm of the immune system to participate in

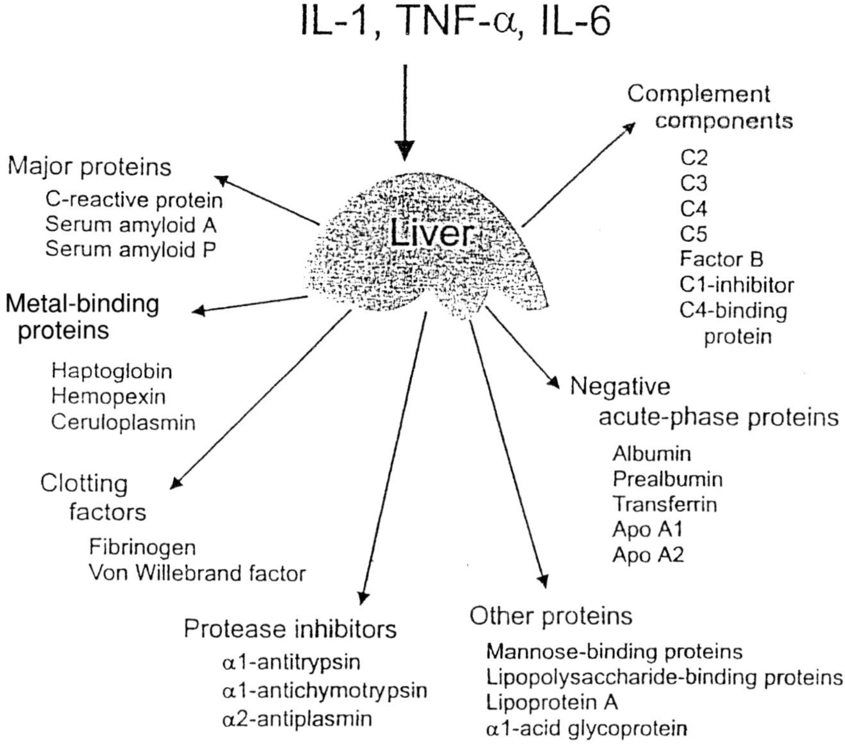

Figure 3.4 Some of the acute phase proteins released by the liver of mammals in response to proinflammatory cytokines. Interleukin 1, tumor necrosis factor α, and interleukin 6 are products of activated macrophages. Although not all of these products have been identified in fish, they are thought to exist (adapted from Tizard, 1996).

specific reactions. This process is best understood in mammals, in which macrophage, and perhaps neutrophil, cytokines released at the site of inflammation circulate to the liver and induce the transcription of these proteins. Serum amyloid protein (SA), lipopolysaccharide-binding protein (LBP), manose-binding protein (MBP), and C-reactive protein (CRP) readily bind to lectins, lipopolysaccharides, carbohydrates, and the C-polysaccharides and phosphorylcholine respectively. Each of these, and others, modify phagocyte activities, including enhanced reactivity to complement components, phagocytosis, chemotaxis, and the respiratory burst. Fish also have a wide variety of agglutinins present in serum, mucus, and bile that appear to be mostly carbohydrate-binding lectins and are either proteins or glycoproteins. Unlike antibodies, these lectins are constitutive. Metal-binding proteins are also produced and include metallothionein (binds copper, zinc, mercury, cadmium, others), apotransferrins (binds iron), and ceruloplasmin. The production of metal-binding proteins is also induced during inflammation, therefore they may be considered as acute phase proteins. In the ongoing battle between pathogen and host, one of the mechanisms that pathogens use to get past the physical and chemical barriers of non-specific immunity is to digest the host's tissues using various proteinases. As a consequence, various antiproteinases have been selected for that inhibit the process. Whether or not antiproteinases are inducible is unknown.

One of the most effective means for dealing with invading pathogens is through the activation of complement. Complement is actually a broader term for several components of an enzyme system. Each step in the cascade results in the formation of a particular product that catalyzes a subsequent reaction, whose products then catalyze the next, and so on. The key to the success of such a reaction is that the intermediate products are present in limiting quantities, have very short half-lives, and are easily inhibited, thus providing for a system that is rapid yet controlled. Its purpose is several fold: to enhance the effectiveness of by-stander components of the immune system, especially cellular activities; to alter vascular permeability; and to foster the destruction of invading organisms. However, left uncontrolled, the complement cascade system may lead to severe tissue destruction or cell death.

The classic and alternate pathways of the complement cascade involve antibodies or microbial membranes respectively. Both lead to the same endproducts: polymerized membrane attack complexes that lead to the destruction of target organisms by forming organized pores, or polyperforins (Figure 3.5). Along the cascade pathways, several 'split' products are formed that are not involved in the subsequent reaction. These split products (C3a, C5a, C2a, C3b, and C5b67) are potent modulators of inflammation, each acting through specific receptors located on target cells. The complement system of fish is still under investigation; however, the components appear to be similar, if not identical, to those of mammals. Complement components are manufactured primarily by macrophages of the RES and the liver, at least in higher vertebrates. It is easy to see that xenobiotics that target macrophages and/or the liver can have wide-ranging effects on immunocompetence through an impaired complement system. White and Anderson (1985) demonstrated that pentachlorophenol and pentachlorinated dibenzo-*p*-dioxin lower both complement levels and activity in mice, but not as a result of hepatocyte damage (White *et al.*, 1986; Lin and White, 1993).

Cellular defenses associated with non-specific defenses have been extensively studied in fish. As with higher vertebrates, fish have granulocytes (primarily neutrophils and eosinophilic granular cells), macrophages, and non-specific cytotoxic cells (Ainsworth, 1992). Neutrophils are involved in phagocytosis and intra- and extracellular killing of invading microorganisms, particularly under inflammatory conditions. At least in mammals, neutrophils are recruited into sites of inflammation as a result of chemokines and cytokines from tissue-resident macrophages and from activated endothelial cells. Adhesion molecules on the neutrophil are first up-regulated and then the cell adheres to activated endothelial cells. The neutrophil then migrates from the vascular compartment, between endothelial cells, across the basement membrane and into the sites of inflammation or tissue damage. Throughout these processes, the neutrophil acquires an enhanced ability to phagocytose tissue debris and microorganisms through activation of the humoral components, namely complement cascade split products. At this time, much of the tissue debris or microbes may be opsonized (coated) with these

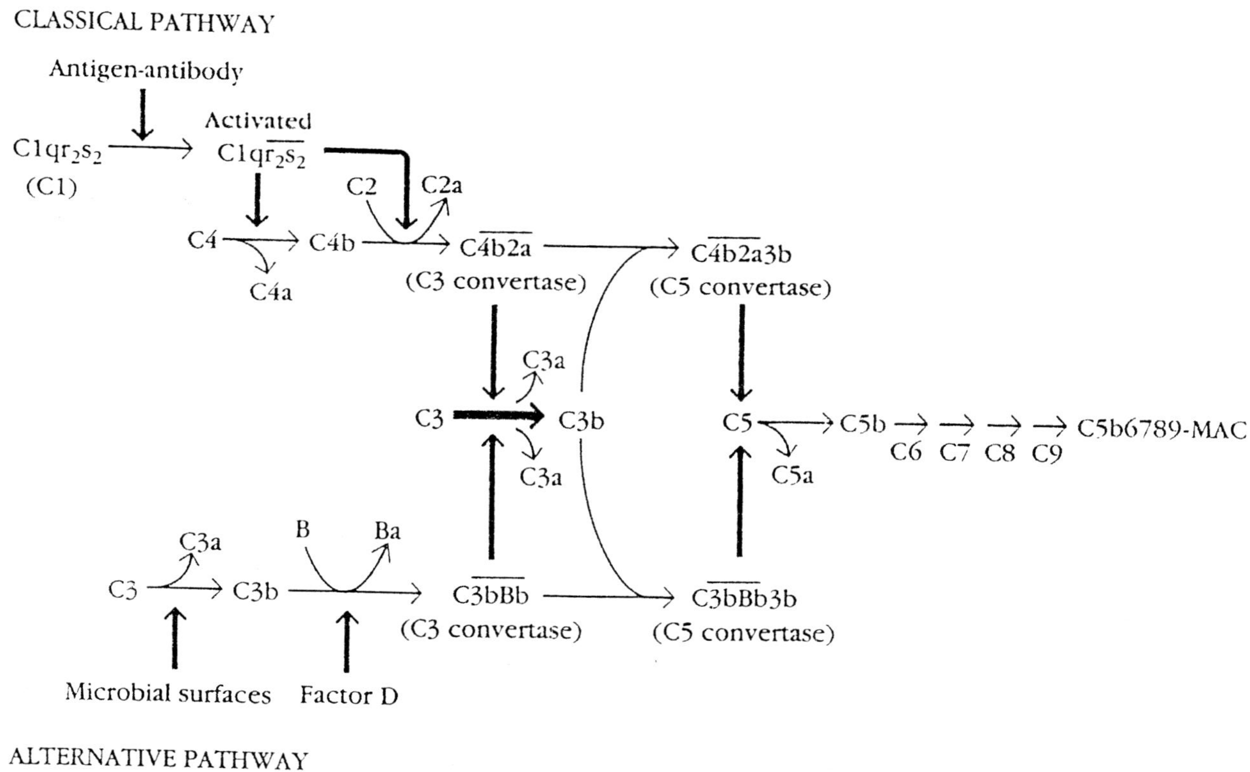

Figure 3.5 The complement activation pathways. The classic pathway is initiated by binding of C1 to antigen–antibody complexes. The alternative pathway is initiated by binding of C3b to activating surfaces such as microbial cell walls. Both pathways generate C3 and C5 convertases and bound C5b, which is converted into a membrane-attacking complex by a common sequence of terminal reactions. Hydrolysis of C3 is the major amplification step in both pathways, generating large amounts of C3b, which forms part of the C5 convertase. C3b can also diffuse away from the activating surface and bind to immune complexes or cell surfaces, where it functions as an opsonin (after Kuby, 1994a).

humoral substances, which enhances neutrophil function. The primary function of neutrophils in all vertebrates is to set the stage for tissue remodeling and wound repair, as well as to promote an environment for specific immune responses. As stated elsewhere, fish have eosinophilic granular cells that may function more as a mast cell than a true eosinophil. These cells are especially abundant in estuarine killifish (*Fundulus* sp.) found along the western Atlantic and northern Gulf of Mexico coastlines (Roszell and Rice, 1998). These marsh inhabitants are heavily parasitized by round worms for most of the year, and during that time the majority of phagocytes from the anterior kidney, spleen, and peritoneal cavity are eosinophilic granular cells (Roszell and Rice, 1998). Vallejo and Ellis (1989) and Reite (1998) have noted that eosinophilic granular cells are also abundant in salmonids, particularly in the intestine. Like mammalian mast cells, these cells readily degranulate upon stimulation with 40:80 poly-L-lysine and substance P. Phagocytosis also initiates degranulation.

Monocytes are circulating cells with a distinctly different lineage from that of granulocytes. Like granulocytes, they are phagocytic and migrate out of the circulating peripheral blood leukocyte compartment and into tissues where they differentiate into tissue macrophages. With proper stimulation, blood monocytes can have enhanced activities, but not to the degree of tissue macrophages. In response to inflammation, tissue macrophages secrete cytokines, various enzymes, reactive oxygen intermediates, antimicrobial peptides, and arachidonate metabolites (prostaglandins and leukotrienes). The key role of macrophages is to bridge non-specific with specific responses, and they do this by acting as professional antigen-presenting cells.

As described earlier, the immune system is sensory in function, and, as such, must communicate the nature of non-self to the rest of the immune system. As the macrophage engulfs and degrades external substances in the lysosomal compartments, smaller fragments (peptides) are generated that are then presented on the surface in a specific groove of class II major histocompatability (MHC) proteins in a process to be discussed in detail in the next section. This information is then forwarded to the responsive side of specific immunity as a signal to T-helper cells. Only antigen-presenting cells express MHC-II molecules and only then in a heightened state of activity, usually after encountering foreign substances. Along with certain dendritic cells, B cells, some glial cells, and activated endothelia, the macrophage serves this critical function. Just after tissue damage or invasion by microbes, tissue macrophages, acting as part of the connective tissue stroma, secrete neutrophil chemotactic cytokines and chemokines. As a result, neutrophils are recruited to establish a milieu that then stimulates the recruitment of monocytes and T cells from the bloodstream into the inflamed area. In effect, the immune system has recognized non-self or altered self (tissue damage), responded in a fashion to remove it, then communicated this information throughout the immune system and the host.

This scenario of inflammation may be another example of how mammalian biomedicine is imposed on lower vertebrates. Inflammation in homeotherms has four classic signs: swelling, raised local temperature, increased blood flow to the area, and redness. Although fish may be able to exhibit two or three of these signs, elevated local temperature is not known to occur. Fish may, however, display behavioral changes such as moving to places in the water column that may provide elevated body temperatures. Inflammation in fish is a poorly understood process and needs more focused research.

Cellular immunologists have given a great deal of attention to the phenomenon of macrophage activation. At least three stages of macrophage activity have been described for mammalian systems and the same probably holds true for teleost macrophages. A resting macrophage may become primed by various factors, the most noted of which is gamma-interferon, a secreted product from type I T-helper cells (Th-1). Additional biologic response modifiers such as lipopolysaccharide from Gram-negative bacteria can then drive the macrophage to a state of heightened activity termed 'activated'. Most immunologists accept an additional state in which hyperactivation is observed. Activated macrophages exhibit enhanced phagocytosis and intracellular killing, enhanced production of reactive oxygen intermediates (superoxide anion, hydrogen peroxide, hydroxyl radicals, and nitric oxide derivatives), enhanced secretion of proinflammatory cytokines [interleukin 1 (IL-1), tumor necrosis factor α (TNF-α), IL-6, IL-12, IL-18], up-regulated MHC-II molecules, and enhanced tumoricidal activity.

One of the most phylogenetically conserved features of macrophages and granulocytes is the respiratory burst. Membrane-bound NADPH oxidase is phosphorylated by a protein kinase C (PKC) that then catalyzes the formation of superoxide from molecular oxygen. The superoxide is dismutated to hydrogen peroxide by superoxide dismutase (SOD). A hydroxyl radical can be formed from both superoxide and hydrogen peroxide in the presence of iron, and perhaps copper, and from the interaction with hypochlorous acid, a by-product of myeloperoxidase and hydrogen peroxide. The respiratory burst can be initiated by phagocytosis, microbial cell wall components such as formylated methionin-leucyl-phenylalanin (f-MLP), glucans, and fatty acids. The most potent initiators of the respiratory burst are the phorbol esters (phorbol myristate acetate, phorbol dibutylrate) which mimic diacylglycerol from the breakdown of phophatidylinositol during cell activation. The other product is inositol triphosphate, which activates intracellular calcium channels in the endoplasmic reticulum, thereby raising intracellular calcium concentrations. Calcium ionophores (ionomycin, A23187) mimic this signal and are often used to stimulate macrophages and neutrophils. Calcium ionophores, however, are very weak stimulators of the respiratory burst, but in fish they do potentiate the effects of phorbol esters (Rice and Weeks, 1989).

Nitric oxide, a product of nitric oxide synthase, is produced by many cell types. This enzyme uses L-arginine as a substrate to generate citrullene and nitric oxide as products. Macrophages have an inducible nitric oxide synthase (iNOS) that is stimulated by lipopolysaccharides (LPS) and cytokines, usually in combination. Nitric oxide is a short-lived reactive molecule that tends to react with oxygen, superoxide, and transition metals, resulting in NOx, peroxynitrite ($OONO^-$), and metal–NO adducts respectively (Stamler *et al.*, 1992; Mohr *et al.*, 1994). Typical thiol targets may include the *N*-methyl-D-aspartic acid (NMDA) receptor, ion channels, NADPH oxidase, protein kinase C, and adenylate cyclase. Cytosolic thiol target sites include glyceraldehyde-3-phosphate dehydrogenase, tissue plasminogen activator, actin, and glutathione, whereas cytosolic metal target sites include guanylate cyclase, hemoglobin, aconitase, oxidoreductase complexes I and II, and ribonucleotide reductase (Mohr *et al.*, 1994).

The presence of macrophage iNOS in fish has been confirmed (Neumann *et al.*, 1995; Campos-Perez *et al.*, 1997; Xiang and Rice, 2000). As with LPS- and cytokine-activated mouse macrophages, similar treatment of the channel catfish monocyte/macrophage cell line 42TA results in the production of an intensely staining 130-kDa protein along with smaller proteins (Xiang and Rice, 2000). It is interesting to note that iNOS, through its NADPH-binding domain, is very similar to the cytochrome P450 reductase component of the cytochrome P (CYP) xenobiotic-metabolizing enzyme systems, although there are no overlapping functions. How, or whether, this similarity manifests itself in inflammatory reactions is unclear at this time. Expression of iNOS protein and activity are suppressed by glucocorticoids, whereas CYP1A activity is enhanced (Celander *et al.*, 1997).

During phagocytosis, an engulfed particle is contained within a phagosome which then fuses with a lysosome. It is here that degradation of the particle occurs, forming both the peptide needed for antigen presentation to T cells and waste products. Some of the most characterized enzymes are summarized in Table 3.1 and include those specific for nucleic acids, nucleotides, proteins, lipids, glycans and glycoproteins, and the sulfate ester and terminal neuraminic acid-specific enzymes for the breakdown of microbial cell walls. Since enzymes are critical targets for many xenobiotics, these phagocyte enzymes may be targeted in some instances of xenobiotic-related phagocyte dysfunction. To date, this possibility is unexplored in fish.

Antigen processing and presentation for specific responses

Through structure and function, the immune system receives information regarding non-self, processes this information and communicates (presents) it to higher levels of the system that dictate the direction and magnitude of a response by the organism as a whole (Figure 3.6). Although our current understanding of these processes in vertebrates has been acquired from rodent models and human biomedicine, these processes are also becoming evident in bony fish. In mammals, the two most important cells in controlling immune responses to exogenous antigens are the professional antigen-presenting cell (e.g. macrophage) and the T-helper cell. The macrophage, as a primary component of the RES, engulfs foreign material and degrades it enzymatically in the lysosomal compartment into smaller peptide fragments. These fragments are then bound to the α and β chains of the MHC-II molecules located in the endosomes that fuse with lysosomes. The entire complex of MHC-II and peptide fragment are transported to the cell surface and presented to cells bearing receptors. T-helper cells serve that purpose in higher vertebrates by having a T-cell receptor (TCR) that recognizes both the MHC-II component

Table 3.1 The enzyme contents of the granules found in most granulated phagocytes and their substrates.

Substrate	*Enzyme*
Nucleic acids	Ribonuclease Deoxyribonuclease
Nucleotides	Acid phosphatase
Proteins	Aminopeptidase Cathepsin Elastase Collagenase
Lipids	Acid triglyceride lipase Phospholipase
Glycans, glycoproteins, proteoglycans	α-Glucosidase Dextranase β-*N*-Acetyl-glucosaminidase α-Fucosidase β-Glucuronidase Lysozyme β-Galactosidase *N*-Acetyl-β-galactosaminidase α-Mannosidase α-Amylase
Sulfate esters	Arylsulfatase
Terminal neuraminic acid	Neuraminidase

Notes
After granulation, the enzymes act on their substrates and thus contribute to the process of digestion of the ingested particles in the phagosome.
After Klein (1996).

and the peptide fragment. The particular sequence of amino acids within the peptide-binding region of MHC-II molecules dictates the ability to bind certain peptide fragments, thus the ability to respond to a given antigen is genetically determined. In other words, what is seen as foreign (non-self) versus self is genetically determined, or MHC restricted. Strategically, MHC-II molecules are expressed only by antigen-presenting cells. Channel catfish monocytes and B cells have been shown to incorporate and degrade antigens and, subsequently, to induce antigen-specific responses (Vallejo *et al.*, 1990, 1991).

Endogenously produced proteins also initiate a vigorous immune response. These antigens may be viral or neoplastic in origin. Evolution has selected for a mechanism to eliminate these threats by simply eliminating the affected cell. This pathway is orchestrated by the MHC-I complex and the TCR on cytotoxic T cells (Figure 3.6). As with the MHC-II molecules, MHC-I molecules (an α chain associates with the β_2-macroglobulin) are generated and assembled in the endoplasmic reticulum (ER). Instead of fusing with peptide fragments in a vesicular

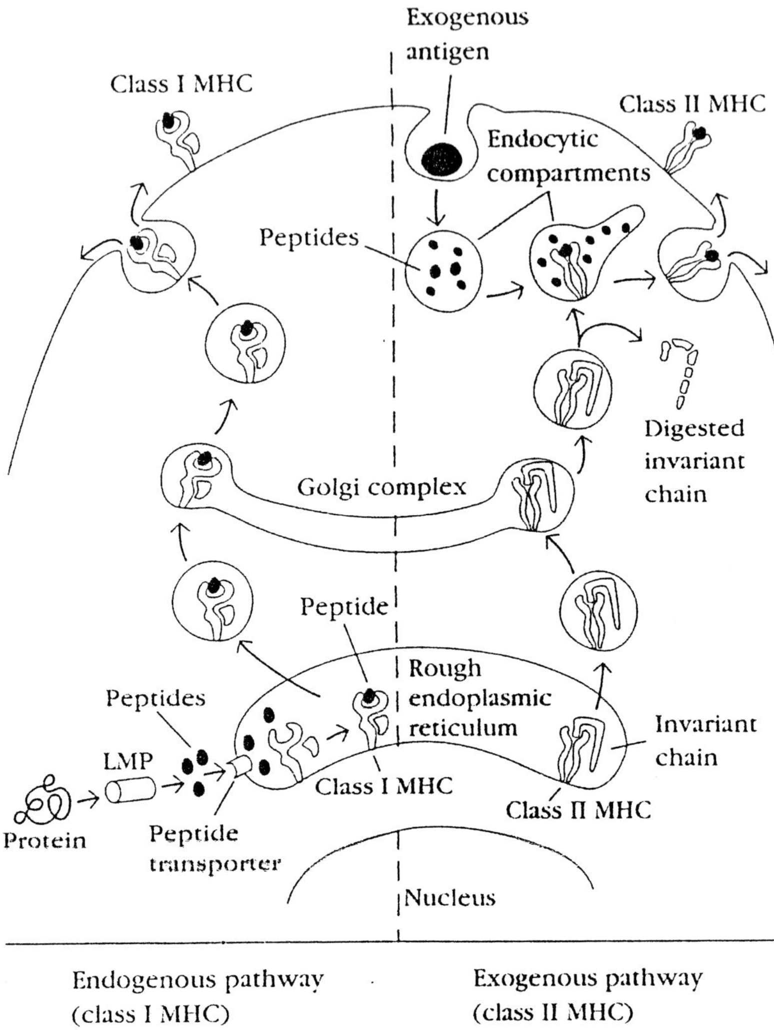

Figure 3.6 Mammalian model of separate antigen-presenting pathways for endogenous and exogenous antigens. The mode of entry into cells and the site of antigen processing appear to determine whether antigenic peptides associate with class I MHC molecules in the rough endoplasmic reticulum or with class II molecules in the endocytic compartments. Not all of these steps have been experimentally demonstrated in fish (after Kuby, 1994b).

compartment, the peptide fragments are generated in the cytosol by proteases, pumped across the ER by specialized membrane transporters where they bind to MHC-I molecules. This complex is then carried to the cell surface in vesicles and presented on the cell surface to cytotoxic T cells that recognize the foreign peptide in the context of MHC-I via the TCR.

Cytosolic proteases and membrane transporters are polymorphic, as are the

MHC-I molecules. Thus, the ability to respond to non-self generated from within the organism is genetically restricted. This endogenous pathway of antigen recognition works because the cytosol is 'sampled' on a continuous basis. In the case of a viral infection, the cytosolic ribosomal machinery manufactures new virons. As cytosolic proteins are sampled, the proteases generate small fragments of this new product of 'self', which are then transported across the ER membrane and assembled into the peptide-binding cleft of MHC-I molecules. In the context of peptide MHC-I, the cell is identified as infected with non-self and is subsequently eliminated by the cytotoxic T cell. Cytotoxic T cells recognize peptide fragments associated with MHC-I, whereas helper T cells recognize the peptide fragments associated with MHC-II molecules. This MHC-I-orchestrated event is ultimately responsible for the rejection of tissues transplanted from other organisms; however, MHC-II/T-helper cell interactions are also involved. Foreign MHC-II molecules on leukocytes of the transplanted tissues trigger the proliferation of host T-helper cells that then release T-cell growth factors, primarily IL-2. Interleukin 2 not only aids in the proliferation of T-helper cells but also promotes the differentiation and proliferation of cytotoxic T-cell precursors that are specific for cells bearing foreign MHC-I. In the end, the vertebrate immune system has mechanisms for recognizing both exogenous and endogenous sources of non-self.

The response to non-self, in the form of peptide fragments bound to MHC-I or MHC-II molecules, requires two signals. The first signal is the cognate recognition of the MHC–peptide complex by the TCR and accessory molecules that facilitate contact (CD4 for T-helper cells and CD8 for cytotoxic T cells). The second signal results from events subsequent to the first signal and includes enhanced expression of adhesion molecules, specific ligands, and signal transduction to the cytosol and nucleus. Concomitant with first and second signaling is the release and response to cytokines, the hormones of the immune system.

The MHC-I- and MHC-II-related responses are not uniform throughout fish. Hagfish and lampreys have a strong mixed lymphocyte reaction but a weak allograft rejection response; the elasmobranchs are weak responders in both reactions, as are the chondrosteans. Teleosts, on the other hand, display very strong mixed lymphocyte reactions and allograft rejection responses. Such disparity in responses between major groups of fish has led to the suggestion that MHC-I and MHC-II genes and products (and functions) are not present in elasmobranchs, cyclostomes, and chondrosteans. This disparity may be due to suboptimal conditions for measuring such responses in these animals (most likely), or it may be that the MHC complex evolved independently or that the system has been lost (Roit *et al.*, 1989).

Once an individual has recognized non-self, this information must be communicated throughout the organism. In higher vertebrates, this is done through the release of cytokines, the hormones of the immune system, and through cell-to-cell contact in the lymphoid or myeloid compartments. Most of the intended targets of cytokines are nearby, therefore the effects are either paracrine (nearby) or autocrine (same cell). Cytokine biology is most advanced in rodents and humans because of its significance in human medicine, but the subject is complicated by the rapidly growing list of new cytokines. Moreover, cytokines are both redundant and pleiotropic, meaning that more than one cytokine can have the same function and a given cytokine can perform more than one function or affect more than one cell type respectively. The list of cytokines that have been identified through cloning, sequencing, and functional analysis is far too long and mammalian in origin to discuss here; however, their functions can be lumped into three categories. The first group of cytokines are proinflammatory and are represented by IL-1, TNF-α, and IL-8. The second category of cytokines is involved in the commitment of stem cells to a defined lineage and is represented by the hematopoietic growth factors IL-3 and granulocyte–macrophage colony-stimulating factor (GM-CSF). The third group of cytokines are lineage-specific hematopoietic differentiation factors represented by IL-4 and IL-6, which promote the differentiation and growth of B cells, IL-2, a T-cell growth factor, and IL-5, which promotes the differentiation of eosinophils. Interleukin 5 is also an isotype switch factor for B cells. Gamma-interferon (γ-IFN) enhances immunoglobulin production and is a switch factor as well. Other cytokines such as IL-9 and IL-11 are produced by T cells and B cells, respectively, and act to enhance T-cell growth. Specific cytokine-like activities in fish have been described by numerous groups (Secombes *et al.*, 1996), but to date no full sequences for either cytokine has been reported. This is one of the fastest growing areas of fish immunology, mostly because of the need to understand more clearly the immunobiology of aquaculture commodities.

Central to assimilating information regarding non-self in the immune system is the production of antigen recognition molecules. The ability to recognize antigens is achieved through the T-cell receptor, a membrane-bound, non-secreted glycoprotein homodimer found on T-helper and cytotoxic T cells, and through the glycoprotein B-cell receptor (BCR), also known as membrane immunoglobulins. The T-cell receptor recognizes an antigen in the form of processed peptide bound to MHC molecules, and the B cell recognizes an antigen in its native form. As a result of activation and differentiation, B cells become either antibody-secreting plasma cells or memory B cells, the latter representing the expanded population, or clones, of the small number of original B cells capable of recognizing the specific antigen. After recognizing native antigens, the antigen–BCR complex is also endocytosed, processed, and presented to the TCR in the context of MHC-II in the same manner as macrophages, therefore B cells also function as antigen-presenting cells. When antigen levels are high, particularly during initial contact, the macrophage is more efficient at antigen processing and presenting; however, on subsequent exposures when antigen levels may be low, the B cell is more efficient.

Vertebrates are endowed with the ability to recognize a wide range of different antigens. To do this requires an ability to generate diversity in the antigen recognition domain of the TCR and B-cell receptor. This is accomplished for the immunoglobulin molecule by the presence of multiple germline variable genes for heavy and light chains, the ability to have multiple recombinations between the heavy and light chain genes, frequent recombination inaccuracies, point mutations, and assorted heavy and light chains. The arrangement of genes and these mechanisms of diversity are conserved throughout vertebrates, but there are differences in how the genes are packaged. For most vertebrates, including teleosts and humans, multiple genes for the immunoglobulin variable regions exist, followed by segments coding for diversity, then joining regions, and finally genes coding for the constant regions of each chain (–V–V–V–V–V–V–D–D–J–J–J–C–). In elasmobranchs, however, the immunoglobulin genes are clustered in that each V, D, and J segment is linked to its own constant gene (–VDJ-C–VDJ-C–VDJ-C–).

Higher vertebrates have different classes, or isotypes, of antibody molecule (i.e. IgM, IgG, IgA, IgD, IgE); however, immunoglobulin class M (IgM) is the only described immunoglobulin in teleosts and elasmobranchs. Pentameric and monomeric forms have been identified in the shark, whereas tetrameric, trimeric, dimeric, and monomeric forms have been identified in teleosts (Tizard, 1996). The circulating immunoglobulin-like molecules of hagfish are more similar to the complement components than immunoglobulin, therefore they may not be inducible by repeated exposure to a particular antigen. The constant regions of the heavy chain of each antibody isotype in mammals are associated with a particular effector function, such as complement activation and Fc receptor binding on leukocytes. In contrast, no Fc receptor for antibody has been found on the leukocytes of fish, although this has been demonstrated to some degree in elasmobranchs (McKinney and Flajnik, 1997). Of particular interest to comparative immunologists is that immunologic memory involves immunoglobulin class switching in higher vertebrates, but not in fish (Kaattari, 1994). Furthermore, the capability to mount a humoral immune response is temperature dependent in fish, but not in higher vertebrates (Clem *et al.*, 1991). Of particular interest to comparative immunotoxicologists is the kinetics of antibody production, and secretion is dramatically different between rodent models and fish. For example, the primary antibody response in B6C3F1 mice reaches a peak at approximately 4 days for splenic antibody-secreting cells and at 6 days for circulating antibody (Holsapple *et al.*, 1991). In channel catfish, numbers of antibody-secreting cells reach their peak at about day 14, and circulating antibody is highest at about day 21 (Waterstrat *et al.*, 1991; Rice and Schlenk, 1995). Similar kinetics are exhibited by rainbow trout (Anderson and Zeeman, 1995). For most species of fish, the highest numbers of antibody-secreting cells are found in the anterior kidney, although they can be found in the spleen (Anderson and Zeeman, 1995) and, in some cases, also in the circulating leukocyte pool (Waterstrat *et al.*, 1991).

Four different sets of genes code for the antigen-binding region of the TCR. Alpha and beta gene products are expressed on most T cells (α/β-TCR) in the peripheral compartment, and the gamma and delta gene products are expressed on a minor portion of T cells (γ/δ-TCR). The distribution of the γ/δ-TCR-bearing cells appears to be more mucosal than for the α/β-TCR-bearing cells. The mechanisms for generation diversity in TCR genes are very similar to the immunoglobulin genes. Both forms of the TCR have been identified in teleosts (Clem *et al.*, 1996).

Fish as models in biomedical research

A significant factor affecting the advancement of fish immunotoxicology is the lack of critical reagents such as recombinant cytokines and various monoclonals to key cellular targets that are available to those studying mammalian systems. None of the cluster of differentiation (CD) antigens described in rodents and humans have been standardized for fish cells. As discussed above, much less is known about the immunobiology of fish than of higher vertebrates. Only recently has the need to understand fish immunology, for the sake of fish, been appreciated at the economic level. Large disease outbreaks and seasonal variations in vaccination success have resulted in large economic losses to the aquaculture industry. Therefore, fish immunology has risen to the level of veterinary biomedical importance recognized by beef, dairy, swine, poultry, and other commodity industries.

Very recent advances in fish immunology are the result of the use of modern molecular biology tools. Consider the following limited examples: T-cell, B-cell, and monocytic cell lines have been developed from channel catfish (Miller *et al.*, 1994a,b; Clem *et al.*, 1996). MHC-II beta and alpha-interferon genes have been cloned and sequenced from Atlantic salmon (Robertsen *et al.*, 1997; Koppang *et al.*, 1998), whereas MHC II beta genes have been cloned from channel catfish (Godwin *et al.*, 1997); rainbow trout iNOS has been partially cloned and sequenced (Campos-Perez *et al.*, 1997; Xiang and Rice, 2000); some proinflammatory cytokines have been characterized in rainbow trout and specific cytokine-like activity has been described in carp and channel catfish (Secombes *et al.*, 1996; Rycyzyn *et al.*, 1998). The basic biology of adhesion processes in inflammatory sequelae in fish has also been under development (Ainsworth and Boyd, 1998), as have signal transduction pathways (Burnett, 1997; Rycyzyn *et al.*, 1998). These and other advances will improve our future ability to understand the molecular and cellular mechanisms of action of immunotoxic xenobiotics.

Compared with the wealth of information regarding immunotoxicology in mammals, very little is known about the mechanisms of action associated with immunotoxic agents in fish. This lack of information remains despite the last three decades of observation and research concerning the contribution of environmental pollutants to the susceptibility of fish to diseases (Arkoosh *et al.*, 1998). To date, we can clearly see immunomodulation in fish as a result of exposure to several classes of xenobiotics, including, but not limited to, pesticides, metals, polyaromatic hydrocarbons, halogenated aromatic hydrocarbons, and pharmaceuticals. The observation that similar immune functions (e.g. antibody production, phagocytosis, natural killer cell-like activity, etc.) in both fish and mammals are modulated as a result of exposure has prompted the use of fish as surrogate models for rodent studies and as models for human immunobiology (Zelikoff *et al.*, 1995; Hart *et al.*, 1997, 1998). As these studies imply, there are similarities in immunotoxic responses between fish and mammals within the context of comparative anatomy, physiology, and immunology. Regardless of phylogeny, two general deviations from normal immunologic homeostasis can be observed following exposure to immunotoxic agents: immune suppression and immune enhancement (Burns *et al.*, 1996). However, it is important to point out that in aquatic immunotoxicology such conclusions are often based on functional endpoints and assays rather than on a whole animal, integrated immunophysiologic approach. In other words, does enhanced or depressed phagocytosis of bacteria or other particles by fish macrophages necessarily equate to immune modulation at the system level, or is this particular endpoint of immune function affected and not the entire immune system? Can any single assay predict immunologic competence?

Historical approaches to immunotoxicology

The above questions prompted pioneering mammalian immunotoxicologists to develop a systematic approach to immunotoxicology based on the need to conserve animals, time, and, especially, costs (Dean *et al.*, 1979). A tier approach to toxicity testing was later validated by Luster *et al.* (1982, 1992) as a means to evaluate immunomodulation in rodents following exposure to chemicals and drugs. In tier I immunoassays, compounds of interest were screened for their potential immunotoxicity, namely immunopathology (hematology, body and organ weights, and lymphoid histology), and simple assays of cell-mediated immunity, and the hemolytic plaque assay to measure humoral immunity. If effects were noted in this battery of endpoints, then a more comprehensive battery of assays (tier II) was performed, which included cell-surface marker profiles, host-resistance models, cytotoxic T-cell functions and cytokine profiles, primary and secondary antibody responses to both T and T-independent antigens, and more detailed evaluations of macrophage and granulocyte functions. Tier II assays were designed to distinguish which cell types were the target for toxicity. This tier approach to testing legitimized immunotoxicology and immunopharmacology as stand alone disciplines within the broader field of immunology. It also ushered in a whole generation of immune function technicians who may or may not have a full appreciation for immunobiology as an integrated subject. The ultimate question for immunotoxicologists, and thus a primary reason for this chapter, is why are certain immune functions modulated only by certain groups of compounds. Because the immune systems of fish are so poorly understood and because of the lack of advanced reagents, most immunotoxicological studies are limited to tier I assays.

Mechanisms of action of some immunotoxicants

The immune system, defined as the lymphoid organs, integument, and circulating leukocytes, is a sensitive target organ for toxicity for several reasons. The fixed tissues of the RES are designed as a filtering system for the circulatory system and as a barrier to the outside of the organism. This barrier may be exposed directly to xenobiotics. In addition to fixed tissues, most immunocompetent cells are found in the peripheral blood compartment. Therefore, both tissues and peripheral blood leukocytes are directly susceptible to toxicants in the parent form and as active metabolites following transformation by xenobiotic-metabolizing enzymes. Moreover, the immune system depends on a high rate of turnover for immunocompetent cells. Any cell type that has a high rate of division, or turnover, is susceptible to toxicity; a fundamental mechanism associated with chemotherapy for the treatment of neoplastic and hyperproliferative diseases. As very little is known about the actual mechanism(s) of action associated with immunotoxic xenobiotics, the following is a discussion on the subject of polycyclic aromatic hydrocarbons, halogenated aromatic hydrocarbons, and metals. The literature base is by no means exhaustive and exclusive, but rather represents those studies that address possible mechanisms of action. The mechanisms of immunotoxicity of pesticides and pharmaceuticals is less clear and the reader is referred to Anderson (1990), Dunier and Siwicki (1993), and Anderson and Zeeman (1995) for reviews on the subject. Of the environmentally persistant pesticides, pentachlorophenol seems to be particularly immunotoxic to phagocytes (Roszell and Anderson, 1993, 1994, 1996), probably through its ability to uncouple oxidative phosphorylation and disrupt electron transport systems involved in the formation of reactive oxygen intermediates.

Polycyclic aromatic hydrocarbons

Polycyclic aromatic hydrocarbons (PAHs) are by-products of fossil fuels and their combustion products. As such, they are practically ubiquitous in distribution, particularly in heavily populated areas as a result of atmospheric deposition, terrestrial run-off, and boating activity. Of the PAHs, benzo(a)pyrene (BAP) has received the most attention in terms of toxicity. As the prototype PAH, it is mutagenic, carcinogenic, and immunotoxic, but only after activation by CYP1A to electrophilic metabolites, namely the 7,8,9,10-diol epoxide intermediate. Induction of CYP1A activity is a classic effect of exposure and is initiated through the Ah receptor, a transcription factor not only for CYP1A genes but also for hormone receptors and other proteins. Adduction of reactive intermediates to both protein and nucleic acids is well described and seems to account for their wide range of toxicity. Other PAHs with similar structures, especially those harboring a 'bay' region are also toxic, including 3-methylcholanthrene (3-MC) and dimethylbenzanthracene (DMBA). Structures lacking a 'bay' region, and thus not forming reactive electrophilic metabolites, are less toxic, especially immunotoxic (White, 1986). White and Holsapple (1984) clearly demonstrated that those PAHs lacking mutagenic properties were also not immunotoxic. In fish, DMBA and BAP are very potent immunotoxic (Faisal *et al.*, 1991; Seeley and Weeks-Perkins, 1991; Hart *et al.*, 1997, 1998; Rose *et al.*, 2000) and carcinogenic (Vogelbein *et al.*, 1990) agents. In recent years, it has become clear that certain populations of fish inhabiting systems heavily contaminated with PAHs, such as the Elizabeth River system, have become resistant to overt toxicity, although the population harbors neoplastic lesions (Vogelbein *et al.*, 1990; Armknect *et al.*,

1998). Earlier work in the Elizabeth River system by Faisal *et al.* (1991) and Kelly-Reay and Weeks-Perkins (1994) demonstrated reduced non-specific cytotoxic cell activity and macrophage function respectively. Although the humoral response in rodents is very sensitive to select PAHs (White, 1986), this has not been examined in the Elizabeth River system. However, Arkoosh *et al.* (1991, 1996) demonstrated that salmon from Puget Sound, Washington, another system heavily contaminated with PAHs, have reduced immunologic memory and reduced peripheral blood mitogenic responses. With regards to suppressed immunologic memory, the work of Kaattari *et al.* (1994) may well underscore the mechanisms of PAH-induced suppression. Aflatoxin B_1, like BAP, is metabolized by the P450 system to genotoxic electrophiles. Early in the embryologic development of the B-cell repertoire, only D-proximal V genes are available for recombination. The authors hypothesize that if fish are exposed to aflatoxin B_1 at an early embryological stage then part of the genotoxicity may be manifested in these regions, thus leading to impaired maturation of the antibody response. Genotoxicity, as a mechanism for PAH-induced immunotoxicity, is supported by the findings of Rose *et al.* (2000). Lymphoid tissues and isolated leukocytes metabolize BAP to structures associated with DNA adducts in those tissues. Although not addressed in fish models, some PAHs (e.g. DMBA) induce apoptosis by elevating intracellular calcium and inducing DNA fragmentation (Burchiel *et al.*, 1992).

Halogenated aromatic hydrocarbons

Halogenated aromatic hydrocarbons (HAHs) include the polychlorinated biphenyls, polychlorinated dioxins, furans, and naphthalenes. TCDD (2,3,7,8-tetrachlorodibenzo-*p*-dioxon) is the prototype congener of this class. It is also the most immunotoxic xenobiotic known to date, at least in mice (Holsapple *et al.*, 1991). Carcinogenesis and reproductive toxicity are also associated with HAHs, but the most toxic congeners are those processing a planar, TCDD-like structure that binds with very high affinity to the Ah receptor (Hahn *et al.*, 1997). There seem to be two types of toxicities associated with HAHs; those associated with TCDD-like structures (e.g. co-planar PCBs) and those with non-planar congeners. Toxicities associated with non planar HAHs are due to their hormone-mimicking, calcium ionophoric, and membrane-perturbing properties (Soto *et al.*, 1992; Ganey *et al.*, 1993; Tithof *et al.*, 1995).

Although TCDD is a very potent immunotoxicant in mice, especially of the primary antibody response, it seems to be much less so in fish. Spitzbergen *et al.* (1986a,b) noted only moderate suppression of the antibody response, but demonstrated enhanced susceptibility to the infectious hematopoietic necrosis virus in rainbow trout. This suggests that non-specific aspects of immunity were targeted, such as non-specific cytotoxic cell activity. This explanation is supported by Rice and Schlenk (1995), who reported that PCB-126 (co-planar PCB) did not affect the antibody response to *Edwardsiella ictaluri*; however, NCC and phagocyte activity were suppressed. This is in contrast to that found in rodent models and is difficult to explain, at least in terms of mammalian immunobiology. Karrass *et al.*

(1996) demonstrated that TCDD elevates basal intracellular calcium levels in B cells, thereby disrupting normal cell signaling and differentiation in these antibody-producing cells. Kerkvliet and Burleson (1994) demonstrated that the developing T cell is the primary target of TCDD. Whether or not co-planar HAHs modulate developing NCC and phagocytes is unknown, but likely.

As with PAHs, HAHs are found in the environment as mixtures of congeners, and often in mixture with PAHs, metals, and pesticides. Most mixtures of HAHs are part of commercial mixtures such as Aroclor 1254 and 1242. Aroclor 1254 has a significant amount of co-planar structures, whereas 1242 does not. The immunotoxic potential of Aroclor 1242 has not been investigated in fish, but it appears to activate neutrophils of higher vertebrates by increasing calcium-mediated events. Dietary Aroclor 1254 suppresses the antibody response in rainbow trout and increases disease susceptibility (Cleland *et al.*, 1988; Thuvander and Carlstein, 1991), but it increases anterior kidney B-cell populations and enhances phagocyte activity in channel catfish, while having no effect on antibody responses (Rice *et al.*, 1998). Clearly, there are species differences in sensitivity of fish to Aroclor 1254.

Metals and organometals

The immunotoxicity of metals in fish has been reviewed extensively by Zelikoff (1993) and Zelikoff *et al.* (1995), while the possible mechanisms of action have been addressed by Burnett (1997). Most metals have dual toxicities: neurologic and immunotoxicologic. Both the nervous system and the immune system rely on intercellular and intracellular signals to perform their function(s). As cadmium is similar to calcium in ionic radius, it may compete for calcium in signal transduction pathways in both neurons and immunocompetent cells. Resting cells have relatively low intracellular and high extracellular calcium levels. Subsequent activation signals elevate intracellular calcium and mediate calmodulin/calcineurin-directed systems. Cadmium can disturb this balance and lead to disruption of intracellular homeostasis. Burnett (1997) demonstrated that low levels of mercury chloride induce an influx of calcium, followed by tyrosine kinase activity and cellular proliferation. In mammals, mercury is associated with autoimmune disease, but this has not been investigated in fish. A similar property is associated with tributyltin (TBT) immunotoxicity. Depending on the concentration, TBT can both activate and inhibit reactive oxygen formation in toadfish macrophages (Rice and Weeks, 1989, 1990, 1991). This activation is associated with an influx of calcium and seems to be consistent with other reports relating to TBT-induced thymic atrophy (Chow *et al.*, 1992; Raffray *et al.*, 1993). In rodent models, lead (Pb) affects immune function by enhancing adenylate cyclase activity and elevating cAMP that inhibits Th1-T cells in favor of Th2-T cells, thereby suppressing cell-mediated immune responses. To date, the effects of lead on adenylate cyclase and cAMP in fish leukocytes has not been investigated.

Autoimmunity and hypersensitivity: a frontier in fish immunotoxicology

Xenobiotic-induced autoimmune and hypersensitivity (allergic) reactions is a major human health problem. Perhaps the best and most common examples are penicillin and poison ivy hypersensitivity in humans. In both cases, the sensitizing agent is too small to stimulate an immune response, but such 'haptens' bind to self proteins (carriers) and together are recognized as non-self in susceptible hosts. As with any carrier–hapten system, antigen processing and presentation to T cells is the critical step in mounting an immune response against the hapten (Enk and Katz, 1992). Because T cells are not educated to recognize xenobiotics during T-cell development in the thymus, these foreign structures are presented as neoantigens once complexed with self proteins. Those xenobiotics that tend to elicit adverse immune reactions do not bind directly to self proteins upon entering the body, but must be altered in order to do so. The mechanisms associated with the development of allergic and autoimmune reactions to xenobiotics is reviewed in detail by Griem *et al.* (1998). Several factors interact in concert to initiate such reactions, including the metabolism of xenobiotics into reactive, haptenic metabolites, the polymorphic nature of metabolizing enzyme systems, and the induction of co-stimulatory signals for antigen presentation and sensitization of T cells. The sensitization of T cells is critical (Goebeler *et al.*, 1993; Rambukkana *et al.*, 1996).

Both organics and metals have the potential of inducing adverse immune responses, albeit through different mechanisms (Lisby *et al.*, 1995; Griem *et al.*, 1998). Reactive electrophilic metabolites of organic compounds bind covalently to nucleophilic groups of proteins such as thiol, amino, and hydroxyl groups. On the other hand, metals form protein–metal complexes by binding to several amino acids or by oxidizing proteins (Figure 3.7). In either case, self proteins act as carriers for their respective organic or metallic xenobiotic.

The generation of xenobiotic–self complexes is referred to as the preimmunologic phase and may involve hepatic metabolism, extrahepatic metabolism, and, in a special case, xenobiotic metabolism by phagocytes. In mammals, autoimmune hepatitis is a common side-effect of long-term treatment with tienilic acid, a diuretic. The prohapten parent compound is metabolized by CYP2C9, which results in covalent CYP2C9–tienilic complexes that then act as carrier–hapten systems. Autoantibodies against CYP2C9 are generated that then initiate the influx of proinflammatory signals, resulting in chronic and destructive hepatitis. Similar situations involving CYP2E and CYP1A and other prohaptens have also been reported (Bourdi *et al.*, 1994; Eliasson and Kenna, 1996). Both CYP2E and CYP1A are known to exist in fish. Furthermore, the lymphoid tissues of the gut are exposed to dietary toxicants before passing into the portal system *en route* to the liver for transformation. As the gut of fish is an active participant in xenobiotic metabolism (Van Veld *et al.*, 1987), gut-associated lymphoid tissues (GALT) and mucosal lymphoid tissues (MALT) are expected to lead to the generation of active metabolites not only of procarcinogens, promutagens, and proimmunotoxicants, but of prohaptens as well.

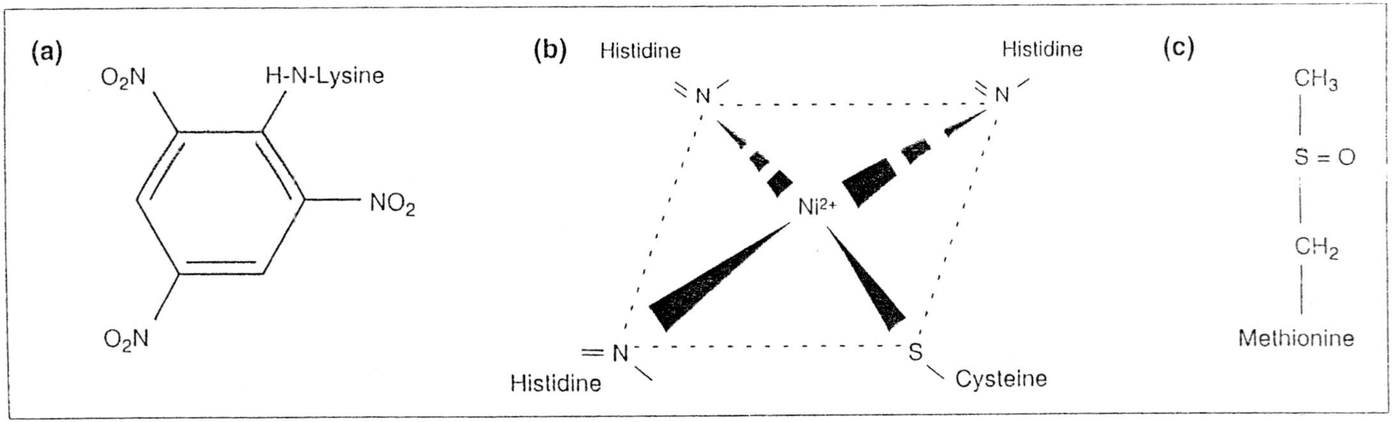

Figure 3.7 Haptens comprise organic compounds and metal ions, and bind to proteins forming either covalent bonds (a) or co-ordinated complexes (b). Organic haptens forming covalent bonds bind to a single amino acid side-chain; the covalent binding of trinitrophenyl (TNP) to lysine is shown. Metal complexes consist of a central metal ion and a set of atoms, ions or small molecules, regarded as ligands. Alternatively, certain reactive chemicals can irreversibly oxidize protein side-chains, such as those of cysteine and methionine (c); a methionine sulfoxide is shown (after Griem *et al.*, 1998).

As mentioned earlier, the integument is the most important immune organ and this is clearly seen in the case of contact hypersensitivity. Skin immunity of higher vertebrates is regulated primarily through the interactions of Langerhans cells and T cells, and, in come cases, macrophages. Langerhans cells are a good source of CYP1A enzymes and are able to metabolize dimethylbenz-(a)-anthracene to haptens. A similar role for fish integument has not been established, but is highly likely.

The importance of phagocytes to both innate and specific immune responses is well known, however their role(s) in generating haptens should be of great interests to fish immunotoxicologists. In what is considered to be a classic study, Ladics *et al.* (1992) demonstrated that mouse splenic macrophages metabolize benzo(a)pyrene to a mutagen and carcinogenic diol-epoxide structures similar to those formed by hepatic parenchymal cells. Rose *et al.* (2000) demonstrated the same for mummichog, *Fundulus heteroclitus*, anterior kidney cells. Furthermore, these metabolites from mummichog were able to form DNA adducts; a possible first step in generating auto-DNA antibodies. However, cells other than macrophages, including lymphocytes, eosinophilic granular cells, endothelial cells, and inter-renal cells, are present in these preparations. Pure populations of macrophages will be needed to equate these findings with those of Ladics *et al.* (1992), but the likelihood of similar responses is good.

Regardless of phylogeny, the ability of neutrophils, macrophages, or eosinophils to generate reactive electrophiles from procarcinogens and prohaptens implies the presence of CYP-like enzymes. These cell types are an abundant source of myeloperoxidase, prostaglandin H synthase, and some cytochrome P450 enzymes. The first two have broad substrate specificity, especially prostaglandin H synthase, a enzyme with co-oxidation–peroxidation as its primary function. It is this enzyme system that Ladics *et al.* (1992) may have been describing during the metabolism of benzo(a)pyrene to reactive intermediates by leukocytes. At least in higher vertebrates, prostaglandin H synthase is inducible by proinflammatory signals and is called cyclooxygenase 2 (COX-2). If and when the presence and function of COX-2 is established in fish, phagocytes will be credited with having a key role in metabolizing certain xenobiotics not only to carcinogenic intermediates but perhaps to haptens as well.

Once the preimmunologic stage has been set by the generation of hapten–self protein carriers, T cells need to be sensitized. In the classic sense, this is carried out via the presentation of hapten to T cells in the context of MHC-II/TCR interactions. Cognate or cell-to-cell contact and generation of primarily two co-stimulatory signals is required. Signal 1 is delivered by the interaction of MCH-II/hapten interaction with the TCR and CD3 components. Signal 2 is a generic term referring to a variety of ligand-initiated signal transduction events between T and antigen-presenting cells and adhesion molecules that contribute to T-cell activation and subsequent events (Table 3.2). As described earlier, fish do have T and B lymphocytes as well as the ability to process and present antigens to T cells (Clem *et al.*, 1996). The ability of fish to generate self protein carrier–hapten systems following hepatic-, extrahepatic-, and phagocyte-associated xenobiotic

Table 3.2 How xenobiotics contribute to co-stimulation of T cells in mammals.

Chemical	*Effects*	*Reference*
TNCB, DNCB	Induction of IL-1 mRNA Langerhans cells; induction of IL-1 and TNF mRNA in keratinocytes	Enk and Katz (1992)
DNFB, nickel sulfate	Induction of IL-1 mRNA Langerhans cells	Rambukkana *et al.* (1996)
Nickel sulfate	Stabilizing TNF mRNA in keratinocytes	Lisby *et al.* (1995)
Nickel sulfate, cobalt sulfate	Expression of adhesion molecules on endothelia	Goebeler *et al.* (1993)
Mercuric chloride	Secretion of IL-1 from macrophages	Zdolsek *et al.* (1994)

Notes
Adapted from Griem *et al.* (1998).
TNCB, trinitrochlorobenzene; DNCB, dinitrochlorobenzene; DNFB, dinitrofluorobenzene; IL-1, interleukin 1; TNF, tumor necrosis factor.

metabolism leading to T-cell activation and subsequent hypersensitive or autoimmune responses has not been established. The likelihood is great and needs focused attention by fish immunologists.

The neuroendocrine–immune connection

Environmentally induced immunomodulation probably occurs most often through indirect mechanisms, especially with long-term chronic exposure to low levels of xenobiotics. This may be due to negative effects on immune regulatory systems such as the neuroendocrine system. Since cross-talk between these two systems is one of the most important components of immune regulation, this is a likely mechanism associated with immunomodulation. The current focus on environmental endocrine disrupters comes at a critical time when the field of neuroendocrine immunology is also growing. Understanding these interactions may help explain hormesis and non-linear dose–response curves often observed in immunotoxicology. This is especially true for biologic rhythms and their associated changes in immune responsiveness to antigens and immunotoxic agents.

The most characterized neuroendocrine interaction that modulates immune function involves hypothalamic–pituitary–intrarenal (HPI) generation of glucocorticoids following CNS recognition of stress. Various forms of stress have been shown to activate the HPI axis in fish (Brown *et al.*, 1989; Anderson, 1990; Van Der Kraak *et al.*, 1992; Weeks *et al.*, 1992). HPI activation, with increases in circulating plasma cortisol (cortisone in humans), is part of the generalized adaptation to stress in fish and other vertebrates (Seyle, 1973; Saad, 1988; Thomas, 1990) and is often associated with immunomodulation.

However, not all vertebrates are glucocorticoid sensitive in terms of immune suppression, lymphocyte lysis, lymphocyte dysfunction, or lymphocyte redistribution. For example, rabbits, mice, rats, and hamsters (main glucocorticoid is cortisone) are very sensitive to glucocorticoids, whereas humans, monkeys, and guinea pigs (main glucocorticoid is cortisol) are relatively insensitive (Claman, 1975; Munck *et al.*, 1984; Munck and Guyre, 1991). Whether or not such differences in sensitivity exist between species of fish is unclear at this time, but the liklihood is good. In addition, both humans and fish secrete cortisol as the main glucocorticoid; a fact that supports the use of fish as immunotoxicologic sentinels in at least understanding some of the effects of xenobiotics on neuroendocrine–immune responses in humans. Interestingly, while handling stress (and other stressors) leads to high levels of circulating cortisol and immunomodulation (Maule *et al.*, 1989), the effects of similar levels of cortisol *in vitro* do not necessarily result in suppressed activity. However, the effects of pharmacologic glucocorticoids (e.g. dexamethasone) are dramatic in both humans and fish. In terms of xenobiotic interactions, those that elevate glucocorticoids may also enhance Ah receptor-mediated CYP1A induction because these steroids are known to enhance CYP1A in rats (Prough *et al.*, 1996) and in fish (Celander *et al.*, 1997). Hontela *et al.* (1992) and Van Der Kraak (1992) have shown that several environmental contaminants alter cortisol responses to stress. Elevated plasma cortisol in response to xenobiotics is common in pharmacologic settings, however, and may be partially responsible for immune dysfunction in many laboratory-based toxicity studies (Pruett *et al.*, 1993).

The physiologic basis for some neuroendocrine–immune interactions may lie in simple anatomy. The significance of neuroendocrine tissue lying directly in or on the anterior kidney of teleosts is often unappreciated and, perhaps, ignored. For example, intrarenal tissue (adrenal equivalent) is located either scattered throughout the anterior kidney or in discrete locations, depending on the species of fish. Simple handling and encroachment upon the holding facility (tank room) can cause a surge in the corticotropin-releasing hormone (CRH)–adrenocorticotropic hormone (ACTH)–cortisol pathway, leading to immunomodulation. Other stress-related secretions, including biogenic amines, prolactin and growth hormone(s), are known to affect fish immune responses. Channel catfish leukocytes were shown to secrete immunoreactive ACTH (Arnold and Rice, 2000), a phenomenon well documented in mammals. Virtually every neuroendocrine hormone is produced by leukocytes (Weigent and Blalock, 1996). Likewise, the receptors for these hormones have been found on mammalian leukocytes (Bardwaj *et al.*, 1997).

Other hypothalamic–pituitary-orchestrated hormones affect immune function as well. For example, growth hormone, which is elevated at rest (Leatherland *et al.*, 1974), increases natural killer cell function (Kajita *et al.*, 1992). Immune function in some lower vertebrates is higher in the evening (Levi *et al.*, 1991); an observation that correlates with elevated circulating growth hormone. Prolactin is also stimulatory to immune function and circulating levels are higher at rest (evening) in diurnally active vertebrates (Nevid and Meier, 1995). On the other hand, cortisol, a hormone typically considered to be immunosuppressive, is higher during the day (Levi *et al.*, 1991). In this regard, prolactin may be a feedback controlling factor for cortisol (Weigent, 1996). Melatonin, which is directly linked to circadian rhythms and possibly reproductive cycles (Zachman *et al.*, 1992; Weaver *et al.*, 1993; Yanez and Meissl, 1996) is also higher in diurnally active vertebrates in the evening (inhibited by light). Melatonin has been reported to enhance immune function (Maestroni, 1995), and to act as a nocturnal antioxidant (Reiter, 1995). Moreover, hormonal release from pineal and hypophyseal tissues is not only pulsatile but also varies dramatically over daily and seasonal cycles (Emata *et al.*, 1991; Nevid and Meier, 1993, 1995). Although these examples are based on lower vertebrates, there are sufficient similarities in the basic aspects of physiology, including immunophysiology, to warrant their use as environmental sentinels for the benefit of other vertebrates. In addition, certain biochemical pathways leading to toxicity are common to all vertebrates.

The relationships between the neuroendocrine–immune system and biologic rhythms are becoming clearer. Seasonal and circadian changes in immune responses and lymphoid tissue structure are well known in lower vertebrates (Schwassmann, 1971; Matty, 1978; Thorp, 1978; White and Fletcher, 1985; Nakanishi, 1986), rodents (Pownall and Knapp, 1980), and humans (Tavida *et al.*, 1975; Kaplan *et al.*, 1976). With regards to environmental contaminants, especially those resembling hormones (in function), significant effects on these interactions resulting in altered NE and biologic rhythms may result from exposure to environmental contaminants. It is well known that lower vertebrates have seasonal responses to contaminants, including the ability to metabolize insecticides (Chambers and Yarbrough, 1979; George *et al.*, 1990). Although seldom published, mammalian immunotoxicologists are well aware of the seasonal variations (times of the year) in assay performance (plaque-forming assays, chromium release assays, etc.). Therefore, the influence of biologic rhythms on immunotoxicologic studies is not necessarily restricted to lower vertebrate models.

Environmental hormones may affect fish immune responses

It is now clear that some very common, if not ubiquitous, environmental contaminants are estrogen mimics (Fry and Toone, 1981; Kubiak *et al.*, 1989; Ellis and Pattisina, 1990; Bishop *et al.*, 1991; Munkittrick *et al.*, 1991; Colborn *et al.*, 1993). For example, nonyl- and octyl-phenolic components of industrial surfactants/detergents mimic estrogen (17β-estradiol; E_2) in stimulating both *in vitro* and *in vivo* vitellogenin production in fish (Purdom *et al.*, 1994; Benson and Nimrod, 1996; Lech *et al.*, 1996). These compounds, and others (kepone, *o,p′*-DDT, some PCB congeners), are also known to mimic estrogens in stimulating the proliferation of estrogen-dependent breast tumor cell lines (Soto *et al.*, 1992). It is important to recognize that the potency of most of these compounds is usually far less than that of either estradiol or the synthetic estrogen diethylstilbestrol (DES). Other studies indicate that many chemical mixtures may contain not only hormonal agonists but also antagonists (Safe and Gaido, 1998). An increase in estrogen/androgen receptor binding by hormone mimics, blocking of hormone receptors, or disruption of enzymes responsible for estrogen/androgen production are the prevailing theories explaining these interactions (Lu *et al.*, 1996).

Environmental estrogens may significantly affect immunophysiologic homeostasis. In mammals, and especially humans, sexual dimorphism exists in immunophysiology. For example, female humans have higher levels of circulating antibody (IgG, IgM, and IgA) than males, and antibody responses to both T-dependent and T-independent antigens are greater in females than males (Butterworth *et al.*, 1967; Terres *et al.*, 1968; Eidinger and Garrett, 1972; Grossman, 1985; see review by Madden and Felten, 1995). These observations, as well as immunogenetics, may partially explain higher incidences of autoimmune diseases [systemic lupus erythematosus (SLE)] in females. This implies a shift from Th1 to Th2 T-cell functions. Whether or not xenoestrogens (or androgens) can tip the scales in this regards is unknown, but Heo *et al.* (1998) show that Pb can differentially suppress Th1 in favor of Th2 T-helper-associated cytokines.

Both estrogen and testosterone may also affect immune function vicariously through the thymus, a primary lymphoid organ. Exposure of normal mice to estrogens induces thymic atrophy; a plausible explanation for dramatic thymic atrophy beginning with the onset of pubescence in humans. If this is the case, then environmental contaminants that mimic estrogen (or androgens) may have serious consequences for immune function of developing vertebrates, regardless of phylogeny. Other endocrine organs are affected by estrogen and testosterone, including the thyroid gland. For example, Leatherland (1985) provided evidence that both estrogen and testosterone suppress thyroid function in rainbow trout. Thyroxine (T_3), and therefore the T_4/T_3 ratio, was significantly lowered and calcium/magnesium levels were elevated as aconsequence of exposure. These findings, along with estrogen-related increases in plasma protein, plasma triglyceride, and phosphoprotein are consistent with an increase in hepatic vitellogenin production (vitellogenesis). It is unknown at this point whether thyroid dysfunction in fish can occur following exposure to environmentally relevant concentrations of xenoestrogens (or xenoandrogens) or whether it can occur only under experimental pharmacologic situations.

Immune function and homeostasis may be affected by thyroid dysfunction because T_3 and T_4 have profound effects in vertebrates (Fabris, 1973; Chatterjee and Chandel, 1983; Scott and Glick, 1987). Thyroidectomy results in decreased lymphoid organ weights, decreased circulating lymphocytes, and lowered antibody responses. Administration of T_3 or T_4 reverses these conditions. Thus, immune function may be indirectly affected by environmental contaminants through the action on neuroendocrine systems. With regards to contaminants, the effects of PCBs on thyroid function is fairly well documented and seems to be related to induction of P450-related enzymes and increased expression/activity of uridine diphosphate glucuronyltransferase (UDP-GT), the rate-limiting enzyme in T_4 metabolism (McClain *et al.*, 1988). T_3/T_4 ratios would then become altered as a result of exposure and effect. Other studies indicate that non-planar PCB congeners may induce monoamine oxidase, thereby decreasing dopamine and other biogenic amine levels. Dopamine secretion leads to prolactin secretion, therefore those systems under prolactin control (T-cell proliferation and differentiation, isomoregulation, and reproduction) may be indirectly impaired by PCBs.

Moreover, epinephrine, norepinephrine, and dopamine are also involved in acute stress responses and vascular lymphoid tonicity. It is generally accepted that co-planar PCBs affect their toxicities primarily through Ah receptor-mediated events (Holsapple *et al.*, 1991), whereas non-planar, ortho-substituted congeners affect membrane and cytosolic-associated events (Ganey *et al.*, 1993; Tithof *et al.*, 1995), including calcium modulation, alterations in mitochondria function, and membrane fluidity. How xenoestrogens modulate PCB effects on immunophysiology and natural biologic rhythms remains unknown.

As previously stated, immune function is tightly coupled with neuroendocrine function (Weigent and Blalock, 1996; Arnold and Rice, 2000) and both are very sensitive targets of many xenobiotics and generalized stress (Pruett *et al.*, 1993). The prototype and most characterized xenobiotics, in terms of immunotoxicity and endocrine disruption, are 2,3,7,8-tetrachlorodibenzo-*p*-dioxin (TCDD) (Holsapple *et al.*, 1991; Rice *et al.*, 1995a) and benzo(a)pyrene (White, 1986). Certain organometallic (OM) compounds such as tributyltin and methylmercury are also classic immunotoxicants and disrupt endocrine functions (Bryan *et al.*, 1986; Weis and Kim, 1988; Ellis and Pattisina, 1990; Zelikoff, 1993, Rice *et al.*, 1995b).

The above few examples represent laboratory studies. Other published studies imply environmentally mediated immunomodulation in fish (Weeks and Warinner, 1984, 1986; Warinner *et al.*, 1988; Arkoosh *et al.*, 1991, 1996; Faisal *et al.*, 1991; Seeley and Weeks-Perkins, 1991; Rice *et al.*, 1996). Although altered immune functions in these studies are associated with contaminated environments, neuroendocrine disruption may have been a significant component of immune dysfunction because seasonality, time of day, and reproductive status were not considered.

Slater and Schreck (1998) demonstrated that in sexually immature salmon leukocyte androgen receptor affinity dropped by a factor of four in the winter but rose during the summer. Confinement stress caused a decrease in androgen receptor affinity on leukocytes, but had no effect on numbers of receptors. Sexual maturity leads to an increase in androgen receptor numbers; the affinity of the receptors decreased after ovulation. As it is well known that sex steroids modulate immune responses, such attention to neuroendocrine–immune interactions (immunophysiology) in immunotoxicology should be focused on all models, regardless of phylogeny. The same is true for cortisol, prolactin, growth hormone, and others. The need to account for natural physiologic fluctuations in environmental immunotoxicology is supported further by the findings of Ottinger and Kaattari (1998). *In vitro* LPS-stimulated leukocyte proliferation and immunoglobulin production are more sensitive to aflatoxin B1 between July and December than between January and June in rainbow trout. The same individual fish were used throughout the study. Neuroendocrine–immune interactions can also influence the interpretation of data resulting from chronic exposure to low levels of contaminants in mixtures. For example, our laboratory developed the gulf killifish, *Fundulus grandis*, as an estuarine model for environmental immunotoxicologic studies (Roszell and Rice, 1998; Rice and Xiang, 2000). We developed monoclonal antibodies against killifish immunoglobulin and vitellogenin that, in turn, allowed us to quantify specific antibody responses and circulating vitellogenin respectively. Killifish were fed a diet containing 10, 1, 1, and 10 parts per million (ppm) of Aroclor 1254, tributyltin chloride, 3-methylcholanthrene, and nonyl-phenol, respectively, for 120 days between January and May. Fish were fed once daily (14:00 h) a ration of 5 percent body weight. At weeks 7 and 13 of the exposure, each fish received an intraperitoneal injection of formalin-killed *Vibrio anguillarium* as the model antigen. Fish were then sacrificed in the morning and again in the evening at 2-week intervals corresponding to the lunar cycle. Serum samples were then taken to measure antigen-specific immunoglobulin and vitellogenin levels. Liver CYP1A protein levels were also determined. At week 10, antibody levels were higher in the evening than in the morning for both control and treated killifish. For fish sampled in the morning, secondary antibody responses were higher in control than in treated fish (weeks 14, 15, and 16). Antibody levels of the controls remained elevated for the remainder of the study, whereas antibody responses of treated animals declined. There were no differences in secondary antibody responses between treated and control fish sampled in the evening. Although each of the dietary xenobiotics are classified as endocrine disrupters, no alteration in vitellogenesis was noted as a result of treatment. Furthermore, CYP1A levels varied from morning to evening, even in controls. As expected, CYP1A levels were higher in treated animals, but fluctuations were the norm for both treated and controls. Taken together, these data indicate the need to incorporate daily and seasonal fluctuations when conducting immunotoxicologic research with fish.

Summary

Fish immunologists and toxicologists will continue to use immune function endpoints as biomarkers of exposure and effect in field studies (Fournier *et al.*, 2000). In the end, however, it will be necessary to distinguish between immunomodulation due to chemical insult and that which is due to extrinsic factors such as neuroendocrine–immune interactions. Neuroendocrine–immune modulations may or may not be the result of exposure to xenobiotics. At this point, there are no immune functional endpoints (phagocytosos, natural killer activity, mitogenic responses, etc.) for which modulation is a signature of a particular class of xenobiotics. Metals and organometallic xenobiotics clearly affect phagocyte responses, with both enhancement and suppression being the end result, depending on the level of exposure (dose). On the other hand, halogenated aromatic hydrocarbons can as well. It becomes a dose–response phenomenon in which at some point in the dose–response relationship one will see immunmodulation. To conclude that the immune system is the target organ for any particular xenobiotic or class of compounds, researchers must show that immune function is affected before other organ systems are affected. This will require integrated approaches to fish immunotoxicology as described for tier I and tier II immunotoxicity testing in rodents.

With regard to the future of fish immunotoxicology, researchers should be forced to have a clearer understanding of their particular fish model. Aquaculture species are the most characterized in terms of immune response kinetics and are those species for which the most reagents are available. In the future, fish immunotoxicologists may find themselves working with only a few species, yet those species (channel catfish, rainbow trout, medaka, mummichog, carp, etc.) will be highly characterized and their immune systems understood. At that point, we may be closer to having systems similar to the few strains of rodents that used in mammalian immunotoxicology.

References

Agius, C. 1985. The melano-macrophage centres of fish. In *Fish Immunology*. Manning, M.J. and Tatner, M.F. (eds), pp. 85–105. Academic Press, London.

Ainsworth, A.J. 1992. Fish granulocytes: Morphology, distribution, and function. *Annual Review of Fish Diseases* 2: 123–148.

Ainsworth, A.J. and Boyd, B. 1998. Intracellular signaling events in superoxide generation and adhesion of channel catfish, *Ictalurus punctatus* rafinseque, neutrophils to the extracellular matrix protein fibrinogen. *Fish and Shellfish Immunology* 22: 173–184.

Anderson D.P. 1990. Immunological indicators: effects of environmental stress on immune protection and disease outbreaks. *American Fisheries Society Symposium* 8: 38.

Anderson, D.P. and Zeeman, M.G. 1995. Immuntoxicology in fish. In *Fundamentals of Aquatic Toxicology*, 2nd edn. Rand, G.M. (ed.), pp. 371–404. Taylor and Francis, Washington, DC.

Anderson, R.S. 1994. Modulation of blood cell-mediated oxyradical production in aquatic species: implications and applications. In *Aquatic Toxicology: Molecular, Biochemical, and Cellular Perspectives*. Malins, D.C. and Ostrander, G.K. (eds), pp. 241–265. CRC Press, Boca Raton.

Armknecht, S.L., Kaattari, S.L. and Van Veld, P.A. 1998. An elevated glutathione S-transferase in creosote-resistant mummichog, *Fundulus heteroclitus*. *Aquatic Toxicology* 41: 1–16.

Arnold, R.E. and Rice, C.D. 2000. Channel catfish, *Ictalurus punctatus*, leukocytes secrete immunoreactive adrenal corticotropin hormone (ACTH). *Fish Physiology and Biochemistry* 22: 303–310.

Arkoosh, M.R., Casillas, E.C., Ethan, B.M. and Varanasi, U. 1991. Suppression of immunological memory in juvenile chinook salmon, *Oncorhyncus tshawytscha*, from an urban estuary. *Fish and Shellfish Immunology* 1: 261–277.

Arkoosh, M.R., Clemons, E., Huffman, P., Sanborn, H.R., Casilla, E. and Stein, J.E. 1996. Leukoproliferative response of splenocytes from English sole (*Pleuronectes vetulus*) exposed to chemical contaminants. *Environmental Toxicology and Chemistry* 15: 1154–1162.

Arkoosh, M.R., Casillas, E., Clemons, E., Kagley, A.N., Olson, R., Reno, P. and Stein, J.E. 1998. Effect of pollution on fish diseases: Potential impacts on salmonid populations. *Journal of Aquatic Animal Health* 10: 182–190.

Bardwaj, R., Becher, E., Mahnke, K., Hartmeyer, M., Schwarz, T., Scholzen, T. and Luger, T.A. 1997. Evidence for the differential expression of the functional alpha-melanocyte stimulating hormone receptor MC-1 on human monocytes. *Journal of Immunology* 158: 3378–3384.

Benson, W.H. and Nimrod, A.C. 1996. Environmental estrogens: potential for reproductive toxicity in fish. Technical Completion Report, G2028-15. US Department of the Interior. WRRI. Mississippi State University, Mississippi.

Bishop, C.A., Brooks, R.J., Carey, J.H., Ng, P., Norstrom, R.J. and Lean, D.R.S.J. 1991. The case for a cause–effect linkage between environmental contamination and development in eggs of the common snapping turtle (*Cheldyra*: *S. serpentina*), from Ontario, Canada. *Journal of Toxicology and Environmental Health* 33: 521–548.

Bourdi, M., Tinel, M., Beaune, P.H. and Pessayre, D. 1994. Interactions of dihydralazine with cytochromes P4501A: a possible explanation for the appearance of anti-cytochrome P4501A2 autoantibodies. *Molecular Pharmacology* 45: 1287–1295.

Bowser, D.H., Frenkel, K. and Zelikoff, J.T. 1994. Effects of in vitro nickel exposure on the macrophage-mediated immune functions of rainbow trout, *Oncorhynchus mykiss*. *The Bulletin of Environmental Contamination and Toxicology* 52: 367–373.

Brown, J.A., Edwards, D. and Whitehead, C.J. 1989. Cortisol and thyroid hormone responses to acid stress in the brown trout (*Salmo trutta* L). *Journal of Fish Biology* 35: 73–81.

Bryan, B.W., Gibbs, P.E., Hummerstone, L.G. and Burt, G.R. 1986. The decline of the gastropod *Nucella lapillus* around south-west England: Evidence for the effect of tributyltin from antifoulant paints. *Journal of the Marine Biology Association, UK* 66: 611–640.

Brousseau, P.Y., Payette, H., Tryphonas, H., Blakely, B., Boermans, H., Flipo, D. and Fournier, M. 1994. *Canadian Network of Toxicology Centre's Handbook of Immunological Methods*, 1st edn.

Burchiel, S.W., Davis, D.P., Ray, S.D., Archuleta, M.M., Thilsted, J.P. and Corcoran, G.G. 1992. DMBA-induced cytotoxicity in lymphoid and nonlymphoid organs of B6C3F1 mice: Relation of cell death to target cell intracellular calcium and DNA damage. *Toxicology and Applied Pharmacology* 113: 126–132.

Burnett, K.G. 1997. Evaluating intracellular signaling pathways as biomarkers for environmental contaminant exposures. *American Zoologist* 37: 585–594.

Burns, L.A., Meade, B.J. and Munson, A.E. 1996. Toxic responses of the immune system. In *Casarett and Doull's Toxicology: The Basic Science of Poisons*, 5th edn. Klaassen, C.D. (ed.), pp. 355–402. McGraw-Hill, New York.

Butterworth, M., McClellan, B. and Allansmith, M. 1967. Influence of sex on immunoglobulin levels. *Nature* (London) 214: 1224–1225.

Camp, K.C. 1997. Disease resistance and immune function in full-sib families of channel catfish, *Ictalurus punctatus*. Master of Science Thesis. College of Veterinary Medicine, Mississippi State University.

Camp, K.C., Rice, C.D. and Wolters, W.R. 2000. Survivability and immune responses after challenge with *Edwardsiella ictaluri* in susceptible and resistant families of channel catfish, *Ictalurus punctatus*. *Fish and Shellfish Immunology* 10: 475–487.

Campos-Perez, J.J., Laing, K.J., Ellis, A.E. and Secombes, C.J. 1997. The role of inducible nitric oxide synthase and nitric oxide in the immune system of rainbow trout. In *Proceedings of the 7th International Congress of the International Society of Developmental and Comparative Immunologists*, Williamsburg, VA, July 1997.

Celander, M., Bremier. J., Hahn, M.E. and Stegeman, J.J. 1997. Glucocorticoid–xenobiotic interactions: Dexamethasone-mediated potentiation of cytochrome P4501A induction by b-naphthoflavone in a fish hepatoma cell line (PLHC-1). *Environmental Toxicology and Chemistry* 16: 900–907.

Chambers, J.E. and Yarbrough, J.D. 1979. A seasonal study of microsomal mixed-function oxidase components in insecticide-resistant and susceptible mosquitofish, *Gambusia affinis*. *Toxicology and Applied Pharmacology* 48: 497–508.

Chatterjee, S. and Chandel, A.S. 1983. Immunomodulatory role of thyroid hormones: in vivo effect of thyroid hormones on the blastogenic response of lymphoid tissues. *Acta Endocrinologica* 103: 95–100.

Chow, S.C., Kass, G.E., McCabe, M.J. and Orrenius, S. 1992. Tributyltin increases cytosolic free Ca^{2+} concentration in thymocytes by mobilizing intracellular Ca^{2+}, activating a Ca^{2+} entry pathway, and inhibiting Ca^{2+} efflux. *Analytical Biochemistry and Biophysics* 298: 143–149.

Claman, H.N. 1975. How corticosteroids work. *Journal of Allergy and Clinical Immunology* 55: 145–151.

Cleland, G.B., McElroy, P.J. and Sonstegard, R.A. 1988. The effect of dietary exposure to Aroclor 1254 and/or mirex on humoral immune expression of rainbow trout, *Salmo gairdneri*. *Aquatic Toxicology* 12: 141–146.

Clem, L.W., Miller, N.W. and Bly, J.E. 1991. Evolution of lymphocyte subpopulations, their interactions, and temperature sensitivities. In *Phylogenesis of Immune Functions*. Warr, G.W. and Cohen, N. (eds), pp. 191–213. CRC Press, Boca Raton.

Clem, L.W., Bly, J.E., Wilson, M., Chinchar, V.G., Stuge, T., Barker, K.S., Luft, C., Rycyzyn, M., Hogan, R.J., van Lopik, T. and Miller, N.W. 1996. Fish Immunology: the utility of immortalized lymphoid cells – a mini review. *Veterinary Immunology and Immunopathology* 54: 137–144.

Colburn, T., VoM Saal, F.S. and Soto, A.M. 1993. Developmental effects of endocrine-disrupting chemicals in wildlife and humans. *Environmental Health Perspectives* 101: 378–384.

Dalmo, R.A., Ingebrigtsen, K. and Bogwald, J. 1997. Non-specific defense mechanisms in fish, with particular reference to the reticuloendothelial system (RES). *Journal of Fish Diseases* 20: 241–273.

Dean, J.H., Padarathsingh, M.L. and Jerrells, T.R. 1979. Assessment of immunobiological effects induced by chemicals, drugs, and food additives. I. Tier testing and screening approach. *Drug Chemistry and Toxicology* 2: 5–17.

Demers, N.E. and Bayne, C.J. 1994. Plasma proteins of rainbow trout, *Oncorhynchus mykiss*: immediate response to acute stress. In *Models for Environmental Toxicology, Biomarkers, Immunostimulants*, Vol. 1. Stolen, J.S. and Fletcher, T.C. (ed.), pp. 1–9. SOS Publications, Fair Haven, NJ.

Dunier, M. and Siwicki, A.K. 1993. Effects of pesticides and other organic pollutants in the aquatic environment on immunity in fish: a review. *Fish and Shellfish Immunology* 3: 423–438.

Eidinger, D. and Garrett, T.J. 1972. Studies of the regulatory effects of the sex hormones on antibody formation and stem cell differentiation. *Journal of Experimental Medicine* 136: 1098–1116.

Eliasson, E. and Kenna, J.G. 1996. Cytochrome P4502E1 is a cell surface autoantigen in halothane hepatitis. *Molecular Pharmacology* 50: 573–582.

Ellis, A.E. 1977. The leukocytes of fish: A review. *Journal of Fish Biology* 11: 453–491.

Ellis, D.V. and Pattisina, L.S. 1990. Widespread neogastropod imposex: a biological indicator of global TBT contamination. *Marine Pollution Bulletin* 21: 248–253.

Emata, A.C., Meier, A.H. and Hsiao, S.M. 1991. Daily variations in plasma hormone concentrations during the semilunar spawning cycle of the gulf killifish, *Fundulus grandis*. *Journal of Experimental Zoology* 259: 343–354.

Enk, A.H. and Katz, S.I. 1992. Early molecular events in the induction phase of contact hypersensitivity. *Proceedings of the National Academy of Sciences of the United States of America* 89: 1398–1402.

Fabris, D.L. 1973. Immune depression in thyroid-deprived animals. *Clinical and Experimental Immunology* 15: 601–611.

Faisal, M., Huggett, R.J. Vogelbein, W.K. and Weeks, B.A. 1991. Evidence of aberration of the natural cytotoxic cell activity in *Fundulus heteroclitus* (Pisces: *Cytprinodontidae*) from the Elizabeth River, Virginia. *Veterinary Immunology and Immunopathology* 29: 339–351.

Faisal, M., Rutan, B.J. and Sami-Demmerle, S. 1995. Development of continuous liver cell cultures from the marine teleost, spot (*Leiostomus xanthurus*, Pisces: Sciaenidae). *Aquaculture* 132: 59–72.

Fange, R. 1994. Blood cells, haemopoiesis and lymphomyeloid tissues in fish. *Fish and Shellfish Immunology* 4: 405–411.

Fletcher, T.C. and White, A. 1973. Lysozyme activity in the plaice, *Pleuronectes platessa* L. *Experientia* 29: 1283–1285.

Fournier, M., Lacroix, A., Voccia, I. and Brousseau, P. 1998. Phagocytic and metabolic activities of macrophages from mummichog naturally exposed to pulp mill effluents in the Miramichi river. *Ecotoxicology and Environmental Safety* 40: 177–183.

Fournier, M., Cyr, D., Blakley, B., Boermans, H. and Brousseau, P. 2000. Phagocytosis as a biomarker of immunotoxicity in wildlife species exposed to environmental xenobiotics. *American Zoologist* 40: 412–420.

Fry, D.M. and Toone, C.K. 1981. DDT-induced feminization of gull embryos. *Science* 231: 919–924.

Ganey, P.E., Sirois, J.E., Denison, M., Robinson, J.P. and Roth, R.A. 1993. Neutrophil function after exposure to polychlorinated biphenyls *in vitro*. *Environmental Health Perspectives* 101: 430–434.

George, S., Young, P., Leaver, M. and Clarke, D. 1990. Activities of pollutant metabolizing and detoxication systems in the liver of plaice, *Pleuronectes platessa*: Sex and seasonal variations in non-induced fish. *Comparative Biochemistry and Physiology* 96C: 185–192.

Godwin, U.B., Antao, A., Wilson, M.R., Chinchar, V.G., Miller, N.W., Clem, L.W. and McConnel, T.J. 1997. MHC class II B genes in the channel catfish, *Ictalurus punctatus*. *Developmental and Comparative Immunology* 21: 13–24.

Goebeler, M., Meinardus-Hager, G., Roth, J., Goerdt, S. and Sorg, C. 1993. Nickel chloride and cobalt chloride, two common contact sensitizers, directly induce expression of intercellular adhesion molecule-1 (ICAM-1), vascular cell adhesion molecule-1 (VCAM-1), and endothelial leukocyte adhesion molecule (ELAM-1) by endothelial cells. *Journal of Investigations in Dermatology* 100: 759–765.

Griem, P., Wulferink, M., Sachs, B., Gonzalez, J.B. and Gleichmann, E. 1998. Allergic and autoimmune reactions to xenobiotics: how do they arise? *Immunology Today* 19: 133–141.

Grossman, D.J. 1985. Interactions between the gonadal steroids and the immune system. *Science* 227: 257–261.

Hahn, M.E., Karchner, S.I., Shapiro, M.A. and Perera, S.A. 1997. Molecular evolution of two vertebrate aryl hydrocarbon (dioxin) receptors (AHR1 and AHR2) and the PAS family. *Proceedings of the National Academy of Sciences of the United States of America* 94: 13743–13748.

Hart, L.J., Smith, S.A., Smith, B.J., Robertson, J. and Holladay, S.D. 1997. Exposure of tilapian fish to the pesticide lindane results in hypocellularity of the primary hematopoietic organ (pronephros) and the spleen without altering activity of phagocytic cells in these organs. *Toxicology* 118: 211–221.

Hart, L.J., Smith, S.A., Smith, B.J., Robertson, J., Besteman, E.G. and Holladay, S.D. 1998. Subacute immunotoxic effects of the polycyclic aromatic hydrocarbon 7,12-dimethylbenzanthracene (DMBA) on spleen and pronephros leukocytic cell counts and phagocytic cell activity in tilapia, *Oreochromis niloticus*. *Aquatic Toxicology* 41: 17–29.

Heo, Y., Lee, W.T. and Lawrence, D.A. 1998. Differential effects of lead and cAMP on development and activities of Th1- and Th2-lymphocytes. *Toxicological Science* 43: 172–185.

Holsapple, M.P., Snyder, N.K., Wood, S.C. and Morris, D.L. 1991. A review of 2,3,7,8-tetrachlorodibenzo-p-dioxin-induced changes in immunocompetence: 1991 update. *Toxicology* 69: 219–255.

Hontela, A., Rasmussen, J.B., Audet, C. and Chevalier, G. 1992. Impaired cortisol stress response in fish from environments polluted by PAHs, PCBs, and mercury. *Archives of Environmental Contamination and Toxicology* 22: 278–285.

Kaattari, S.L. 1994. Development of a piscine paradigm of immunological memory. *Fish and Shellfish Immunology* 4: 447–457.

Kaattari, S.L., Adkison, M., Shapiro, D. and Arkoosh, M.R. 1994. Mechanisms of immunosuppression by aflatoxin B1. SOS Publications, NJ. *Modulators of Fish Immune Responses* 1: 151–156.

Kajita, Y., Sakai, M., Kobayashi, M. and Kawauchi, H. 1992. Enhancement of non-specific cytotoxic activity of leucocytes in rainbow trout, *Oncorhynchus mykiss*, injected with growth hormone. *Fish and Shellfish Immunology* 2: 155–157.

Kaplan, M.S., Byers, V.S., Levin, A.S., German, D.F., Fudenburg, H.H. and Lecam, L.N. 1976. Circadian rhythms of stimulated lymphocyte blastogenesis. *Journal of Allergy and Clinical Immunology* 58: 180–189.

Karrass, G.J., Morris, D.L., Matulka, R.A., Kramer, C.M. and Holsapple, M.P. 1996. 2,3,7,8-tetrachlorodibenzo-*p*-dioxin (TCDD) elevates basal B-cell intracellular calcium concentration and suppresses surface Ig- but not CD40-induced antibody secretion. *Toxicology and Applied Pharmacology* 137: 275–284

Kelly-Reay, K. and Weeks-Perkins, B.A. 1994. Determination of the macrophage chemiluminescent response in *Fundulus heteroclitus* as a function of pollution stress. *Fish and Shellfish Immunology* 4: 95–105.

Kenedy-Stoskopf, S. 1993. Immunology. In *Fish Medicine*. Stoskopf, M.K. (ed.), pp. 149–159. W.B. Saunders Company, Philadelphia.

Kerkvliet, N.I. and Burleson, G.R. 1994. Immunotoxicity of TCDD and related halogenated aromatic hydrocarbons. In *Immunotoxicology and Immunopharmacology*, 2nd edn. Dean, J.H., Luster, M.I., Munson, A.E. and Kimber, I. (eds), pp. 97–121. Raven Press, New York.

Klein, J. 1996. Defence reactions mediated by phagocytes. In *Immunology*. Blackwell Scientific Publications, Boston, USA.

Koppang, E.O., Lundin, M., Press, C., Ronningen, K. and Lie, O. 1998. Differing levels of Mhc class II b chain expression in a range of tissues from vaccinated and non-vaccinated Atlantic salmon, *Salmo salar* L. *Fish and Shellfish Immunology* 8: 183–196.

Kubiak, T.J., Harris, H.J., Smith, L.M., Schwartz, T.P., Stalling, D.L., Trick, J.A., Sileo, L., Docherty, D.E. and Erdman, T.C. 1989. Microcontaminants and reproductive impairment of the Forster's tern on Green Bay, Lake Michigan – 1983. *Archives of Environmental Contamination and Toxicology* 18: 706–727.

Kuby, J. 1994a. The complement system. In *Immunology*, 2nd edn. Kuby, J. (ed.), p. 396. W.H. Freeman and Company, New York.

Kuby, J. 1994b. Antigen processing and presentation. In *Immunology*, 2nd edn. Kuby, J. (ed.), p. 247. W.H. Freeman and Company, New York.

Ladics, G.S., Kawabata, T.T., Munson, A.E. and White, K.L. 1992. Metabolism of benzo-a-pyrene by murine splenic cell types. *Toxicology and Applied Pharmacology* 116: 248–257.

Leatherland, J.F. 1985. Effects of 17-estradiol and methyl testosterone on the activity of the thyroid gland in rainbow trout, *Salmo gairdneri* Richardson. *General and Comparative Endocrinology* 60: 343–352.

Leatherland, J.F., McKeown, B.A. and John, T.M. 1974. Circadian rhythm of plasma prolactin, growth hormone, glucose and free fatty acid in juvenile kokanee salmon, *Oncorhynchus nerka*. *Comparative Biochemistry and Physiology* 47A: 821–828.

Lech, J.J., Lewis, S.K. and Ren, L. 1996. In vivo estrogenic activity of nonylphenol in rainbow trout. *Fundamental Applied Toxicology* 30: 229–232.

Levi, F., Canon, C., Dipalma, M., Florentin, I. and Misset, J.L. 1991. When should the immune clock be reset? From circadian pharmacodynamics to temporally optimized drug delivery. *Annals of the New York Academy of Science* 618: 312–329.

Lin, W. and White, K.L. 1993. Production of complement component C3 in vivo following 2,3,7,8-tetrachlorodibenzo-*p*-dioxin exposure. *Journal of Toxicology and Environmental Health* 39: 273–285.

Lisby, S., Muller, K.M., Jongeneel, C.V., Saurat, J.H. and Hauser, C. 1995. Nickel and skin irritants up-regulate tumor necrosis factor-alpha mRNA in keratinocytes by different but potentially synergistic mechanisms. *International Immunology* 7: 343–352.

Lu, Y., Wang, X. and Safe, S. 1996. Interaction of 2,3,7,8-tetrachlorodibenzo-p-dioxin and retinoic acid in MCF-7 human breast cancer cells. *Toxicology and Applied Pharmacology* 127: 1–8.

Luer, C.A., Walsh, C.J., Bodine, A.B., Wyffels, J.T. and Scott, T.R. 1995. The elasmobranch thymus: Anatomical, histological, and preliminary functional characterization. *Journal of Experimental Zoology* 273: 342–354.

Luster, M.I., Dean, J.H. and Moore, J.A. 1982. Evaluation of immune functions in toxicology. In *Methods in Toxicology*. Hayes, W. (ed.), pp. 561–586. Raven Press, New York.

Luster, M.I., Portier, C., Pait, D.G., White, K.L., Gennings, C., Munson, A.E. and Rosenthal, G.J. 1992. Risk assessment in immunotoxicology. I. Sensitivity and predictability of immune tests. *Fundamental Applied Toxicology* 18: 200–210.

McClain, R.M., Posch, R.C., Bosakowski, T. and Armstrong, J.M. 1988. Studies on the mode of action for thyroid gland tumor promotion in rats by phenobarbital. *Toxicology and Applied Pharmacology* 99: 254–265.

McKinney, E.C. 1992. Shark lymphocytes: Primitive antigen reactive cells. *Annual Review of Fish Diseases* 2: 43–51.

McKinney, E.C. and Flajnik, M.F. 1997. IgM-mediated opsonization and cytotoxicity in the shark. *Journal of Leukocyte Biology* 61: 141–146.

Madden, K.S. and Felten, D.L. 1995. Experimental basis for neural–immune interactions. *Physiology Review* 75: 77–106.

Maestroni, G.J.M. 1995. The immunoneuroendocrine role of melatonin. *Journal of Pineal Research* 14: 1–10.

Marc, A.M., Quentel, C., Severe, A., Le Bail, P.Y. and Boeuf, G. 1995. Changes in some endocrinological and non-specific immunological parameters during seawater exposure in the brown trout. *Journal of Fish Biology* 46: 1065–1081.

Matty, A.J. 1978. Pineal and some pituitary hormone rhythms in fish. In *Rhythmic Activity in Fishes*. Thorpe, J.E. (ed.), pp. 21–30. Academic Press, London.

Maule, A.G., Tripp, R.A., Kaattari, S.L. and Schreck, C.B. 1989. Stress alters immune function and disease resistance in chinook salmon, *Oncorhynchus tshawytscha*. *Journal of Endocrinology* 120: 135–142.

Miller, N.W., Rycyzyn, M.A., Wilson, M.R., Warr, G.W., Naftel, J.P. and Clem, L.W. 1994a. Development and characterization of channel catfish long term B cell lines. *Journal of Immunology* 152: 2180–2189.

Miller, N.W., Chinchar, V.G. and Clem, L.W. 1994b. Development of leukocyte cell lines from channel catfish, *Ictalurus punctatus*. *Journal of Tissue Culture Methods* 16: 117–123.

Mohr, S., Stamier, J.S. and Brune, B. 1994. Mechanism of covalent modification of glyceraldehyde-3-phophate dehydrogenase at its active thiol site by nitric oxide, peroxynitrite, and related nitrosating agents. *FEBS Letters* 348: 223–227.

Muhvich, A.G., Jones, R.T., Kane, A.S., Anderson, R.S. and Reimscheussel, R. 1995. Effects of chronic copper exposure on the macrophage chemiluminescent response and gill histology in goldfish, *Carassius auratus*, L. *Fish and Shellfish Immunology* 5: 251–264.

Munck, A. and Guyre, P.M. 1991. Glucocorticoids and immune function. In *Psychoneuroimmunology*, Vol. II. Ader, R., Felten, D.L. and Cohen. N. (eds), pp. 447–474. Academic Press, San Diego.

Munck, A., Guyre, P.M. and Holbrook, N.J. 1984. Physiological functions of glucocorticoids in stress and their relation to pharmacological actions. *Endocrinology Reviews* 5: 25–44.

Munkittrick, K.R., Port, C.B., Van Der Kraak, G.J., Smith, I.R. and Rokosh, D.A. 1991. Impact of bleached kraft mill effluent on population characteristics, liver MFO activity, and serum steroids of a Lake Superior white sucker (*Catostomus commersoni*) population. *Canadian Journal of Aquatic Science* 48: 1–10.

Nakanishi, T. 1986. Seasonal change in the humoral immune response and the lymphoid tissues of the marine teleost, *Sebastiscus marmoratus. Veterinary Immunology and Immunopathology* 12: 228–234.

Neumann, N.F., Fagan, D. and Belosevic, M. 1995. Macrophage activating factor(s) secreted by mitogen stimulated goldfish kidney leukocytes synergize with bacterial lipopolysaccharide to induce nitric oxide production in teleost macrophages. *Developmental and Comparative Immunology* 19: 473–482.

Nevid, N.J. and Meier, A.H. 1993. A day–night rhythm of immune activity during scale allograft rejection in the gulf killifish, *Fundulus garndis. Developmental and Comparative Immunology* 17: 221–228.

Nevid, N.J. and Meier, A.H. 1995. Timed daily administration of hormones and antagonists of neuroendocrine receptors alter day–night rhythms of allograft rejection in the gulf killifish, *Fundulus grandis. General and Comparative Endocrinology* 97: 327–339.

Ottinger, C.A. and Kaattari, S.L. 1998. Sensitivity of rainbow trout leucocytes to Aflatoxin B_1. *Fish and Shellfish Immunology* 7/8: 515–530.

Payne, J.F. and Fancey, L.F. 1989. Effect of polycyclic aromatic hydrocarbons on immune responses in fish: Change in melanomacrophage centers in flounder, *Pseudopleuronectes americanus*, exposed to hydrocabon-contaminated sediments. *Marine Environmental Research* 28: 431–444.

Porter, W.P., Hinsdill, R., Fairbrother, A., Olson, L.J., Jaeger, J., Yuill, T.L., Bisgaard, S., Hunter, W.G. and Nolan, K. 1984. Toxicant–disease environment interactions associated with suppression of immune system, growth, and reproduction. *Science* 224: 1014–1017.

Pownall, R. and Knapp, M.S. 1980. Immune responses have rhythms; are they important? *Immunology Today* 1: 7–10.

Prough, R.A., Linder, M.W., Pinaire, J.A., Xiao, G.H. and Falkner, K.C. 1996. Hormonal regulation of the hepatic enzymes involved in foreign compound metabolism. *FASEB Journal* 10: 1369–1377.

Pruett, S.B., Ensley, D.K. and Crittenden, P.L. 1993. The role of chemical-induced stress responses in immunosuppression: A review of quantitative associations and cause–effect relationships between chemical-induced stress responses and immunosuppression. *Journal of Toxicology and Environmental Health* 39: 163–192.

Purdom, C.E., Hardiman, P.A., Bye, V.J., Eno, N.C., Tyleer, C.R. and Sumpter, J.P. 1994. Estrogenic effects of effluents from sewage treatment works. *Chemistry and Ecology* 8: 275–285.

Raffray, M., McCarthy, D., Snowden, R.T. and Cohen, G.M. 1993. Apoptosis as a mechanism of tributyltin cytotoxicity to thymocytes: Relationship of apoptotic markers to biochemical and cellular effects. *Toxicology and Applied Pharmacology* 119: 122–130.

Rambukkana, A., Pistoor, F.H., Bos, J.D., Kapsenberg, M.L. and Das, P.K. 1996. Effects of contact allergens on human Langerhans cells in skin organ culture: Migration, modulation of cell surface molecules, and early expression of interleukin 1β protein. *Laboratory Investigations* 74: 422–436.

Reite, O.B. 1998. Mast cells/eosinophilic granule cells of teleostean fish: A review focusing on staining properties and functional responses. *Fish and Shellfish Immunology* 8: 489–513.

Reiter, R.J. 1995. Functional pleiotropy of the neurohormone melatonin: Antioxidant protection and neuroendocrine regulation. *Frontiers in Neuroendocrinology* 16: 383–415.

Rice, C.D. and Schlenk, D. 1995. Immune function and cytochrome P4501A activity after acute exposure to 3,3′,4,4′,5-pentachlorobiphenyl (PCB 126) in channel catfish. *Journal of Aquatic Animal Health* 7: 195–204.

Rice, C.D. and Weeks, B.A. 1989. Influence of tributyltin on in vitro activation of oyster toadfish macrophages. *Journal of Aquatic Animal Health* 1: 62–68.

Rice, C.D. and Weeks, B.A. 1990. The influence of in vivo exposure to tributyltin on reactive oxygen formation in oyster toadfish macrophages. *Archives of Environmental Contamination and Toxicology* 19: 854–857.

Rice, C.D. and Weeks, B.A. 1991. Tributyltin stimulated reactive oxygen formation in toadfish macrophages. *Developmental and Comparative Immunology* 15: 431–436.

Rice, C.D. and Xiang, Y. 2000. Immune function, hepatic CYP1A, and reproductive biomarker responses in the gulf killifish, *Fundulus grandis*, during dietary exposures to endocrine disrupters. *Marine Environmental Research* 50: 161–166.

Rice, C.D., Merchant, R.E., Jeong, T.C., Karras, J.B. and Holsapple, M.P. 1995a. The effects of acute exposure to 2,3,7,8-tetrachlorodibenzo-p-dioxin on glioma-specific cytotoxic T-cell activity in Fischer 344 rats. *Toxicology* 95: 177–185.

Rice, C.D., Banes, M.M. and Ardelt, T.C. 1995b. Immunotoxicity in channel catfish, *Ictalurus punctatus*, following acute exposure to tributyltin. *Archives of Environmental Contamination and Toxicology* 28: 464–470.

Rice, C.D., Kergosien, D.H. and Adams, M.S. 1996. Innate immune function as a bioindicator of pollution stress in fish. *Ecotoxicology and Environmental Health Safety* 33: 186–192.

Rice, C.D., Roszell, L.E., Banes, M.M. and Arnold, R.E. 1998. Effects of dietary PCBs and nonyl-phenol on immune function and CYP1A activity in channel catfish, *Ictalurus punctatus*. *Marine Environmental Research* 46: 351–354.

Robertsen, B., Trobridge, G. and Leong, J.-A. 1997. Molecular cloning of double stranded RNA inducible MX genes from Atlantic salmon, *Salmo salar* L. *Developmental and Comparative Immunology* 21: 397–412.

Roit, I., Brostoff, J. and Male, D. 1989. Evolution of immunity. In *Immunology*, 2nd edn. Roit, I., Brostoff, J. and Male, D. (eds), Ch. 15. Gower Medical Publishers, London.

Rose, W.L., French, B.L., Reichert, W.L. and Faisal, M. 2000. DNA adducts in hematopoietic tissues and blood of the mummichog, *Fundulus heteroclitus*, from a creosote-contaminated site in the Elizabeth River, Virginia. *Marine Environmental Research* 50: 581–589.

Roszell, L.E. and Anderson, R.S. 1993. Immunosuppressive effect of pentachlorophenol on phagocytes from an estuarine teleost, *Fundulus heteroclitus*, as measured by chemiluminescence activity. *Archives of Environmental Contamination and Toxicology* 25: 492–489.

Roszell, L.E. and Anderson, R.S. 1994. Inhibition of phagocytosis and superoxide production by pentachlorophenol in two leukocyte populations from *Fundulus heteroclitus*. *Marine Environmental Research* 38: 195–206.

Roszell, L.E. and Anderson, R.S. 1996. Effect of in vivo pentachlorophenol exposure on *Fundulus heteroclitus* phagocytes: modulation of bactericidal activity. *Diseases of Aquatic Organisms* 26: 205–211.

Roszell, L.E. and Rice, C.D. 1998. Innate cellular immune function of anterior kidney leucocytes in the gulf killifish, *Fundulus grandis*. *Fish and Shellfish Immunology* 8: 129–142.

Rycyzyn, M.A., Wilson, M.R., Clems, L.W. and Miller, N.W. 1998. Evidence for Jak-stat interaction in channel catfish lymphoid cells. *Developmental and Comparative Immunology* 22: 401–406.

Saad A.H. 1988. Corticosteroids and immune systems of non-mammalian vertebrates: A review. *Developmental and Comparative Immunology* 12: 481–494.

Safe, S.H. and Gaido, K. 1998. Phytoestrogens and anthropogenic estrogenic compounds. *Environmental Toxicology and Chemistry* 17: 119–126.

Salo, H.M., Aaltonen, T.M., Markkula, E. and Jokinen, E.I. 1998. Ultraviolet B irradiation modulates the immune system of fish (*Rutilus rutilus*, Cyprinidae). I. Phagocytes. *Photochemistry and Photobiology* 67: 433–437.

Schwassmann, H.O. 1971. Biological rhythms. In *Fish Physiology*, Vol. 6. Hoar, W.S. and Randall, D.J. (eds), pp. 371–428. Academic Press, New York.

Scott, T. and Glick, B. 1987. Organ weights, T-cell proliferation, and graft vs. host capabilities of hypothyroidic chickens. *General and Comparative Endocrinology* 67: 270–276.

Secombes, C.J. 1992. NADPH oxidase-like activity in rainbow trout, *Oncorhynchus mykiss* (Walbaum) macrophages. *Developmental and Comparative Immunology* 16: 405–413.

Secombes, C.J., Hardie, L.J. and Daniels, G. 1996. Cytokines in fish: an update. *Fish and Shellfish Immunology* 6: 291–304.

Secombes, C.J., Tahir, A. and Stagg, R.M. 1997. Immunocompetence in flatfish as a measure of the biological effects of exposure to sewage sludge or hydrocarbon contaminated sediments. In *Ecotoxicology: Responses, Biomarkers and Risk Assessment*. Zelikoff, J.T., Lynch, J.M. and Shepers J. (eds), pp. 281–292. OECD by SOS Publications, Fair Haven, NJ.

Seeley, K.R. and Weeks-Perkins, B.A. 1991. Altered phagocytic activity of macrophages in oyster toadfish from a highly polluted subestuary. *Journal of Aquatic Animal Health* 3: 224–227.

Seyle H. 1973. The evolution of stress concept. *American Scientist* 61: 692.

Sinderman, C.J. 1979. Pollution associated diseases and abnormalities of fish and shellfish: a review. *Fish Bulletin* 76: 717–720.

Sinderman, C.J., Esser, S.C., Gould, E., McCain, B.B., McHugh, J.L., Morgan, R.P., Murchelano, R.A., Sherwood, M.J. and Spitzer, P.R. 1982. Effects of pollutants on fishes. In *Proceedings of the Symposium of Ecological Stress*. Estuarine Res. Fed. Columbia, pp. 23–38.

Slater, C.H. and Schreck, C.B. 1998. Season and physiological parameters modulate salmonid leucocyte androgen receptor affinity and abundance. *Fish and Shellfish Immunology* 8: 379–391.

Snieszko, S.F. 1974. The effects of environmental stress on outbreaks of infectious diseases of fishes. *Journal of Fish Biology* 6: 374–383.

Soto, A.M., Lin, T.M., Justicia, H., Slivia, R.M. and Sonnenschein, C. 1992. An 'in culture' bioassay to access the estrogenicity of xenobiotics (E-screen). In *Chemically Induced Alterations in Sexual and Functional Development; the Wildlife/Human Connection*. Colborn, T. and Clement, C. (eds), pp. 295–309. Princeton Scientific Publishing, Princeton, HJ.

Spitzbergen, J.M., Schat, K.A., Kleeman, J.M. and Peterson, R.E. 1986a. Interactions of 2,3,7,8-tetrachlorodibenzo-p-dioxin (TCDD) with immune responses of rainbow trout. *Veterinary Immunology and Immunopathology* 12: 263–280.

Spitzbergen, J.M., Schat, K.A., Kleeman, J.M. and Peterson, R.E. 1986b. Effects of 2,3,7,8-tetrachlorodibenzo-p-dioxin (TCDD) or Aroclor 1254 on the resistance of rainbow trout, *Salmo gairdneri* R, to infectious haematopoietic necrosis virus. *Journal of Fish Diseases* 11: 73–83.

Stamler, J.S., Singei, D.J. and Loscalzo, J. 1992. Biochemistry of nitric oxide and its redox-activated forms. *Science* 258: 1898–1902.

Tavida, H.B., Fleming, K.A., Hume, P.D. and Simpson, H.W. 1975. Circadian rhythmicity of human plasma cortisol and PHA-induced lymphocyte transformation. *Clinical and Experimental Immunology* 22: 190–193.

Terres, G., Morrison, S.L. and Habicht, G.S. 1968. A quantitative difference in the immune response between male and female mice. *Proceedings of the Society for Experimental Biology and Medicine* 127: 664–667.

Thomas P. 1990. Molecular and biochemical responses of fish to stressors and their potential use in environmental monitoring. *American Fisheries Society Symposium* (Adams, S.M., ed.) 8: 9–28.

Thorp, J.E. 1978. *Rhythmic Activity of Fishes*. Academic Press, London.

Thuvander, A. and Carlstein, M. 1991. Sublethal exposure or rainbow trout, *Oncorhynchus mykiss*, to polychlorinated biphenyls: effects on the humoral immune response to *Vibrio anguillarum*. *Fish and Shellfish Immunology* 1: 77–86.

Tithof, P.K., Contreras, M.L. and Ganey, P.E. 1995. Aroclor 1242 stimulates the production of inositol phosphates in polymorphonuclear neutrophils. *Toxicology and Applied Pharmacology* 131: 136–143.

Tizard, I.R. 1996. *Veterinary Immunology, An Introduction*, 5th edn. W.B. Saunders Co. Publishers, Philadelphia.

Vallejo, A.N. and Ellis, A.E. 1989. Ultrastructural study of the response of eosinophilic granule cells to *Aeromonas salmonicida* extracellular products and histamine liberators in rainbow trout, *Salmo gairdneri* Richardson. *Developmental and Comparative Immunology* 13: 133–148.

Vallejo, A.N., Miller, N.W., Jorgensen, T. and Clems, L.W. 1990. Phylogeny of immune recognition: antigen processing/presentation in channel catfish immune responses to hemocyanins. *Cellular Immunology* 130: 364–377.

Vallejo, A.N., Miller, N.W. and Clem, L.W. 1991. Phylogeny of immune recognition: processing and presentation of structurally defined proteins in channel catfish immune responses. *Developmental and Comparative Immunology* 1: 137–148.

Van Der Kraak, G.J., Munkittrick, K.R., McMaster, M.E., Portt, C.B. and Chang, J.P. 1992. Exposure to bleached kraft pulp mill effluent disrupts the pituitary–gonadal axis of white sucker at multiple sites. *Toxicology and Applied Pharmacology* 115: 224–234.

Van Veld, P.A., Vetter, R.D., Lee, R.F. and Patton, J.S. 1987. Dietary fat inhibits the intestinal metabolism of the carcinogen benzo-(*a*)-pyrene in fish. *Journal of Lipid Research* 28: 810–817.

Vogelbein, W.K., Fournie, J.W., Van Veld, P.A. and Huggett, R.J. 1990. Hepatic neoplasms in the mummichog, *Fundulus heteroclitus*, from a Creosote-contaminated site. *Cancer Research* 50: 5978–5986.

Warinner, J.E., Mathews, E.S. and Weeks, B.A. 1988. Preliminary investigations of the chemiluminescent response in normal and pollutant-exposed fish. *Marine Environmental Research* 24: 281–284.

Waterstrat, P.R., Brazil, J. and Ainworth, A.J. 1991. Use of an ELISA-based assay for the detection of antibody-secreting cells in channel catfish, *Ictalurus punctatus* (Rafinesque). *Journal of Fish Diseases* 14: 669–675.

Weaver, D.R., Stehle, J.H., Stopa, E.G. and Reppert, S.M. 1993. Melatonin receptors in human hypothalamus and pituitary: Implications for circadian and reproductive responses to melatonin. *Journal of Clinical Endocrinology and Metabolism* 76: 295.

Weeks, B.A. and Warinner, J.E. 1984. Effects of toxic chemicals on macrophage phagocytosis in two estuarine fishes. *Marine Environmental Research* 14: 327–335.

Weeks, B.A. and Warinner, J.E. 1986. Functional evaluation of macrophages in fish from a polluted estuary. *Veterinary Immunology and Immunopathology* 12: 313–320.

Weeks, B.A., Warinner, J.E. and Mathews, E.S. 1988. Influence of toxicants on phagocytosis, pinocytosis, and melanin accumulation by fish macrophages. *Aquatic Toxicology* 11: 395–438.

Weeks, B.A., Anderson, D.P., DuFour, A.P., Fairbrother, A., Goven., A.J., Lahvis, G.P. and Peters, G. 1992. Immunological biomarkers to assess environmental stress. In *Biomarkers: Biochemical, Physiological, and Histological Markers of Anthropogenic Stress*. Hugget, R.J., Kimerle, R.A., Mehrle, P.M. and Bergman, H.L. (eds). Lewis Publishers, Boca Raton.

Weigent, D.A. 1996. Immunoregulatory properties of growth hormone and prolactin. *Pharmacology and Therapeutics* 69: 237–257.

Weigent, D.A. and Blalock, J.E. 1996. Neuroendocrine peptide hormones and receptors in the immune response and infectious diseases. In *Psychoneuroimmunology, Stress, and Infection*. Friedman, H., Klein, T.W. and Friedman, A.L. (eds), pp. 47–78. CRC Press, Boca Raton.

Weis, J.S. and Kim, K. 1988. Tributyltin is a teratogen in producing deformities in limbs of the Fiddler Crab, *Uca pugilator*. *Archives of Environmental Contamination and Toxicology* 17: 583–587.

Wester, P.W., Vethaak, A.D. and van Muiswinkel, W.B. 1994. Fish as biomarkers in immunotoxicology. *Toxicology* 86: 213–232.

White, A. and Fletcher, T.C. 1985. Seasonal changes in serum glucose and condition of the plaice, *Pleuronectes platessa* L. *Journal of Fish Biology* 26: 755–764.

White, Jr, K.L. 1986. An overview of immunotoxicology and carcinogenic polycyclic hydrocarbons. *Environmental Carcinogenic Reviews* (*Journal of Environmental Health Science*) C4: 163–202.

White, K.L. and Anderson, A.C. 1985. Suppression of mouse complement activity by contaminants of technical grade pentachlorophenol. *Agents Actions* 16: 385.

White, K.L. and Holsapple, M.P. 1984. Direct suppression of in vitro antibody production by mouse spleen cells by the carcinogen benzo-*a*-pyrene, but not by the noncarcinogenic congener benzo-*e*-pyrene. *Cancer Research* 44: 3388–3393.

White, K.L., Lysy, H.H., McCay, J.A. and Anderson, A.C. 1986. Modulation of serum complement levels following exposure to polychlorinated dibenzo-p-dioxins. *Toxicology and Applied Pharmacology* 84: 209–219.

Xiang, Y. and Rice, C.D. 2000. Expression of fish iNOS is increased by pro-inflammatory signals and xenobiotics. Tenth International Symposium on Pollutant Responses in Marine Organisms (PRIMO-10), Williamsburg, VA, 25–29 April 1999 – abstracts. 50: 61–84.

Yanez, J. and Meissl, H. 1996. Secretion of methoxyindoles melatonin, 5-methoxytryptophol, 5-methoxyindolacetic acid, and 5-methoxytryptamine from trout pineal organs in superfusion culture: Effects of light intensity. *General and Comparative Endocrinology* 101: 165–172.

Yang, C.Z. and Albright, L.J. 1994. Anti-phytoplankton therapy of finfish: the mucolytic agent L-cysteine ethyl ester protects coho salmon *Oncorhynchus kisutch* against the harmful phytoplankter *Chaetoceros concavicornis*. *Diseases of Aquatic Organisms* 20: 197–202.

Zachman, A., Falcon, J., Knijff, S.C.M., Bolliet, V. and Ali, M.A. 1992. Effects of photoperiod and temperature on rhythmic melatonin secretion form the pineal organ of the white sucker (*Catostomus commersoni*), *in vitro*. *General and Comparative Endocrinology* 86: 26–33.

Zdolsek, J.M., Soder, O. and Hultman, P. 1994. Mercury induces *in vivo* and *in vitro* secretion of interleukin-1 in mice. *Immunopharmacology* 28: 201–208.

Zeeman, M. 1986. Modulation of the immune response in fish. *Veterinary Immunology and Immunopathology* 12: 235–241.

Zeeman, M. 1994. Comparative immunotoxicology and risk assessment. In *Modulators of Fish Immune Responses*. Stolen, J.S. (ed.), pp. 317–328. SOS Publications, Fair Haven, NJ.

Zeeman, M.G. and Brindley, W.A. 1981. Effects of toxic agents upon fish immune systems: a review. In *Immunologic Considerations in Toxicology*, Vol. 2. Sharma, R.P. (ed.), pp. 1–60. CRC Press, Boca Raton.

Zelikoff, J.T. 1993. Metal pollution-induced immunomodulation in fish. *Annual Review of Fish Diseases* 3: 305–325.

Zelikoff, J.T. 1994. Fish immunotoxicology. In *Immunotoxicology and Immunopharmacology*, 2nd edn. Dean, J.H., Luster, M.I., Munson, A.E. and Kimber, I. (eds). Raven Press, New York.

Zelikoff, J.T., Bowser, D., Squibb, K.S. and Frenkel, K. 1995. Immunotoxicology of low level cadmium exposure in fish: An alternative animal model for immunotoxicological studies. *Journal of Toxicology and Environmental Health* 45: 235–248.

4 Neurobehavioral toxicity in fish

Edward E. Little and Sandra K. Brewer

Introduction

Behavior is a sequence of quantifiable actions that operate through the central and peripheral nervous system (Keenleyside, 1979). These patterns are the culmination of genetic, biochemical, and physiologic processes and as such are sensitive to alterations in the steady state of the organism (Warner *et al.*, 1966; Beitinger, 1990). Most behavioral patterns are the result of adaptation to environmental variables, and a normal behavioral repertoire is required for an animal to avoid unfavorable conditions or to actively seek favorable environmental situations. Ultimately, adaptive behavioral responses are required for the growth and survival of individuals, as well as reproduction, to ensure viability of the population. Behavior, then, is not a random event, but rather is a structured and predictable sequence of activities of significant adaptive value that is essential to the organism's existence (Little *et al.*, 1993a).

Behavior provides a unique perspective between the organism and its environment; between physiology and ecology. Behavior is the cumulative manifestation of physiologic processes critical for the performance of essential life functions. Behavior enables continuous adjustments to external stimuli and internal cues to maintain the organism under beneficial circumstances to grow and to reproduce. Thus, behavior is a selective response that ensures adaptation to environmental conditions – to a myriad of physical, chemical, and social cues. In terms of evolution, behavior is highly selective, and in concert with morphologic and physiologic adaptations permits the best opportunity for survival and reproductive success by enabling organisms to exploit efficiently resources within the defined parameters of their habitat.

Because adaptive behavioral function is crucial, behavioral investigations are particularly relevant when evaluating the effects of contaminants on fish. Behavioral toxicity occurs when a contaminant or other stressful condition induces a behavioral change that exceeds the normal range of variability (Marcucella and Abramson, 1978). The effects of contaminants on fish behavior have received increased attention over the past decade and several reviews have resulted (Little *et al.*, 1985, 1993a; Atchison *et al.*, 1987; Beitinger, 1990; Henry and Atchison, 1991; Blaxter and Hallers-Tjabbes, 1992; Scherer, 1992; Birge *et al.*, 1993).

A diversity of quantifiable behavioral responses can be affected by changes in water quality. These responses vary in complexity, ranging from reflexive responses to complex social interactions. Behavioral responses most useful in contaminant assessment should be (1) well-defined endpoints that are practical to measure, (2) well understood relative to environmental factors that cause variation in the response, (3) sensitive to a range of contaminants and adaptable to different species, and (4) ecologically relevant (Rand, 1985). The accumulated evidence described in several reviews (Olla *et al.*, 1980; Westlake, 1984; Little *et al.*, 1985; Rand, 1985; Atchison *et al.*, 1987; Beitinger, 1990) indicates that behavioral functions are usually quite sensitive to contaminant exposure and are often the first visually noticeable responses exhibited by fish that are exposed to toxic substances. In addition to being directly affected by environmental stressors, behavioral responses, such as avoidance–attractance or feeding, may also significantly influence the level or duration of exposure actually experienced by an organism through contact with their environment or diet (Beitinger, 1990).

A range of behaviors may be affected during contaminant exposure, depending on the type of chemical, its concentration, and the duration of exposure. For example, during a 28-day exposure of juvenile rainbow trout to dioxin (2,3,7,8-TCDD), five different behavioral responses were altered (Figure 4.1; Mehrle *et al.*, 1988). Reduced feeding was evident after 7 days of exposure to 789 pg L^{-1}; hypoactivity and diminished responsiveness to external stimuli were evident by day 10 of exposure; abnormal swimming postures were noted by day 12; and severe lethargy occurred by day 19. The appearance of a particular abnormality varied with concentration. For example, reduced feeding appeared after 7 days of exposure at 789 pg L^{-1}; by day 14 at 176 pg L^{-1}; and by day 17 at 38 pg L^{-1}. Thus, as with other toxicologic endpoints, the magnitude or severity of behavioral response varies with the duration and concentration of exposure.

Although there are many ways that behavior can be used to determine effects of exposure to toxicants, the use of behavior as an index for toxicant identification is somewhat more limited because there have been few systematic studies examining behavioral responses that are indicative of exposure to specific toxic agents. In an approach developed by Drummond *et al.* (1986), behavioral and morphologic responses were divided into ten categories containing a total of forty unique descriptors. Responses were monitored during acute exposures of fathead minnows (*Pimephales promelas*) to 139 single compounds. Statistical pattern recognition, based on the types of behavioral changes induced by exposure, was used to identify three general responses (types I, II, and III) which correlated with three classes of contaminants. Type I responses were indicative of narcosis-producing chemicals such as ethers, alcohols, ketones, and phthalates. These chemicals depress central and peripheral nervous system activities. Exposed fish exhibit depressed locomotory activity, loss of startle responses, rapid shallow opercular rates, darkened coloration, and tetany. Type II responses were indicative of chemicals such as rotenone, benzene, and phenol, which disrupt metabolic activity. Exposed fish exhibit heightened locomotory activity, hyperactivity to stimulation, increased rate and amplitude of opercular activity, slight darkening, and edema. Type III responses were indicative of neurotoxic chemicals such as

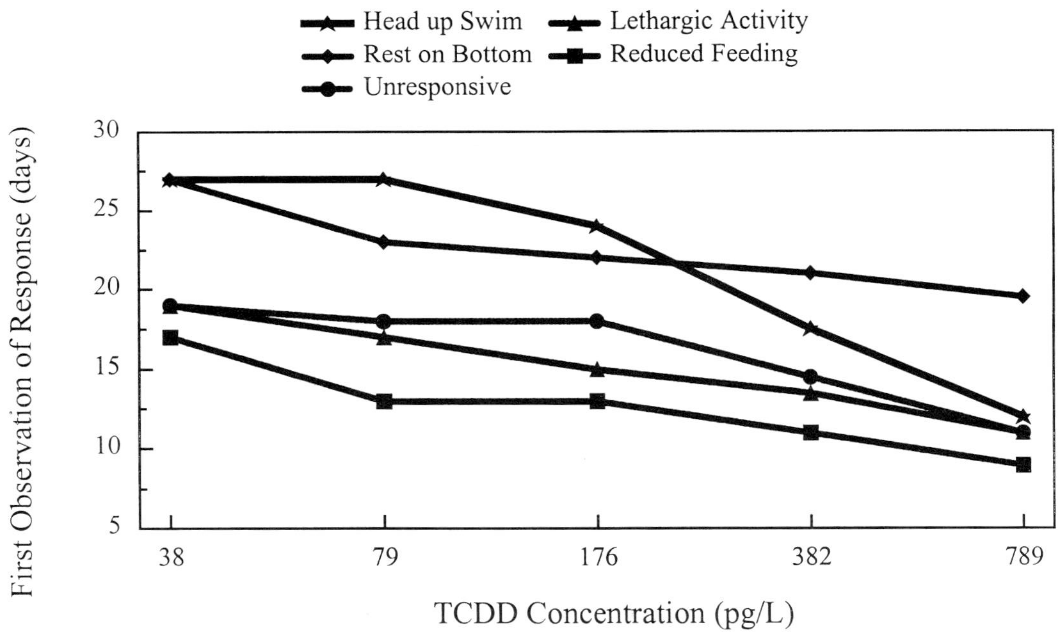

Figure 4.1 Days of exposure to the dioxin TCDD to induce behavioral changes in rainbow trout exposed as free-swimming juveniles for 28 days (redrawn from Merhle *et al.*, 1988).

carbamates, organophosphates, caffeine, and strychnine. Exposed fish exhibit depressed locomotory activity with hyperactivity to stimulation, convulsions, spasms, tetany, scoliosis, lordosis, or vertebral hemorrhage. Thus, behavioral changes can be induced through direct effects to the nervous system as well as through indirect physiologic alterations.

Recently, the approach by Drummond *et al.* (1986) of using behavioral alterations as an index for toxicant identification was used by Rice *et al.* (1997), who exposed 30-day-old Japanese medaka (*Oryzias latipes*) to five single compounds with different toxicologic mechanisms. Unique behavioral and morphologic abnormalities were observed for each chemical except 2,4-dinitrophenol (2,4-DNP), an uncoupler of oxidative phosphorylation. 2,4-DNP displayed the fewest behavioral symptoms, with a loss of equilibrium being the most common. These results were similar to the work of Drummond *et al.* (1986) and Drummond and Russom (1990). Because behavioral changes can be induced through direct effects to the nervous system as well as through indirect physiologic alterations, the development of 'behavioral toxicity syndromes' is a promising tool for assessing both mechanisms of toxicity and, possibly, identifying toxicants in complex environmental samples.

Neural basis of behavioral toxicity

In the simplest form, behavior results when a stimulus from the environment is encoded by the sensory cell as neural impulses which induce a muscle to respond (Figure 4.2). The synaptic connection between sensory cell and interconnecting neuron and between neuron and muscle fiber is the focus of communication between environmental stimuli and behavioral response. Contained within that neural chain is a network of chemical reactions that include: the synthesis and transport of neurosecretory substances to the presynaptic membrane (Figure 4.2A); the development and population of receptor sites on the post-synaptic dendrites (Figure 4.2B); the release of neurosecretory substances across the synaptic gap (Figure 4.2C); the binding of neurotransmitter to receptors on the post-synaptic membrane (Figure 4.2D); the removal of excitatory and inhibitory neurosecretory substances from those receptors through diffusion, enzyme degradation, or reuptake; and the generation of the action potential and its propagation down the axon of the nerve to the next synapse with nerve, secretory cell, or muscle (Purves *et al.*, 1992). Thus, there are numerous sites of action for toxicants to affect the nervous system, and interference at any of these sites can block or alter the sequence of neural responses and inhibit or alter behavioral function. Although the relationship between nervous system and behavior would appear to be intuitively obvious, remarkably the physiologic and behavioral aspects of neurotoxicity are seldom examined and, as a result, are poorly understood. The complexity of fish nervous systems seldom allows a direct causal linkage between physiology and behavior as the simplified example (Figure 4.2) would imply. There is sufficient redundancy in the nervous system to allow organisms to compensate (Olla *et al.*, 1980). Thus, it is possible to have seemingly normal behavioral responses even

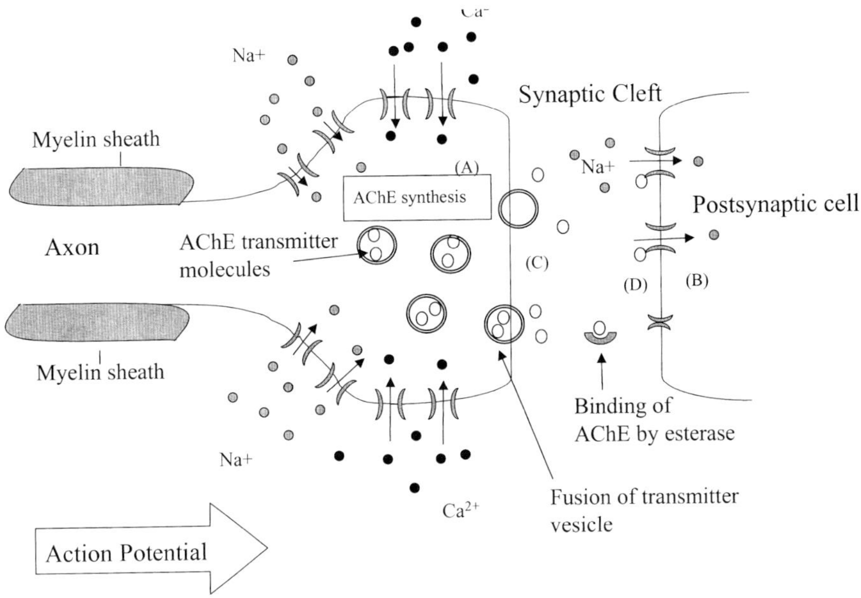

Figure 4.2 Chemical synapse. Action potential spreads along the axon, which terminates next to another neuron. Information flows from the presynaptic cell to the post-synaptic cell by way of transmitter substances (redrawn from Purves *et al*., 1992).

as toxicosis approaches lethality (Carlson *et al*., 1998; Brewer *et al*., 1999). In other cases, the response may be very subtle. Available information reveals that changes in behavioral responses induced by toxicosis include impacts to sensory-mediated responses as well as changes in neuromotor performance.

Neurotoxicity through sensory-mediated impairments

Aquatic organisms are awash in a constant stream of environmental stimuli. The brain and sensory systems must have sufficient acuity to detect relevant cues and must sort through this array of stimuli to detect those that have an immediate bearing on survival. In this section, we are concerned with behavioral alterations resulting from the action of contaminants on receptor function and perception. Most behaviors are the result of sensory integration of environmental cues. Although significant advances have been made in understanding the sensory systems of fish and sensory-mediated behavioral responses of fish, information on contaminant impacts to such systems is remarkably limited, perhaps because of the technical and experimental problems with conducting such studies.

Chemical senses

More is known about toxicant impacts on the physiomorphology of chemosensory systems than for any other sense in fish, perhaps because the sensory receptors are bathed in the aqueous medium and are in immediate contact with contaminants. Most studies fall into one of three general areas: (1) histologic investigations of lesions in the receptor organs; (2) monitoring changes in nerve impulses originating in specific receptors during or following toxicant exposure; and (3) recording summed electrical activity from a stimulated receptor organ (Hara *et al.*, 1983). Chemoreception plays a major role in mediating the basic behavioral functions critical to survival of the individual and species: foraging, reproduction (including homing behavior), and the recognition and avoidance of predators (Little, 1983). Fish detect chemicals through two primary modes of chemoreception: olfaction (smell) and gustation (taste). Solitary chemosensory cells also provide additional chemoreceptor function (Kotrschal, 1991; Whitear, 1992). In fish, olfaction and gustation are mediated by molecules dissolved in water and their distinction is mainly based on differences in anatomy, their central connections to the central nervous system, and the behavioral responses they mediate (Atema *et al.*, 1969). Olfactory chemoreceptors are limited to the epithelium of the olfactory rosette located within the olfactory chambers and olfactory organs are innervated by the first cranial nerve. In most fish, the gustatory organs are generally concentrated within the buccal cavity, but may also be found on sensory barbels, specialized fin rays, or, in the case of the catfish, covering the surface of the body. Depending on their location, taste buds are innervated by the seventh, ninth or tenth cranial nerves.

CONTAMINANTS AS DISRUPTING STIMULI

When the contaminant is perceived as a stimulus, the resulting behavioral response, such as avoidance–attractance or feeding, can significantly influence the level or duration of exposure experienced by the organism through contact with water, sediment, or diet (Beitinger, 1990). If contaminants are detected by fish as noxious stimuli, the organism responds in a manner that minimizes contact. This avoidance response is adaptive in that it effectively removes the organism from the contaminated areas, thus minimizing exposure. In contrast, a chemical-inducing attractance would increase exposure and probability of injury or death. Avoidance–attractance responses depend on: (1) the substance activating the receptor; (2) the subsequent stimulation being interpreted by the fish as sufficiently noxious to induce locomotory responses to escape; and (3) sufficient directional information from the chemical concentration gradient to lead the organism from the contaminant plume. Chemosensory receptors are highly evolved structures relative to the types of molecular structures that can bind with them to initiate stimulation (Sutterlin, 1974). Organisms are unlikely to have sufficient exposure history with many anthropogenic substances to evolve sensory receptors that can bind with them, thus any sensory stimulation may simply be a generalized response to the molecule – an accidental fit. Stimulation will not necessarily result in aversive reactions either; there may not be sufficient exposure history of the species to evolve adaptive responses nor sufficient experience by the organism to acquire an aversive response to this stimulation. Thus, avoidance responses may also be accidental. This is suggested because some contaminants such as certain petroleum products induce attraction and stimulate feeding as if the substance is confused with a natural food stimulus (Atema, 1976).

Many contaminants induce avoidance responses and several reviews have been made of this extensive literature (Cherry and Cairns, 1982; Giattina and Garton, 1983; Hara *et al.*, 1983; Beitinger, 1990). Avoidance data from studies with rainbow trout are summarized in Table 4.1. The sensitivity of avoidance responses ranged from less than 3 percent of the LC_{50} (concentration lethal to 50 percent of the test population) for the herbicide 2,4-D to more than a 1000 times the LC_{50} for the chlorinated biphenol Arochlor 1254 (Little *et al.*, 1985). DeLonay *et al.* (1995) conducted avoidance tests with brown trout (*Salmo trutta*) using a metals mixture representative of the Clark Fork River (Montana) to assess the impact of metal contamination from mining. Concentrations as low as 10 percent of the average metals concentration measured in the river induced avoidance. Similar avoidance responses were observed in laboratory and field tests of metals characteristic of the Coeur d'Alene River (Idaho) downstream from a large mining extraction operation. Telemetry studies conducted at the confluence with an uncontaminated tributary of the river revealed a similar avoidance of the contaminated water within the concentration range that induced avoidance responses in laboratory studies (Woodward *et al.*, 1997). Gray (1990), using telemetry, documented the avoidance of oil-contaminated water and gas-supersaturated water by free-ranging fish. Hartwell *et al.* (1987) conducted integrated laboratory–field studies of avoidance and showed that fathead minnows (*Pimephales promelas*) avoided a blend of heavy metals (copper, chromium, arsenic, and selenium) that are representative of effluent from fly ash settling basins of coal-burning electrical plants. Fish avoided a 73.5 $\mu g\ L^{-1}$ mixture of these metals in a natural stream, and similarly would avoid 34.3 $\mu g\ L^{-1}$ in an artificial stream.

Generalizations regarding the avoidance of aquatic contaminants by fish are

Table 4.1 Avoidance reactions of rainbow trout to contaminants.

Toxicant	*Lowest concentration avoided (mg L^{-1})*	*Lethal concentration (mg L^{-1})*
Aquathol[a]	No response	150.0
Chlorine[b]	0.050	1.000
Copper[c]	0.002	48.000
Diquat[a]	No response	12.000
2,4-D[a]	1.000	1.000
Xylene[a]	0.01	10.000
Zinc[d]	0.006	0.600[e]

Notes
a Folmar (1976).
b Schumacher and Ney (1980).
c Sprague (1964a).
d Sprague (1968).
e Sprague (1964b).

difficult to make because of the variety of species and experimental designs used to test behavioral responses, as well as variations in the modes and sites of action of the chemicals studied (Giattina and Garton, 1983). In his review of published avoidance data for over seventy-five different chemicals, Beitinger (1990) found that roughly one-third of the chemicals were avoided, whereas the others either failed to elicit a response or induced inconsistent responses. Many contaminants cause avoidance reactions but some may attract aquatic organisms, including detergents (Hara and Thompson, 1978), certain metals (Timms *et al.*, 1972; Kleerekoper *et al.*, 1973; Black and Birge, 1980) and petroleum hydrocarbons (Lawrence and Scherer, 1974; Atema, 1976). In the case of attractance, some compounds may be confused with natural chemical cues (Sutterlin, 1974), or damage to chemoreceptors (Hansen *et al.*, 1999a). Generically similar organic compounds may not induce similar reactions (Little *et al.*, 1985). Different species of fish may have different avoidance responses. Largemouth bass (*Micropterus salmoides*) have been shown to be insensitive to 50 μg L^{-1} copper sulfate, whereas goldfish (*Carassius auratus*) and channel catfish (*Ictalurus punctatus*) were attracted to this concentration (Timms *et al.*, 1972). Black and Birge (1980) demonstrated that largemouth bass and bluegill (*Lepomis macrochirus*) were less sensitive than rainbow trout to zinc, copper and cadmium.

Avoidance reactions are also influenced by variables such as contaminant concentration gradient and temperature (Kleerekoper, 1976). Most studies use steep gradient test systems in which the chemical concentration increases sharply over a short distance: fish must choose between clean water or contaminated water (Figure 4.3). Kleerekoper *et al.* (1973) have shown that when gradients were shallow goldfish were attracted to copper concentrations between 11 and 17 μg L^{-1}, but when the gradient was steep fish avoided concentrations as low as 5 μg L^{-1} (Westlake *et al.*, 1974). Because numerous factors are responsible for the spatial distribution of aquatic animals, one must be careful in evaluating laboratory results of single-parameter laboratory tests as reflective of field conditions (Jones *et al.*, 1985).

Sullivan and Fisher (1942, 1954) conducted early basic research on avoidance responses of fish to natural environmental conditions. They simultaneously offered fish darkened (preferred habitat) or light areas and temperatures that fish had avoided or preferred in single-factor trials. There was a 'competing gradient' between light and temperature with fish remaining in preferred darkened areas even though the temperatures in those areas had been avoided under uniform light conditions.

Kleerekoper *et al.* (1973) examined the interaction of temperature and copper. Goldfish preferred water temperatures of 21.5°C and avoided copper concentrations of 10 μg L^{-1} in single-factor trials. However, 10 μg L^{-1} copper was no longer avoided when the water temperature of this treatment was increased to 21.5°C. Giattina *et al.* (1981) studied the interactive effects of heated and intermittently chlorinated effluents on spotfin shiners (*Notropis spilopterus*) downstream from a power plant. They found that the threshold chlorine concentration for avoidance increased from 0.24 mg L^{-1} to a maximum of

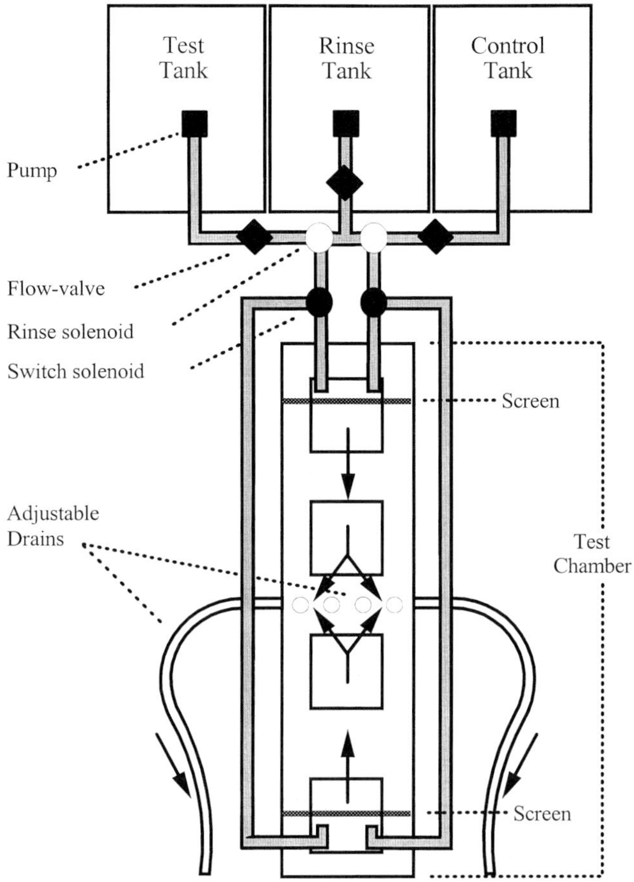

Figure 4.3 Countercurrent avoidance apparatus used to determine avoidance response of brown trout to Clark Fork River (Montana) metals contamination. Reference and contaminated water are introduced at opposite ends of the chamber, which drains from the center to form a steep contaminant gradient (redrawn from DeLonay *et al.*, 1995).

0.38 mg L^{-1} when fish acclimated to 12°C had the opportunity to select simultaneously a preferred temperature (21°C). Apparently, the preference of cold-acclimated fish for warmer water overrode the avoidance response and kept fish close to the LC_{50} values (0.41–0.65 mg L^{-1}) for residual chlorine. Vertical spatial preferences of juvenile chinook salmon (*O. tshawytscha*) within vertical gradients of pulp mill effluents, oxygen, and temperature were significantly correlated with *in situ* temperature, pH, and effluent color (Birtwell and Kruzynski, 1989). Such studies suggest the adaptive value of avoidance may be lost for species that overwinter in discharge areas where effluents may be warmer than ambient stream or lake conditions.

Light intensity and habitat cover are also important variables in avoidance behavior. Lawrence and Scherer (1974) found that infrared light reduced the attraction response of rainbow trout and whitefish (*Coregonus clupeaformis*) to oil well drilling mud. Apparently, the 'shade' provided by the turbid test solutions was a causal factor. Walleye (*Stizostedion vitreum vitreum*) abandoned normally preferred shade when respiratory stress was applied (Scherer, 1971) or after ingestion of mercury (Scherer *et al.*, 1975). Similarly, lake whitefish avoided copper (1 $\mu g\ L^{-1}$) as well as lead and zinc ($\geq 10\ \mu g\ L^{-1}$) which were at or below Canadian water quality guidelines; however, when shade was provided in the test apparatus, higher test concentrations (72 $\mu g\ L^{-1}$ copper; 3.2 $mg\ L^{-1}$ lead; 1000 $\mu g\ L^{-1}$ zinc) were required to induce avoidance (Scherer and McNicol, 1998). The addition of shelter to breeding male fathead minnows increased their threshold for avoiding zinc by more than sixfold, from 0.28 $mg\ L^{-1}$ to 1.83 $mg\ L^{-1}$ (Korver and Sprague, 1989).

CONTAMINANT-INDUCED ALTERATIONS OF CHEMOSENSORY SYSTEMS

Chemical contamination affects chemosensory systems through binding interactions with the chemosensory receptors, morphologic changes in the chemosensory organs and receptors, and changes in central integration of sensory input (Sutterlin, 1974). When a fish encounters a contaminant, it will either detect the chemical or not. If the fish detects the chemical but the resultant stimulation is below a critical threshold, or if the chemical is not detected, then toxicosis will occur and affect perception of the environment as well as the ability of the fish to respond to environmental cues.

Contaminants that are not avoided may impair the perception of natural stimuli by masking them or by altering the ability of the receptor or sensory organ to receive sensory input. Lemly and Smith (1987) found that the responsiveness of fathead minnow to feeding stimuli was abolished following exposure to acidic water (pH $\leq$ 6.0). Smith and Lawrence (1988), however, found that responsiveness of this species to the alarm pheromone, *shreckstoffe,* was not abolished by exposure to water at a pH of 5.0. These studies demonstrate that different stimulus–response systems have different thresholds for acidification and possibly for other types of ecologic stressors. Perfusion of the olfactory rosettes of the whitefish with cadmium (10 μM) was found to induce electrical activity in the olfactory bulb identified as electro-olfactogram (EOG) responses which reflect the summed generator potentials of the olfactory receptors (Hara *et al.*, 1983). This correlated with avoidance reactions observed for cadmium. However, such exposure resulted in diminished physiologic responses to amino acid stimuli. Single exposures to cadmium, silver, copper, mercury, nickel, and zinc were found to depress the olfactory bulbar response to L-serine, an olfactory stimulant (Brown *et al.*, 1982). *In vitro* and *in vivo* studies determined that olfactory receptor sites for L-serine sites were competitively bound by cadmium as low as 25 $\mu g\ L^{-1}$, other metals have been shown to have a similar effect on the olfactory response (Hara *et al.*, 1983). In recent studies, Hansen *et al.* (1999b) found simultaneous alterations in chemosensory-mediated behavior, in the physiologic responsiveness of the olfactory system to L-serine, and evidence of damage to the tissues of the olfactory mucosa and olfactory receptor cells. Hansen and co-workers found that even brief exposure of chinook salmon and rainbow trout to copper (25 µg L^{-1}) caused a significant reduction in EOG responses that recovered over several days; however, exposure to higher copper concentrations (44 µg L^{-1}, chinook salmon; 180 µg L^{-1}, rainbow trout) abolished behavioral avoidance. Furthermore, physiologic recording revealed these higher copper concentrations diminished both the EOG responses recorded on the mucosa as well as electroencephalogram (EEG) responses to L-serine recorded from the olfactory tract. Necrosis of epithelial structures and reduced density of olfactory receptors were evident injuries to the olfactory epithelium.

Similarly, McNicol and Scherer (1991) determined that whitefish avoided cadmium concentrations of 1 µg L^{-1} and less and also avoided cadmium at 8 µg L^{-1} and greater, but showed little response to concentrations between this range. In most recent physiologic and behavioral studies (McNichol and Hara, personal communication), EOG responses to the lower concentrations were shown to be mediated by the olfactory system. The olfactory system apparently became injured at concentrations greater than 1 µg L^{-1}, when avoidance responses ceased to occur and EOG responsiveness to L-serine was abolished. The renewed avoidance response to 8 µg L^{-1} was likely induced by generalized irritation.

Octavolateralis system

Fish must be constantly vigilant of predators and respond immediately to avoid capture. Detection of the predator includes not only visual and olfactory cues but also lateral line input. In studies with the petroleum hydrocarbon naphthalene, the lateral line appeared to be one of the first systems to be adversely affected. An exposure of 0.02 mg L^{-1} caused lesions in the lateral line, and gill epithelia were affected at a concentration of only 0.002 mg L^{-1}; concentrations that did not affect other organs (DiMichelle and Taylor, 1978).

The motion of attack by a predator with its attendant near-field sound pressure waves, other vestibular sensations (e.g. gravity, acceleration), electric fields, and visual stimuli gives rise to a rapid startle response in many fish, comprising a 30–100° rotation of the body followed by axial acceleration that propels the organism from its initial position (Eaton and Hackett, 1984). This C-type rapid escape response, completed within 200 ms, is directed by Mauthner cells (M cells), which are paired interneurons located in the fish hind brain with axons that project the length of the spinal column.

As summarized by Carlson *et al.* (1998), when stimulated by sensory input from the octavolateralis and visual systems, an impulse from the M cell excites primary and secondary motorneurons and interneurons which, in turn, excite white muscle on the side opposite to the stimulus while simultaneously inhibiting motor neurons and interneurons on the side of the fish receiving the stimulation. This gives rise to a severe flexure of the body which reorients and propels the organism from its original location (Fetcho and Faber, 1988). Carlson and co-workers found that the neural activity associated with this sequence was of sufficient magnitude to enable external monitoring of the neural events leading up to the escape response (Figure 4.4). Their routine recording procedures discerned a T1 interval between the peak M-cell depolarization to peak motor interneuron depolarization; followed by a second interval, T2, which occurred from the peak interneuron depolarization to the onset of the neuromuscular response indicated by the electromyogram (EMG). A third interval that they measured, T3, equaled the total interval from M-cell depolarization to the onset of the EMG. The escape response was elicited during this study by touching the caudal surface of the fish with a fine glass rod. This was continued until five responses were elicited. The number of times that the organism was stimulated to induce five M-cell responses was calculated as the R/S ratio. The ratio was determined for each organism before exposure and after 24 and 48 h of exposure. Several toxicants were studied, each having different modes of action. They found that carbaryl (9.4 mg L^{-1}) and phenol (15.1 mg L^{-1}) increased the T1 interval, thus delaying the stimulation of the next interneuron. Carbaryl (5.1 mg L^{-1}), chlorpyrifos (0.03 mg L^{-1}) and 2,4 dinitrophenol (10 mg L^{-1}) significantly increased the T2 interval or the time necessary to initiate the neuromuscular response, whereas phenol exposure shortened this interval. Chlorpyrifos (0.03 mg L^{-1}), carbaryl (7 mg L^{-1}), phenol (15.1 mg L^{-1}), 2,4-DNP (11.2 mg L^{-1}), endosulfan (0.5 µg L^{-1}), and l-octanol (7.8 mg L^{-1}) significantly increased the R/S ratio, reflecting the greater magnitude of stimulation required to induce an M-cell response. The significance of these alterations in the M-cell and escape responses were confirmed in tests that showed the fish were more

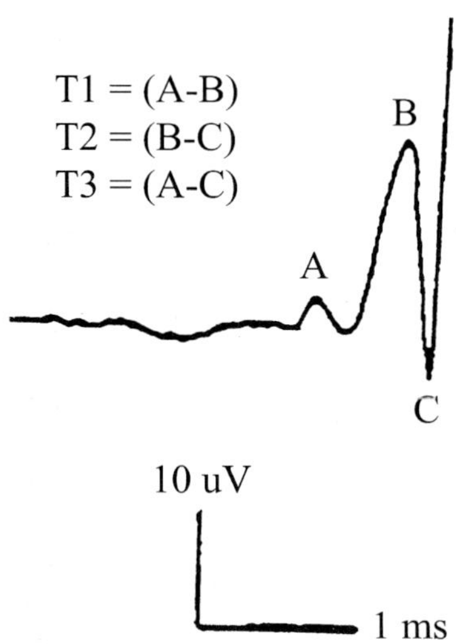

Figure 4.4 Electrophysiologic response of the Mauthner cell of medaka showing the Mauthner cell action potential (A), motor neuron compound biopotential (B), and onset of muscle fiber biopotentials (C). Time measures include T1 = A–B; T2 = B–C; T3 = A–C (redrawn from Carlson *et al.*, 1998).

susceptible to predation following exposure to chlorpyrifos, carbaryl, fenvalerate, endosulfan, phenol, l-octanol and 2,4-DNP at 10 percent of the LC_{50} concentration.

Somatosensory

Many aquatic organisms control body temperature behaviorally by actively seeking preferred temperatures and avoiding others (Reynolds, 1977). Preferred temperatures differ widely among species (Coutant, 1977) and conspecifics (Reynolds and Thomson, 1974). Preferred temperatures are often correlated with physiologic optima for various metabolic functions (Beitinger *et al.*, 1975), and temperature is likely a major variable in habitat selection. Some contaminants affect temperature selection by fish. Atlantic salmon (*Salmo salar*), for example, exposed to 10 μg L^{-1} DDT preferred temperatures that were 3–5°C lower than unexposed fish; whereas fish exposed to 30 and 50 μg L^{-1} preferred temperatures 3–6°C warmer than controls (Ogilivie and Anderson, 1965). In screening assays, Peterson (1976) found that fish preferred significantly lower temperatures following exposures to sodium pentachlorophenate, guthion, malathion, dursban, and dibrom. In a review of temperature selection, Beitinger (1990) concluded that cholinergic stimulation and increased L-dopamine concentrations in the brain raised preferred temperatures, whereas histamine, thyroxine and testosterone lowered them.

Vision

The maximum distance from which a fish responds to prey (reactive distance) is an important visual variable in predator–prey interactions and may be affected by contaminant exposure through direct histologic changes in the eye and associated structures. Unfortunately, there have been few studies examining histologic changes in the visual systems of fish. The bulk of research occurred prior to the mid-1980s and was focused on inorganic contaminants. The cornea of larval striped bass (*Morone saxatilis*) was found to be damaged by copper (Bodammer, 1985). The visual system in mammals is extremely sensitive to mercury poisoning and the few studies conducted with fish suggest the same may be true for teleosts. Adult Indian catfish (*Anabes scandens*) exposed to inorganic mercury (3 μg L^{-1}) for 4 weeks were found to be suffering from blindness and exophthalmia by Panigrahi and Misea (1978), who attributed this to the effects of mercury on the brain and optic lobe, although they did not specifically test this. Hawryshyn *et al.* (1982) found that rainbow trout injected with methylmercury lost both rod and cone sensitivity. Dial (1978) found effects related to retinal organization, pigment distribution, and corneal thickening among others in developing medaka exposed to methylmercury (80 μg L^{-1}).

Vision in fish is often evaluated through the measurement of an optokinetic response to visual stimuli. Optokinetic responses may include movement of the eye to visual stimuli (Brockerhoff *et al.*, 1995) or whole body movements (optomotor) relative to a moving visual stimuli (Scherer and Harrison, 1979).

The optomotor response is usually tested in the laboratory by rotating a background of black and white vertical lines past the fish, which are enclosed in a circular chamber (Figure 4.5). Fish are observed to determine how well they follow the background (Heath, 1995). The response involves vision, integration of information from multiple sensory inputs by the CNS, co-ordinated motor activity, and, in the case of optomotor responses, sufficient swimming performance to maintain position in relation to visual stimuli. Each of these variables are potentially affected by toxicosis. For example, fish that developed cataracts as a result of diphenol phosphate exposure (0.025 mg L^{-1}) lost the optomotor response as well as the ability to capture daphnid prey (Little *et al.*, 1985). Richmonds and Dutta (1992) tested bluegill exposed to the cholinesterase inhibitor malathion by rotating a striped drum around the fish tank. They recorded frequency of turns either in the direction of the drum movement or in the opposite direction (reversals). Hypersensitivity appeared to exist at 0.016 mg L^{-1} because there was a distinct increase in orientations and reversals, whereas at 0.032 mg L^{-1} exposed fish were similar to controls and fish exposed at 0.048 mg L^{-1} were hyposensitive. The increase in response induced by low-exposure concentrations might be attributed to an increase in sensitivity to stimuli, as seen with other cholinesterase inhibitors (McKim *et al.*, 1987). Dutta *et al.* (1992) observed a dose–response loss of the optomotor response following exposure to diazinon, another cholinesterase inhibitor.

The response to environmental stimuli is also altered in early life stage fish. Young juvenile rainbow trout are negatively phototaxic for several days prior to the onset of free swimming, a response that confines the fish to the protective reed. Fish exposed to 6.25 mg L^{-1} planar PCBs showed a positive phototactic response and made no effort to avoid light (Fisher *et al.*, 1993). Furthermore,

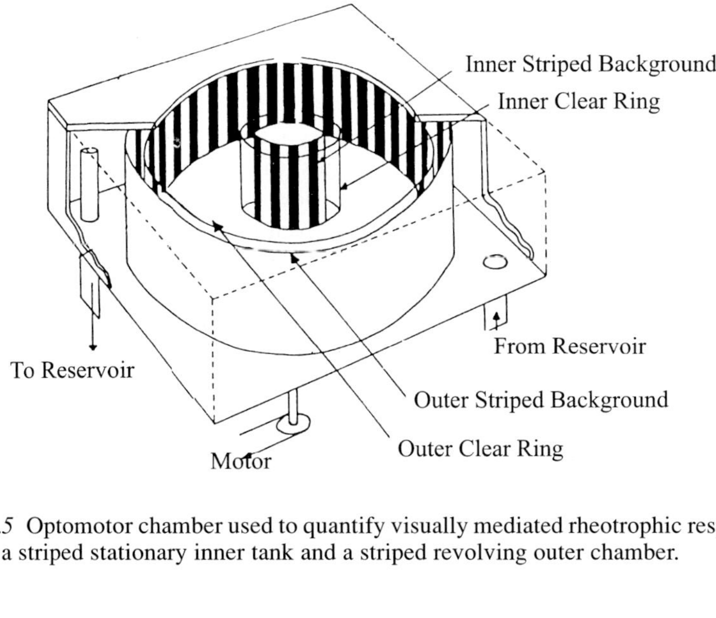

Figure 4.5 Optomotor chamber used to quantify visually mediated rheotrophic responses includes a striped stationary inner tank and a striped revolving outer chamber.

when rainbow trout fry initiate free swimming, they normally orient into the current. Fish exposed to 25 mg L^{-1} benzo(a)pyrene during late embryonic development appeared normal after hatching but did not orient into the current (Ostrander *et al.*, 1990).

Differences in experimental design can dramatically affect results. For example, Dodson and Mayfield (1979) found that rainbow trout exposed to diquat at field concentrations (0.5 and 1.5 mg L^{-1}) lost the ability to maintain position relative to background. De Peyster and Long (1993), using 9.2 mg L^{-1} diquat, found reductions in swimming capacity as well; however, the optomotor response of fathead minnows was enhanced. Heath (1995) suggested this discrepancy resulted because the Dodson and Mayfield (1979) study measured swimming capacity, as reflected by the inability of tested fish to maintain their position relative to the moving stimuli. Recent procedures used by Brockerhoff *et al.* (1995) eliminate the physical performance variable in optomotor response work by measuring movement of the eye.

Contaminant impacts to neuromuscular responses of fish

By design, many pesticides target the nervous system and cause mortality by inhibiting neuromotor responses such as respiration and cardiac output. In non-target organisms, such as fish, pesticides and other substances affect a number of neuromotor responses ranging from respiration to locomotion. Alterations of these responses by sublethal exposure can have a major impact on the ability of the organism to feed, avoid predators and reproduce. Disruption of neuron to neuron and neuron to muscle events in the nervous system can affect behavioral performance through a loss of co-ordination, control, or reaction time. The neurotransmitter chemical released across the synapse can be affected: (1) by reducing its production in the presynaptic membrane, (2) by being bound by the toxicant, or (3) through a competitive binding with the post-synaptic receptor.

Respiration is a rhythmic neuromuscular sequence regulated by endogenous sensory cues as well as by external environmental stimuli. Acute contaminant exposure can induce reflexive cough and gill purge responses to clear the opercular chamber of the irritant, and can also increase rate and amplitude of the respiratory cycle as the fish adjusts the volume of water in the respiratory stream. As the exposure continues, the respiratory cycle can become irregular, largely through diminished modulating input as well as alterations in the endogenous pacemaker. Diamond *et al.* (1990) found that the frequency and amplitude of bluegill opercular rhythms and cough responses were altered following exposure to different contaminants (Figure 4.6). For example, zinc at 300 μg L^{-1} reduced the amplitude of the respiratory response. Dieldrin, an organochlorine insecticide, increased ventilatory frequency at concentrations above 24 μg L^{-1} and also caused cough responses and erratic movements. The use of this approach has proven effective in detecting the presence of metals in complex mixtures (J. M. Diamond, personal communication).

Contaminant impacts to serotonin, a synaptic neurotransmitter in certain brain

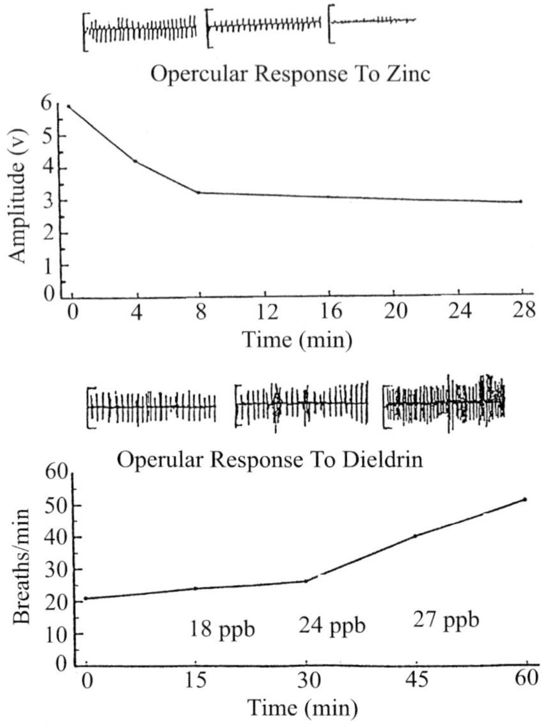

Figure 4.6 Amplitude of opercular movements of juvenile bluegill during exposure to 300 μg L^{-1} zinc and frequency of opercular movements induced in bluegill during exposure to three concentrations of dieldrin and a control. Oscillograph of opercular movements is shown above each graph, each vertical deflection represents one full respiratory cycle of gill opening and closing (redrawn from Diamond *et al.*, 1990).

pathways, have been observed. Heightened rates of locomotory activity and concomitant decreases in brain serotonin concentrations were observed in gulf killifish (*Fundulus grandis*) following exposure to 0.0004 percent Arochlor 1254 (Fingerman and Russell, 1980). Unfortunately, the mode of inhibition between serotonin and toxicant was not established. Weber *et al.* (1991) found that exposure to lead resulted in an increase in brain serotonin and norepinephrine levels in the fathead minnow. Exposures to 0.5 mg L^{-1} lead induced significant increases in brain norepinephrine concentrations which were associated with significant decreases in the frequency of feeding. Exposures to 1.0 mg L^{-1} lead also induced significant increases in serotonin which were associated with impaired reaction distances in response to prey.

Similarly, the neurotransmitter GABA was found to be bound by contaminants such as endosulfan, which blocks chloride channels (Coats, 1990; Carlson *et al.*, 1998). This was found to cause hyper-reactivity to physical stimuli following exposure to endosulfan (0.0015 mg L^{-1}) in medaka (Carlson *et al.*, 1998) and fathead minnow (Drummond and Russom, 1990). It also caused neurophysiologically apparent changes in the nervous system consistent with the convulsive effects resulting from the disruption of post-synaptic inhibition by GABA and glycine neurotransmitters (Faber and Korn, 1988; Carlson *et al.*, 1998).

The most commonly measured physiologic biomarker used to monitor pesticide exposure is cholinesterase (ChE) inhibition. This enzyme is required to prevent ongoing reactions between the neurotransmitter acetylcholine and its receptor. Cholinesterase inhibition is the intended mode of action of carbamate and organophosphorous insecticides which competitively bind with this enzyme to prevent the deactivation of the neurotransmitter with its receptor. Failure to deactivate the neurotransmitter prevents further stimulation of the post-synaptic membrane by preventing full recovery of the post-synaptic membrane from depolarization and desensitizing the receptors. At neuromuscular junctions, this results in behavioral manifestations of tetany and subsequent death. Recent studies in our laboratory have attempted to understand the extent to which behavior is altered by acetyl cholinesterase (AChE) inhibitions. Rainbow trout were exposed to a concentration series of various insecticides, and locomotory responses were observed at 24 and 96 h of exposure as well as after 48 h of recovery in uncontaminated water. Immediately after the behavioral observation, fish were removed from exposure and brain tissues were taken for AChE measurements. Following 20 $\mu g\ L^{-1}$ exposure to the organophosphorous compound malathion, alterations of several locomotory responses occurred. Both speed of movement and distance traveled declined about 50 percent as a result of a 24-h exposure. As shown in Figure 4.7 tortuosity of path (e.g. frequency of turns or ratio of turns to linear movement) also decreased as the organisms became more stereotypic in their movements (Brewer *et al.*, 1999). Brain AChE concentrations significantly declined as a result of inhibition and correlated strongly with the behavioral responses (Figure 4.8). When carbaryl was tested, similar correlations between AChE and behavior were observed. Acetylcholine concentrations were not affected by exposure; however, receptors in the post-synaptic membrane declined in response to acetylcholine left in the synapse. Transmission electron microscopy of synaptic areas of the forebrain and olfactory bulb of the brain also indicated cellular changes in the post-synaptic membrane consistent with the down-regulation of receptors (N.N. Ruzshinskaya, unpublished).

Locomotory responses as an indicator of neurotoxicity

Behavioral neurotoxicity is frequently evident as changes in the form, frequency or posture of locomotory behavior. When bluegill received pulsed doses of the pyrethroid insecticide ES-fenvalerate (0.025 $\mu g\ L^{-1}$), the first indication of toxicity was a tremorous tail movement (Figure 4.9) as fish initiated movement (Little *et al.*, 1993b). Exposure of rainbow trout to sublethal concentrations of 40 $\mu g\ L^{-1}$ malathion resulted in convulsive movements in rainbow trout (Brewer *et al.*, 1999). Changes in swimming behavior resulting from sublethal contaminant exposure may impair the performance of essential life functions. Swimming behavior is well studied as a measure of toxicant effect (Little and Finger, 1990) and has

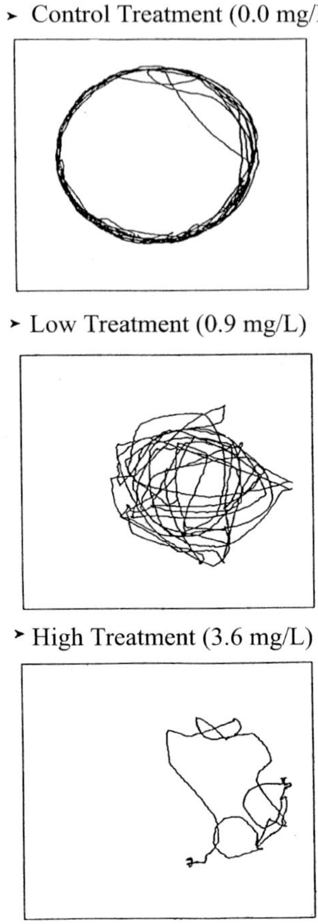

Figure 4.7 Dissolution of swimming paths following exposure of juvenile rainbow trout to carbaryl.

become one of the most common behavioral measures in toxicity investigations. Swimming behavior is affected by many chemical stressors, including metals, organochlorines, and industrial contaminants. Swimming responses have been used in automated biomonitoring systems because of their consistent sensitivity to numerous contaminants (Miller *et al.*, 1982; Smith and Bailey, 1988; Baatrup and Bayley, 1998). Swimming can be measured in many ways, including the frequency and duration of movements as well as the form and pattern. Descriptions of swimming form and pattern include frequency and angle of turns, distance of movement, linear and angular velocity, position in the water column, and swimming posture (Kleerekoper, 1974). The measurement of these swimming responses is usually limited to the laboratory. In contrast, assessment of fish at contaminated field sites currently is not possible as species-typical responses have not been defined to permit the evaluation of behavioral function except for the most extreme aberrations. In the laboratory, the subtle changes which arise from

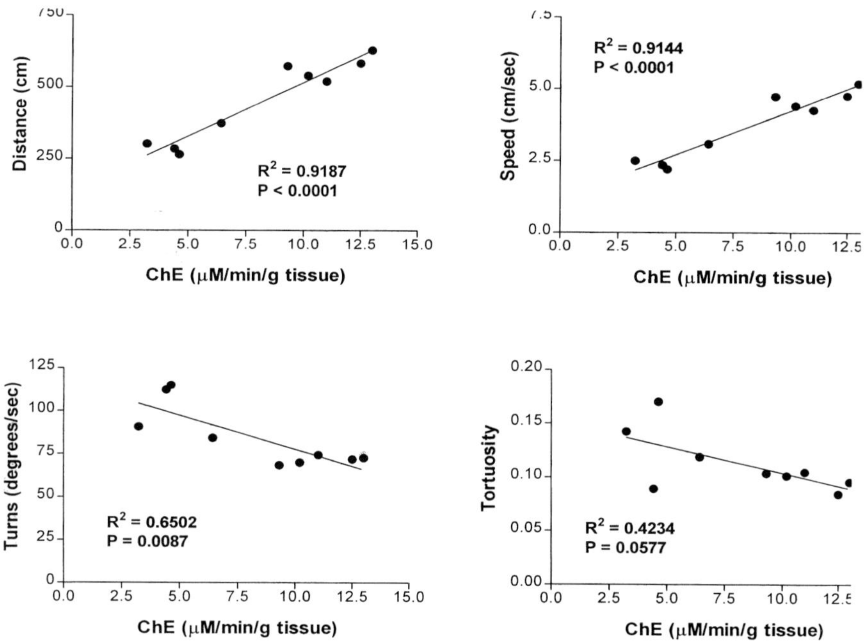

Figure 4.8 Correlations between locomotor behaviors and cholinesterase levels for rainbow trout exposed to malathion (from Brewer *et al.*, 1999).

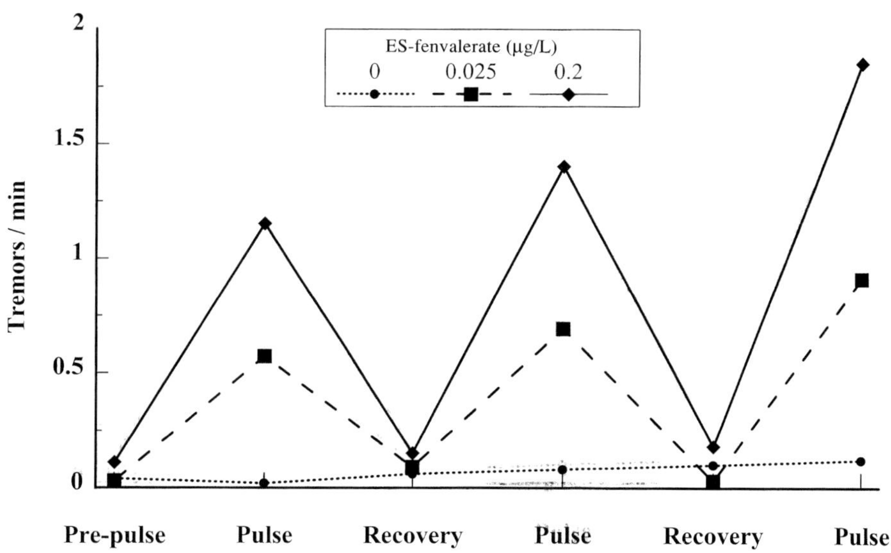

Figure 4.9 ES-fenvalerate exposure induced tremorous swimming responses in bluegill (redrawn from Little *et al.*, 1993b).

sublethal exposures may be confirmed through comparisons with controls or with responses observed during a pre-exposure period.

Behavioral changes, such as altered swimming behavior, usually occur much earlier than mortality when fish are exposed to contaminants. A review by Little and Finger (1990) revealed that the lowest behaviorally effective toxicant concentration that induced changes in swimming behavior of fish ranged from 0.1 percent to 5.0 percent of the LC_{50} (Table 4.2 summarizes results from tests with rainbow trout). When observations were made over time, behavioral changes commonly occurred 75 percent earlier than the onset of mortality. Often, swimming behavior is also affected before reductions in growth are detected. For example, swimming activity was significantly reduced in rainbow trout after a 96-h exposure to 0.005 mg L^{-1} of an organophosphorous defoliant; a concentration that affected growth after 30 d (Little *et al.*, 1990). Development of locomotory responses, frequency of movements, and duration of activity were significantly inhibited in brook trout (*Salvelinus fontinalis*) alevins at lower aluminum concentrations (300 μg L^{-1}) under acidic conditions (pH $\leq$ 6.1) than those that affected survival or growth (Cleveland *et al.*, 1991).

Endocrine impacts to neurotoxicity

Endocrine function is a fundamental physiologic requirement for the nervous system and behavior, thus contaminant alterations to the endocrine system can affect the nervous system and also behavioral performance. There are few studies specifically examining endocrine impacts on neurotoxicity. Further studies investigating contaminant effects on the linkage of hormonally induced behaviors need to be carried out. In this section, we review the sparse literature on toxicity

Table 4.2 Effects of contaminants on the pattern or frequency of swimming behavior and survival of rainbow trout.

Toxicant	*Lowest behaviorally effective concentration (mg L^{-1})*	*Percent of lethal exposure*
Aluminum[a]	0.510	6.0
Chlordane[b]	0.001	2.5
Copper[c]	0.005	12.0
Tribufos (DEF)[b]	0.100	15.0
2,4-D[b]	2.500	2.5
Diquat[d]	0.500	0.5
Lindane[e]	0.060	25.0
Methyl-parathion[b]	0.010	0.2
Penta-chlorophenol[b]	0.007	2.5

Notes
a Freeman and Everhart (1971).
b Little *et al.* (1990).
c Waiwood and Beamish (1978).
d Dodson and Mayfield (1979).
e Poels (1977).

to endocrine-mediated reproductive behaviors, hormonal control of parr–smolt transformation, and stress hormones.

Reproductive hormones induce a number of changes in the nervous system and behavior to initiate spawning. The effects of environmental stressors and contaminants on reproductive processes in fish have been reviewed by Donaldson (1990). Initially, studies on toxicant effects on reproduction were geared towards estimating a maximum acceptable toxicant concentration (MATC) for establishing water quality criteria. Schröder and Peters (1988) examined the effects of sublethal concentrations of lindane and riverwater on specific courtship behaviors of competing male guppies; however, the sequence of behaviors was not examined.

Changes in the physiologic mechanisms of behavior may explain the alterations in fish reproductive behavior, even after short-term exposures. For example, male fathead minnows must prepare a quality nest site to attract gravid females. After spawning, males assume parental responsibilities for maintaining the nest site and guarding the eggs. Fathead minnows have been shown to reduce or even cease these spawning behaviors after exposure to 1.0 mg L^{-1} lead (Weber, 1993) and greater than 37 μg L^{-1} copper (Pickering *et al.*, 1977) immediately prior to spawning. What is interesting about these studies is that fish at different stages of sexual development (approximated by exposing fish at different times of the year) were unequally affected by metal exposure. It is reasonable to assume that specific behaviors may be induced by sex steroids. Concurrent or later exposure to toxicants might inhibit further behavioral development and expression by stopping gonadal development or blocking receptor sites for various neurotransmitters or hormones (Govoni *et al.*, 1980; Wiebe *et al.*, 1983).

Toxicant-induced behavioral dysfunctions may be passed on to subsequent generations. 'Behavioral teratology' was observed among the offspring of Atlantic croaker (*Micropogonias undulatus*) that had received a 30-day dietary exposure to *o,p′*-DDT (2 and 10 μg L^{-1} per 100 g food per day) (Faulk *et al.*, 1999). Larvae exhibited reduced mean and maximum burst speeds in response to visual stimuli. DDT exposure appeared to have its greatest effects on routine swimming activity prior to complete oil globule absorption. Based on this study, it is possible that parental exposure may decrease offspring survival by altering important behaviors, thus lowering the genetic fitness of the current generation. 'Behavioral teratology' is a promising new area that relates the genetic fitness of an organism to behavioral alterations. Contaminant body burdens have long been suspected to affect fitness, but it has been difficult to prove in the field or the laboratory. The extent to which other chemical classes that do not readily bioaccumulate or concentrate have similar subsequent effects in later generations remains unknown.

Parr–smolt transformation of anadromous salmonids involves a series of hormonal, physiologic, and behavioral changes that stimulate out-migration and prepare the fish for the transition from fresh to seawater. Environmental perturbations can disrupt smoltification and prevent out-migration of juvenile salmon. Reduced or delayed smolt migration may result in reduced growth, poor recruitment, lowered survival, and, ultimately, fewer returning adult salmon. Studies with coho salmon (*Oncorhynchus kisutch*), chinook and sockeye

(*Oncorhynchus kisutch nerka*) salmon have shown that metals and other aquatic contaminants inhibit smoltification and impair the ability of fish to survive in seawater (Lorz and McPherson, 1976; Nichols *et al.*, 1984; Hamilton *et al.*, 1986). Coho salmon exposed to sublethal concentrations of copper (5–30 μg L^{-1}) exhibited reduced downstream migration, lowered gill ATPase activity, and poor survival in sea water (Lorz and McPherson, 1976).

The general adaptation response (Selye, 1950) of fish to stress is generally characterized by elevations in cortisol and catecholamines (Pickering, 1981). Hormones, especially the corticosteroids, mediate the stress response by acting as a stimulus between the neural transducer and physiologic responses (Mazeaud *et al.*, 1977; Strange *et al.*, 1977). These hormones not only aid in maintaining homeostasis under stressful conditions but also they may have a strong effect on immunologic functions. Secondary stress responses such as a change in the composition of blood cells and immunosuppression are believed to be in response to elevation in corticoid hormones or catecholamines (Schreck and Lorz, 1978; Thomas and Lewis, 1987; Pickering and Pottinger, 1989; Ainsworth *et al.*, 1991). Exogenous cortisol has been shown to make trout more susceptible in a dose-dependent way to bacterial and fungal diseases (Pickering and Pottinger, 1989). This relationship may also have a negative feedback aspect. Schreck and Bradford (1990) reported that fish lymphokines can suppress the secretion of cortisol from inter-renal cells, the opposite of what occurs in mammals (Sapolsky *et al.*, 1987). Clearly, cortisol can have strong effects on the immune system, but this hormone is most likely not solely responsible for all of the effects of toxicants and other stressors (Schreck and Lorz, 1978; Maule and Schreck, 1990).

Immune function and neurotoxicity

Fish immunotoxicology is a relatively new field and our understanding of fish immunology is far behind that of mammals, despite several good reviews of fish immunology (Satchell, 1991; Weeks *et al.*, 1992). Because fish are exposed to a number of environmental stressors such as predators or pollutants, they frequently become more susceptible to disease. It is commonly thought that such stressors adversely affect immunologic function. The stress hormones, especially cortisol and the catecholamines, have been shown to be involved in immune response of both fish and mammals, although they may have different actions (Sapolsky *et al.*, 1987; Schreck and Bradford, 1990). Clearly, we are only beginning to understand the relationship between neurotoxicity and immunotoxicology. The reader is referred to Chapter 3 for a fuller treatment of the toxic responses of the immune system.

Toxicant impacts on higher behavioral function

Information about neurotoxicity is especially useful in support of hazard assessment and environmental regulation. Although physiologic data provide an indication of exposure and effect, and behavioral data provide indications of effect, it is also important to show a causal linkage with the population in order to provide a predictive index of population-level effects. The ecologic consequences of behavioral change resulting from sensory or neuromuscular neurotoxicity are presumably manifested through the impairment of other adaptive behaviors such as migration, predator avoidance, or predatory success. Hypoactivity and hyperactivity, as well as deviations in adaptive diurnal rhythmicity, may disrupt feeding and increase vulnerability to predation (Steele, 1983). Laurence (1972) found that heightened activity may increase an organism's vulnerability to predation, whereas reductions in swimming activity lessen the chance of encountering prey by reducing effective search areas.

Behavioral measures of injuries include behavioral changes that clearly represent a harmful response that would directly limit the individual's survival and long-term viability. Predator–prey interactions measure the ability of fish to feed, as well as their ability to avoid being eaten by other organisms. Such studies provide a good example of a behavioral measure of injury because deficits in either response will have immediate implications for the organism's survival (Brown *et al.*, 1987). Feeding and prey vulnerability have been used to examine sublethal contamination because predator and prey may be differentially affected by toxicants (Sandheinrich and Atchison, 1990).

Survival of early life stage fish will decline if the development of foraging behavior is delayed or inhibited. Impaired feeding behavior also correlates with reduced growth, which may lengthen a fish's period of vulnerability to predation (Werner and Hall, 1974) and impair overwinter survival (Oliver *et al.*, 1979). Because changes in the ability of a fish to detect, pursue, capture, and consume prey will affect growth and survival, these behavioral variables may provide an ecologically relevant measure of contaminant-induced injury. Contaminants have been shown to affect numerous aspects of the feeding sequence, including detection of prey (Lemly and Smith, 1987), prey capture (Little *et al.*, 1990), handling time, and ingestion of prey (Sandheinrich and Atchison, 1990), as well as general motivation to feed (Little *et al.*, 1990). Thus, several variables of feeding behavior can be measured in toxicity studies. The experimental environment for such studies can vary in complexity depending on the amount of ecologic realism one hopes to achieve, but it can also be readily adapted to standard toxicity testing procedures (Mathers *et al.*, 1985).

A number of studies have shown that feeding behavior is impaired by sublethal exposure to a diversity of contaminants. Responses summarized in Table 4.3 for rainbow trout are typical of those reported in the literature for other species. Reductions in feeding have been observed after sublethal exposure to metals (Atchison *et al.*, 1987), organophosphates (Bull and McInerney, 1974), dioxins (Mehrle *et al.*, 1988), and petroleum hydrocarbons (Woodward *et al.*, 1987). Feeding, measured as frequency of prey capture, was sensitive in defining no-effect concentrations of aluminum in acid-exposed brook trout and was among the most sensitive indices of sublethal pH and aluminum exposure (Cleveland *et al.*, 1989). The sensitivity of feeding behavior, as an index of sublethal exposure among rainbow trout exposed to various agricultural chemicals, ranged from less

Table 4.3 Effects of contaminants on the feeding behavior of juvenile rainbow trout.

Toxicant	*Lowest behaviorally effective concentration (mg L^{-1})*	*Percent of lethal concentration*
Aluminum[a]	0.052	2.0
Chlordane[b]	0.001	2.5
Copper[c]	0.100	10.0
Tribufos (DEF)[b]	0.056	8.0
2,4-D[b]	25.000	25.0
Methyl parathion[b]	0.100	20.0
Pentachlorophenol[b]	0.007	2.5

Notes
a Freeman and Everhart (1971).
b Little *et al.* (1990).
c Waiwood and Beamish (1978).

than 0.3 percent to 50 percent of the LC_{50} (Little *et al.*, 1990). Inhibited motivation to feed seemed to be the predominant effect at higher concentrations, whereas reduced feeding efficiency and reduced strike frequencies were the predominant impairments at exposure to lower concentrations.

Predation tests measure the ability of prey to escape predation. Practically all organisms are vulnerable to predation during some portion of their life cycle. Increased vulnerability to predation may occur when a toxicant alters the ability of fish to detect or respond to predators (Brown *et al.*, 1985). A common laboratory approach for measuring predation combines predators with contaminant-exposed and unexposed prey. Census of the surviving control and exposed prey are made when approximately 50 percent of the prey population have been captured (Little *et al.*, 1985). Structural complexity of the experimental setting, including the addition of refuges for prey, is used to increase the environmental realism of the test. A range of contaminants have been shown to increase predation-induced mortality (Table 4.4). Little *et al.* (1990) reported that different types of toxicants at concentrations as low as 2 percent of the lethal concentration significantly increased the predation of exposed prey. The response did not consistently follow a dose–response relationship, however, because the behavioral aberrations resulting from the exposure sometimes made the prey less conspicuous to the predators because of inactivity or reduced mobility. In tests with methyl parathion, E. E. Little (unpublished) found that a 50 percent decrease in brain AChE was associated with a 40 percent decrease in swimming activity and heightened rates of predation (Figure 4.10). Assessment of predator–prey behaviors during toxicity tests provides not only a sensitive measure of toxicant effect but also a measure of injury as growth and predation-induced mortality are affected by exposure.

Contaminant exposure may also cause aberrations or inhibitions of other behaviors such as schooling, shelter seeking, and locomotory responses which may influence predator–prey interactions. For example, schooling declined following exposure of yearling carp to 0.05 mg L^{-1} DDT (Besch *et al.*, 1977) and

Table 4.4 Effects of contaminants on the vulnerability of early juvenile rainbow trout to predation.

Toxicant	*Lowest behaviorally effective concentration (mg L^{-1})*	*Percent of lethal exposure*
Carbaryl[a]	0.010	0.5
Chlordane[a]	0.02	5.0
Tribufos (DEF)[a]	0.05	5.0
2,4-D[a]	50.0	50.0
Methyl parathion[a]	0.1	2.7
Pentachlorophenol[a]	0.0002	0.5

Note
a Little *et al.* (1990)

of fathead minnows to 7.43 mg L^{-1} of the herbicide 2,4-dinitrophenol at a pH of 7.57 (Holcombe *et al.*, 1980).

Ecologic relevance of behavior as a predictor of contaminant effects in the natural environment

Impaired behavioral performance may be predictive of contaminant effects in the field when ecologic consequences can be linked to impaired behavioral performance. Disruption of essential functions such as habitat selection, competition, predator–prey relationships, or reproduction can become ecologically

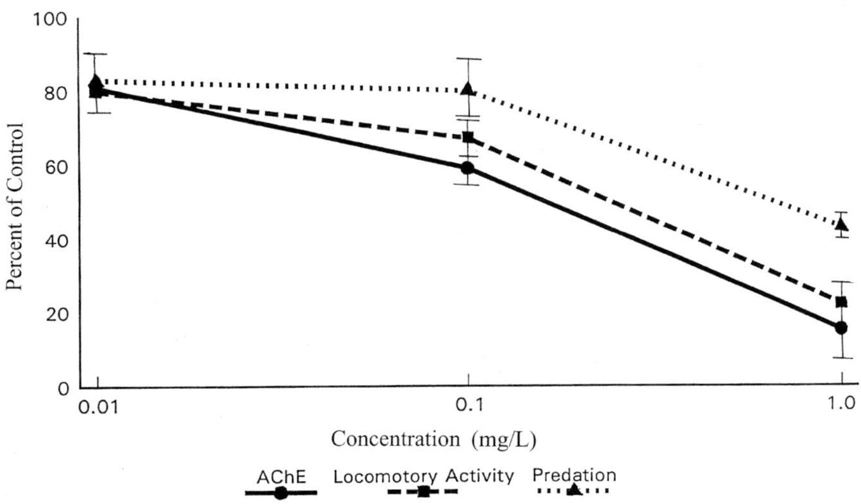

Figure 4.10 Effects of methyl parathion on acetylcholinesterase (AChE) and behavior of bluegill. Means ± standard error. Predation is reported as percent survival as percent of control.

apparent through loss of populations or changes in year-class strength when enough individuals are affected. Verification of behavioral effects in the field is an important step in understanding the causal relation between observed behavioral changes and the impact of contaminants on natural populations and communities (Sandheinrich and Atchison, 1990). However, relating behavioral change to higher ecologic organization in the field poses significant technical challenges for behavioral toxicology. Attempts to verify behavioral responses in the field have been few and usually rely on indirect methods, such as mark and recapture or telemetry, to infer habitat selection or avoidance responses or on stomach content analysis to support hypotheses about foraging behavior. Experimental designs are difficult because of the mobility of the organisms, physical constraints of the environment, verification of exposure, and manpower required to accomplish such determinations (Little, 1990).

Appropriate habitat is critical to the survival of individuals and their offspring. Many fish species require different habitats for essential life functions such as feeding, reproduction, and reducing predator contact. Adaptive behavior allows fish to avoid areas containing unfavorable conditions and to occupy acceptable areas. Multimodal sensory information is often used to determine the suitability of a potential habitat, including vision and chemoreception as well as thermo- and mechanoreception. The selection of appropriate habitat is a complex series of interactions, with many factors potentially affecting the outcome. Contaminants can produce behavioral alterations through interactions with receptors of various sensory systems.

Fish may respond to noxious, aversive habitat conditions in a manner that minimizes exposure to them. On the other hand, the contamination may impair reception of important environmental stimuli or alter the physiologic function of the sensory system such that the organism is less sensitive or unresponsive to natural stimuli. Such deficits in behavioral responses are caused by lesions or alterations in the sense organ or its receptors, changes in the neural response following stimulation of the sensory system, or changes in higher integrative processes of the central nervous system.

Avoidance reactions to contaminants provide an example of the type of behavioral measure that is useful as a predictor of ecologic effects in contaminant assessment. Avoidance responses can be detrimental because organisms may be displaced from preferred habitats to suboptimal areas where they may face greater competition and predation pressure or inadequate resource availability (Atchison *et al.*, 1987). This can result in locally reduced biomass, loss of species diversity, diminished year-class strength, and loss of productivity and diversity in that area.

Because the avoidance response has been verified as a response to contaminants in the field, it has been legally accepted as evidence of injury for Natural Resource Damage Assessments under proceedings of the Comprehensive Environmental Response, Compensation, and Liability Act of 1980 (NRDA, 1986). However, with the exception of avoidance responses, the link between behavioral impairment and injurious population or community effects has not been well documented in the natural environment. It is this lack of well-established causal links between behavioral responses and population effects, not the biologic significance of the response, that limits the utility of behavioral data as predictive indices of field effects and measures of injury.

Although most behavioral effects observed in laboratory studies have not yet been directly linked to ecologic responses, they can still provide protective criteria which correspond to no-effect concentrations documented in the field. For example, in laboratory studies of bluegill exposed to the hydrocarbon fluorene, feeding response was the most sensitive laboratory endpoint (Finger *et al.*, 1985). This is particularly significant since these feeding response data were the only laboratory observations that accurately predicted an adverse impact on growth and survival of bluegills in companion pond ecosystem studies (Boyle *et al.*, 1985). The behavioral response observed in the laboratory was not causally linked to an ecologic impact, but it was predictive of field effects because it indicated a concentration that affected the mesocosm population. Many of the ecologic measurements obtained during field studies do not provide the resolution to link causal relationships between behavioral changes and the response of natural fish populations. Inhibited growth among fish from field populations, for example, may be due to many factors; this is where laboratory behavioral studies could be useful in interpreting and defining the mechanisms of ecotoxicity. In other words, more effort is required to link neurobehavioral endpoints with changes in growth, survival, and reproduction.

Conclusions

Behavioral measurements may be useful as indicators of sublethal contamination because they frequently occur below concentrations that are chronically lethal and at lower concentrations than those that affect growth. Consequently, behavioral tests provide definable, interpretive endpoints that could be used for regulatory purposes in product registration, damage assessments and in the formulation of water quality criteria. Regardless of their sensitivity, utility, or biologic significance, behavioral responses have not been routinely included in standard aquatic toxicity assessment programs (Little, 1990). The development of water quality criteria often relies heavily on measured concentrations of chemicals causing mortality in laboratory exposures with limited data on the chronic effects of exposure, such as growth, on aquatic organisms. Acute mortality, however, may not necessarily be the best predictor of survival or toxicant effect in the field where more pervasive and subtle sublethal exposures are prevalent. Most natural fish populations are exposed to sublethal concentrations of toxicants that can diminish long-term adaptability and survival by altering or impairing critical behavioral functions (Kleerekoper, 1976). Investigation of the sublethal behavioral effects of contaminants on early life stage fish is particularly relevant to the toxicity assessment process because these rapidly developing life stages are typically more sensitive to environmental contamination than juveniles or adults (Mayer and Ellersieck, 1986), and are less tolerant of long-term sublethal alterations of critical organism function. In developing behavioral test methods for contaminant assessment, responses used in behavioral toxicology must address one or more of three assessment objectives, including (1) how well does the response measure effect or injury arising from exposure, (2) does the response aid in the identification of the toxic agent, and (3) does the use of the behavioral response increase the capability of contaminant assessments to predict the ecologic consequences of exposure?

Certain behavioral indices have been used to measure contaminant effects and injury, to provide diagnostic identification of contaminant classes, and to predict ecologic impact. The extent to which a behavioral variable is used in these aspects of contaminant assessment depends upon the complexity of the behavioral variable and the extent to which the response is understood within different levels of biologic organization. Certain elemental responses, such as frequency of movements, underlie more complex responses, such as feeding. Changes in the elemental responses may be excellent signals of change in environmental quality yet be of little apparent significance to the organism's long-term survival. Likewise, growth inhibitions reflecting reduced feeding efficiency, although deleterious for the organisms, may not be directly evident in the status of the population or the community.

The challenges for behavioral neurotoxicity have been explicitly defined by Little *et al.* (1993a): 'More and better behavioral testing procedures, especially those using sensitive early life stage fish, must be developed and refined into effective tools which can be readily and clearly applied in contaminant assessments. Secondly, more work must be done to establish links between behavioral effects observed in the laboratory and ecologic effects observed in the field to aid in the development of contaminant assessments that are adequately predictive of population and community response. Finally, behavioral toxicologists must continue to work to dispel the erroneous paradigm prevalent in aquatic toxicology that suggests that behavioral responses are not suitable for inclusion in contaminant evaluation and assessment because they are either unquantifiable, too complex, too variable, extraordinarily difficult to measure, not biologically significant, or lacking in ecologic relevance.'

Acknowledgments

Funding was provided by the US Army Medical Research Command Project no. MIPR 97MM7721. We thank Sheryl Beauvais, Paulo Carvalho, James Fairchild and an anonymous reviewer for their comments and Aaron DeLonay for technical assistance with figures.

References

Ainsworth, A.J., Dexiang, C. and Waterstrat, P.R. 1991. Changes in peripheral blood leukocyte percentages and function of neutrophils in stressed channel catfish. *Journal of Aquatic Animal Health* 3: 41–47.

Atchison, G.J., Henry, M.G. and Sandheinrich, M.B. 1987. Effects of metals on fish behavior: A review. *Environmental Biology of Fishes* 18: 11–25.

Atema, J. 1976. Sublethal effects of petroleum fractions on the behavior of the lobster, *Homarus americanus*, and the mud snail *Nassarius obsoletus*. In *Estuarine Processes: Uses, Stresses and Adaptation to the Estuary*, Vol. I. Wiley, M. (ed.). Academic Press, New York.

Atema, J., Todd, H. and Bardach, J.E. 1969. Olfaction and behavioral sophistication in fish. In *Olfaction and Taste*, Vol. III, pp. 241–251. Rockefeller University Press, New York.

Baatrup, E. and Bayley, M. 1998. Animal locomotor behavior as a health biomarker of chemical stress. *Archives of Toxicology* 20 (Suppl.): 163–178.

Birtwell, I.K. and Kruzynski, G.M. 1989. *In situ* and laboratory studies on the behavior and survival of Pacific salmon (genus *Oncorhynchus*). *Hydrobiologica* 188/189: 543–560.

Black, J.A. and Birge, W.J. 1980. An avoidance response bioassay for aquatic pollutants. Water Resources Research Institute Research Report 123. University of Kentucky, Kentucky.

Beitinger, T.L. 1990. Behavioral reactions for the assessment of stress in fishes. *Great Lakes Research* 16: 495–528.

Beitinger, T.L., Magnuson, J.J., Neill, W.H. and Schaffer, W.R. 1975. Behavioral thermoregulation and activity patterns in the green sunfish, *Lepomis cyanellus*. *Animal Behavior* 23: 203–210.

Besch, W.K., Kemball A., Meyer-Waarden K. and Scharf, B. 1977. A biological monitoring system employing rheotaxis of fish. In *Biological Monitoring of Water and Effluent Quality*, ASTM STP 607. Cairns, Jr., J., Dickson, K.L. and Westlake, G.F. (eds), pp. 56–74. American Society for Testing and Materials, Philadelphia, PA.

Birge, W., Shoyt, R., Black, J., Kercher, M. and Robinson, W. 1993. Effects of chemical stresses on behavior of larval and juvenile fishes and amphibians. In *American Fisheries Society Symposium 14 Water Quality and the Early Life Stages of Fishes*. Fuiman, L. (ed.), pp. 55–65. American Fisheries Society, Bethesda.

Blaxter, J. and Hallers-Tjabbes, C.T. 1992. The effect of pollutants on sensory systems and behaviour of aquatic animals, Netherlands. *Journal of Aquatic Ecology* 26: 43–52.

Bodhammer, J. 1985. Corneal damage in larval striped bass, *Morone saxatilis* exposed to copper. *Transactions of the American Fisheries Society* 114: 577–583.

Boyle, T.P., Finger, S.E., Paulson, R.L. and Rabeni, C.F. 1985. Comparison of laboratory and field assessment of fluorene. Part II. Effects on the ecological structure and function of experimental pond ecosystems. In *Validation and Predictability of Laboratory Methods for Assessing the Fate and Effects of Contaminants in Aquatic Ecosystems*, ASTM STP 865. Boyle, T.P. (ed.), pp. 134–151. American Society for Testing and Materials, Philadelphia, PA.

Brewer, S.K., Little, E.E., DeLonay, A.J., Beauvais, S.L. and Jones, S.B. 1999. The use of automated monitoring to assess behavioral toxicology in fish: Linking behavior and physiology. In *Environmental Toxicology and Risk Assessment. Eighth Volume* ASTM STP 1364. Henshel, D.S., Black, M.C. and Harrass, M.C. (eds), pp. 370–386. American Society for Testing and Materials, West Conshohocken, PA.

Brockerhoff, S.E., Hurley, J.B., Janssen-Bienhold, U., Neuhauss, S.C.F., Driever, W. and Dowling, J.E. 1995. A behavioral screen for isolating zebrafish mutants with visual system defects. *Journal of the National Academy of Sciences* 92: 10545–10549.

Brown, S.B., Evans, R.E., Thompson, B.E. and Hara, T.J. 1982. Chemoreception and aquatic pollutants. In *Chemoreception in Fishes*. Hara, T.J. (ed.), pp. 363–393. Elsevier Scientific Publishing, Amsterdam.

Brown, J.A., Johansen, P.H., Colgan, P.W. and Mathers, R.A. 1985. Changes in the predator-avoidance behavior of juvenile guppies (*Poecilia reticulata*) exposed to pentachlorophenol. *Canadian Journal of Zoology* 63: 2001–2005.

Brown, J.A., Johansen, P.H., Colgan, P.W. and Mathers, R.A. 1987. Impairment of early feeding behavior of largemouth bass by pentachlorophenol exposure: A preliminary assessment. *Transactions of the American Fisheries Society* 116: 71–78.

Bull, C.J. and McInerney, J.E. 1974. Behavior of juvenile coho salmon (*Oncorhynchus kisutch*) exposed to Sumithion (fenitrothion), an organophosphate insecticide. *Journal of the Fisheries Research Board of Canada* 31: 1867–1872.

Carlson, R.W., Bradbury, S.P., Drummond, R.A. and Hammermeister, D.E. 1998. Neurological effects on startle response and escape from predation by medaka exposed to organic chemicals. *Aquatic Toxicology* 43: 51–68.

Cherry, D.S. and Cairns, Jr, J. 1982. Biological monitoring. V. Preference and avoidance studies. *Water Research* 16: 263–301.

Cleveland, L., Little, E.E., Wiedmeyer, R.H. and Buckler, D.R. 1989. Chronic no-observed-effect concentrations of aluminum for brook trout exposed in low calcium dilute acidic water. In *Environmental Chemistry and Toxicology of Aluminum*. Lewis, T.E. (ed.), pp. 229–246. Lewis Publishers, Chelsea, MI.

Cleveland, L., Little, E.E., Ingersoll, C.G., Wiedmeyer, R.H. and Hunn, J.B. 1991. Sensitivity of brook trout to low pH, low calcium and elevated aluminum concentrations during laboratory pulse exposures. *Aquatic Toxicology* 19: 303–317.

Coats, J.R. 1990. Mechanisms of toxic action and structure–activity relationships for organochlorine and synthetic pyrethroid insecticides. *Environmental Health Perspectives* 87: 255–262.

Coutant, C.C. 1977. Compilation of temperature preference data. *Journal of the Fisheries Research Board of Canada*. 34: 739–745.

DeLonay, A.J., Little, E.E., Lipton, J., Woodward, D.F. and Hansen, J.A. 1995. Behavioral avoidance as evidence of injury to fishery resources: Applications to natural resource damage assessments. In *Environmental Toxicology and Risk Assessment Fourth Volume*, ASTM STP 1262. LaPoint, T.W., Price, F.T. and Little, E.E. (eds), pp. 268–280. American Society for Testing and Materials, Philadelphia, PA.

De Peyster, A. and Long, W. 1993. Fathead minnow optomotor response as a behavioral endpoint in aquatic toxicity testing. *Bulletin of Environmental Contamination and Toxicology* 51: 88–95.

Dial, N.A. 1978. Some effects of methylmercury on development of the eye in medaka fish. *Growth* 42: 309–318.

Diamond, J.M., Parson, M.J. and Gruber, D. 1990. Rapid detection of sublethal toxicity using fish ventilatory behavior. *Environmental Toxicology and Chemistry* 9: 3–12.

DiMichelle, L. and Taylor, M.H. 1978. Histopathological and physiological responses of *Fundulus heteroclitus* to naphthalene exposure. *Journal of the Fisheries Research Board of Canada*. 35: 1060–1065.

Dodson, J.J. and Mayfield, C.I. 1979. Modification of the rheotropic response of rainbow trout (*Salmo gairdneri*) by sublethal doses of the aquatic herbicides diquat and simazine. *Environmental Pollution* 18: 147–157.

Donaldson, E.M. 1990. Reproductive indices as measures of the effects of environmental stressors in fish. *American Fisheries Society Symposium* 8: 109–122.

Drummond, R.A. and Russom, C.L. 1990. Behavioral toxicity syndromes: a promising tool for assessing toxicity mechanisms in juvenile fathead minnows. *Environmental Toxicology and Chemistry* 9: 37–46.

Drummond, R.A., Russom, C.L., Geiger, D.L. and DeFoe, D.L. 1986. Behavioral and morphological changes in fathead minnows, *Pimephales promelas*, as diagnostic endpoints for screening chemicals according to modes of action. In *Aquatic Toxicology*, ASTM STP 921. 9th Aquatic Toxicity Symposium. Poston, T.M. and Purdy, R. (eds), pp. 415–435. American Society for Testing and Materials. Philadelphia, PA.

Dutta, H., Marcelino, J. and Richmonds, C. 1992. Brain acetylcholinesterase activity and optomotor behavior in bluegills, *Lepomis macrochirus* exposed to different concentrations of diazinon. *Archives Internationales de Physiologie de Biochimie et de Biophysique* 100: 331–334.

Eaton, R.C. and Hackett, J.T. 1984. The role of the Mauthner cell in fast-starts involving escape in teleost fishes. In *Neural Mechanisms of Startle Behavior*. Eaton, R.C. (ed.), pp. 213–266. Plenum Press, New York.

Faber, D.S. and Korn H. 1988. Synergism at central synapses due to lateral diffusion of transmitter. *Proceedings of the National Academy of Sciences of the United States of America* 85: 8708–8712.

Faulk, C.K., Fuiman, L.A. and Thomas, P. 1999. Parental exposure to *ortho*, *para*-dichlorodiphenyltrichloroethane impairs survival skills of Atlantic croaker (*Micropogonias undulatus*) larvae. *Environmental Toxicology and Chemistry* 18: 254–262.

Fetchco, J.R. and Faber, D.S. 1988. Identification of motorneurons and interneurons in the spinal network for escapes initiated by the Mauthner cell in goldfish. *Journal of Neuroscience* 8: 4192–4213.

Finger, S.E., Little, E.E., Henry, M.G., Fairchild, J.F. and Boyle, T.P. 1985. Comparison of laboratory and field assessment of fluorene. I. Effects of fluorene on the survival, growth, reproduction, and behavior of aquatic organisms in laboratory tests. In *Validation and Predictability of Laboratory Methods for Assessing the Fate and Effects of Contaminants in Aquatic Ecosystems*, ASTM STP 865. Boyle, T. (ed.), pp. 120–133, American Society for Testing and Materials, Philadelphia, PA.

Fingerman, S.W. and Russell, L.C. 1980. Effects of polychlorinated biphenyl Arochlor 1254 on locomotor activity and on the neurotransmitters dopamine and norepinephrine in the brain of gulf killifish, *Fundulus grandis*. *Bulletin of Environmental Contamination and Toxicology* 25: 682–687.

Fisher, J.P., Spitsbergen, J.M., Bush, B. and Parwar, B.J. 1993. Effect of embryonic PCB exposure on hatching success, survival, growth and developmental behavior in landlocked Atlantic salmon, Salmo salar. In *Environmental Toxicology and Risk Assessment*, Vol. II. Gorsuch, J.W., Dwyer, F.J., Ingersoll, C.G. and LaPoint, T.W. (eds). American Society for Testing and Materials, Philadelphia PA.

Folmar, L.C. 1976. Overt avoidance reaction of rainbow trout fry to nine herbicides. *Bulletin of Environmental Contamination and Toxicology* 15: 509–514.

Freeman, R.A. and Everhart, W.H. 1971. Toxicity of aluminum hydroxide complexes in neutral and basic media to rainbow trout. *Transactions of the American Fisheries Society* 100: 644–658.

Giattina, J.D. and Garton, R.R. 1983. A review of the preference-avoidance responses of fishes to aquatic contaminants. *Residue Reviews* 87: 43–90.

Giattina, J.D., Cherry D.S., Cairns, J. and Larrick S.R. 1981. Comparison of laboratory and field avoidance behavior of fish in heated chlorinated water. *Transactions of the American Fisheries Society* 110: 526–535.

Govoni, S., Memo, M., Lucchi, L., Spano, P.F. and Trabucchi, T. 1980. Brain neurotransmitter systems and chronic lead intoxication. *Pharmacology Research Communication* 12: 447–458.

Gray, R.H. 1990. Fish behavior and environmental assessment. *Environmental Toxicology and Chemistry* 9: 53–68.

Hamilton, S.J., Palmisano, A., Wedemeyer, G.A. and Yatsutake, W.T. 1986. Impacts of selenium on early lifestages and smoltification of fall chinook salmon. In *Transactions of the 51st Wildlife and Natural Resources Conference*, pp. 343–356. Wildlife Management Institute, Washington, DC.

Hansen, J.A., Marr, J.C.A., Lipton, J., Cacela, D. and Bergman, H.L. 1999a. Differences in neurobehavioral responses of chinook salmon (*Oncorhynchus tshawytscha*) and rainbow trout (*O. mykiss*) exposed to copper and cobalt: 1. Behavioral avoidance. *Environmental Toxicology and Chemistry* 18: 1972–1978.

Hansen, J.A., Rose, J.D., Jenkins, R.A., Gerow, K.G. and Bergman, H.L. 1999b. Chinook salmon (*Oncorhynchus tshawytscha*) and rainbow trout (*O. mykiss*) exposed to copper: 2. Neurophysiological and histological effects on the olfactory system. *Environmental Toxicology and Chemistry* 18: 1979–1991.

Hara, T.J. and Thompson, B.E. 1978. The reaction of whitefish, *Coregonus clupeaformis*, to the anionic detergent sodium lauryl sulphate and its effects on their olfactory responses. *Water Research* 12: 893–897.

Hara, T.J., Brown, S.B. and Evans, R.E. 1983. Pollutants and chemoreception in aquatic organisms. In *Aquatic Toxicology*, Nriagu, J.O. (ed.), pp. 248–306. John Wiley and Sons, New York.

Hartwell, S.I., Cherry, D.S. and Cairns, Jr, J. 1987. Field validation of avoidance of elevated metals by fathead minnows (*Pimephales promelas*) following in situ acclimation. *Environmental Contamination and Toxicology* 6: 189–200.

Hawryshyn, C.W., Mackay, W.C. and Nelson, T.H. 1982. Methylmercury induced visual deficits in rainbow trout. *Canadian Journal of Zoology* 60: 3127–3137.

Heath, A.G. 1995. *Water Pollution and Fish Physiology*, 2nd edn. Lewis Publishers, Boca Raton.

Henry, M.G. and Atchison, G.J. 1991. Metal effects on fish behavior-advances in determining the ecological significance of responses. In *Ecotoxicology of Metals: Current Concepts and Applications*. Newman, M.C. and McIntosh, A. (eds), pp. 131–144. Lewis Publishers, Boca Raton.

Holcombe, G.W., Fiandt, J.T. and Phipps, G.L. 1980. Effects of pH increases and sodium chloride additions on the acute toxicity of 2,4-dichlorophenol to the fathead minnow. *Water Research* 14: 1073–1077.

Jones, K.A., Hara, T.J. and Scherer, E. 1985. Locomotor responses by Arctic char (*Salvelinus alpinus*) to gradients of H^+ and CO_2. *Physiology and Zoology* 58: 413–420.

Keenleyside, M.H.A. 1979. Diversity and adaptation in fish behavior. In *Zoophysiology*, Vol. XI. Springer-Verlag, Berlin.

Kleerekoper, H. 1974. Effects of exposure to a subacute concentration of parathion on the interaction between chemoreception and water flow in fish. In *Pollution and Physiology in Marine Organisms*. Vernburg, F.J. and Vernberg, W.B. (eds), pp. 237–245. Academic Press, New York.

Kleerekoper, H. 1976. Effects of sublethal concentrations of pollutants on behavior of fish. *Journal of the Fisheries Research Board of Canada* 33: 2036.

Kleerekoper, H., Waxman, J.B. and Matis, J. 1973. Interaction of temperature and copper ions as orienting stimuli in the locomotor behavior of the gold fish (*Carassius auratus*). *Journal of the Fisheries Research Board of Canada* 30: 725–728.

Korver, R.M. and Sprague, J.B. 1989. Zinc avoidance by fathead minnows (*Pimephales promelas*): Computerized tracking and greater ecological relevance. *Canadian Journal of Fisheries and Aquatic Science* 46: 494–502.

Kotrschal, K. 1991. Solitary chemosensory cells – taste, common chemical sense or what? *Reviews in Fish Biology and Fisheries* 1: 3.

Laurence, G.C. 1972. Comparative swimming abilities of fed and starved larval largemouth bass (*Micropterus salmoides*). *Journal of Fish Biology* 4: 73.

Lawrence, M. and Scherer, E. 1974. Behavioural responses of whitefish and rainbow trout to drilling fluids. Fisheries Marine Service Research Developmental Technical Report 502, Freshwater Institute, Winnipeg, Manitoba.

Lemly, A.D. and Smith, R.J.F. 1987. Effects of chronic exposure to acidified water on chemoreception of feeding stimuli by fathead minnows, *Pimephales promelas*: Mechanisms and ecological implications. *Environmental Toxicology and Chemistry* 6: 225–238.

Little, E.E. 1983. Behavioral function of olfaction and taste in fish. In *Fish Neurobiology and Behavior*, Vol. I. Northcutt, G. (ed.), pp. 351–371 University of Michigan Press, Ann Arbor.

Little, E.E. 1990. Behavioral toxicology: Stimulating challenges for a growing discipline. *Environmental Toxicology and Chemistry* 9: 1–2.

Little, E.E. and Finger, S.E. 1990. Swimming behavior as an indicator of sublethal toxicity in fish. *Environmental Toxicology and Chemistry* 9: 13–19.

Little, E.E., Flerov, B.A. and Ruzhinskaya, N.N. 1985. Behavioral approaches in aquatic toxicity: A review. In *Toxic Substances in the Aquatic Environment: An International Aspect*. Mehrle, Jr, P.M., Gray, R.H. and Kendall, R.L. (eds), pp. 72–98. American Fisheries Society, Water Quality Section, Bethesda.

Little, E.E., Archeski, R.D., Flerov, B.A. and Kozlovskaya, V.I. 1990. Behavioral indicators of sublethal toxicity in rainbow trout. *Archives of Environmental Contamination and Toxicology* 19: 380–385.

Little, E.E., Fairchild, J.F. and DeLonay, A.J. 1993a. Behavioral methods for assessing the impacts of contaminants on early life stage fishes. In *American Fisheries Society Symposium 14 Water Quality and the Early Life Stages of Fishes*. Fuiman, L. (ed.), pp. 67–77. American Fisheries Society, Bethesda.

Little, E.E., Dwyer, F.J. and Fairchild, J.F., DeLonay, A.J. and Zajicek, J.L. 1993b. Survival and behavioral response of bluegill during continuous and pulsed exposures to the pyrethroid insecticide, ES-fenvalerate. *Environmental Toxicology and Chemistry* 12: 871–878.

Lorz, H.W. and McPherson, B.P. 1976. Effects of copper or zinc in freshwater on the adaptation to sea water and ATPase activity, and the effects of copper on migratory disposition of coho salmon (*Oncorhynchus kisutch*). *Journal of the Fisheries Research Board of Canada* 33: 2023–2030.

McKim, J., Bradbury, S. and Niemi, G. 1987. Fish acute toxicity syndromes and their use in the QSAR approach to hazard assessment. *Environmental Health Perspectives* 71: 171–186.

McNicol, R.E. and Scherer, E. 1991. Behavioral responses of lake whitefish (*Coregonus clupeaformis*) to cadmium during preference-avoidance testing. *Environmental Toxicology and Chemistry* 10: 225–234.

Marcucella, H. and Abramson, C.I. 1978. Behavioral toxicology and teleost fish. In *The Behavior of Fish and other Aquatic Animals*. Mostofsky, D.I. (ed.), pp. 33–77. Academic Press, New York.

Mathers, R.A., Brown, J.A. and Johansen, P.H. 1985. The growth and feeding behavior responses of largemouth bass (*Micropterus salmoides*) exposed to PCP. *Aquatic Toxicology* 6: 157–164.

Maule, A.G. and Schreck, C.B. 1990. Changes in numbers of leukocytes in immune organs of juvenile coho salmon after acute stress or cortisol treatment. *Journal of Aquatic Animal Health* 2: 298–304.

Mayer, Jr, F.L. and Ellersieck, M.R. 1986. Manual of acute toxicity: Interpretation and data base for 410 chemicals and 66 species of freshwater animals. United States Fish and Wildlife Service Resource Publication 160. Department of the Interior, Washington, DC.

Mazeaud, M.M., Mazeaud, F. and Donaldson, E.M. 1977. Stress resulting from handling in fish: primary and secondary effects. *Transactions of the American Fisheries Society* 106: 201.

Mehrle, P.M., Buckler, D.R., Little, E.E., Smith, L.M., Petty, J.D., Peterman, P.H., Stalling, D.L., DeGraeve, G.M., Coyle, J.J. and Adams, W.J. 1988. Toxicity and bioconcentration of 2,3,7,8-tetrachlorodibenzodioxin and tetrachlorodibenzofuran in rainbow trout. *Environmental Toxicology and Chemistry* 7: 47.

Miller, D.C., Lang, W.H., Graeves, J.O.B. and Wilson, R.S. 1982. Investigations in aquatic behavioral toxicology using a computerized video quantification system. In *Aquatic Toxicology and Hazard Assessment: Fifth Conference*, STP 766. Pearson, J.G., Foster, R.B. and Bishop, W.E. (eds), pp. 206–220. American Society for Testing and Materials, Philadelphia, PA.

Nichols, J.W., Wedemeyer, G.A., Mayer, F.L., Dickhoff, W.W., Gregory, S.V., Yatsutake, W.T. and Smith, S.D. 1984. Effects of freshwater exposure to arsenic trioxide on the parr–smolt transformation of coho salmon. *Water Research* 1: 143–149.

NRDA (Natural Resource Damage Assessments). 1986. Final rule. Federal Register 51: 27674–27753.

Ogilivie, D.M. and Anderson, J.M. 1965. Effect of DDT on temperature selection by young Atlantic salmon, *Salmo salar*. *Journal of the Fisheries Research Board of Canada* 22: 503–512.

Oliver, J.D., Holeton, G.D. and Chua, K.E. 1979. Overwintering mortality of fingerling smallmouth bass in relation to size relative energy stores and environmental temperature. *Transactions of the American Fisheries Society* 108: 130–136.

Olla, B.L., Pearson, W.H. and Studholme, A.L. 1980. Applicability of behavioral measures in environmental stress assessment. *Rapports et Procès-Verbaux des Réunions Commission Internationale pour l'Exploration Scientifique de la Mer Mediterranée Monaco* 179: 162–173.

Ostrander, G., Anderson, J., Fisher, J., Landholt, M. and Kocan, R. 1990. Decreased performance of rainbow trout, *Oncorhynchus mykiss* emergence behaviors following embryonic exposure to benzo [a] pyrene. *Fisheries Bulletin* 85: 551–555.

Panigrahi, A.K. and Misra, B.N. 1978. Toxicological effects of mercury on a freshwater fish, *Anabas scandens* and their ecological implications. *Environmental Pollution* 16: 31–39.

Peterson, R.H. 1976. Temperature selection of juvenile Atlantic salmon, *Salmo salar*, as influenced by various toxic substances. *Journal of the Fisheries Research Board of Canada* 33: 1722–1730.

Pickering, A.D. (ed.) 1981. *Stress and Fish*. Academic Press. New York.

Pickering, A.D. and Pottinger, T. 1989. Stress responses and disease resistance in salmonid fish: effect of chronic elevation of plasma cortisol. *Fish Physiology and Biochemistry* 7: 253–258.

Pickering, Q., Brungs, W. and Gast, M. 1977. Effect of exposure time and copper concentration on reproduction of the fathead minnow (*Pimephales promelas*). *Water Research* 11: 1079–1083.

Poels, C.L.M. 1977. An automatic system for rapid detection of acute high concentrations of toxic substances in surface water using trout. *American Society for Testing and Materials Special Technical Publication* 607: 85–95.

Purves, W.K., Orians, G.H. and Heller, H.C. 1992. *Life: The Science of Biology*, pp. 815–844. Sinauer Associates, Sunderland, MA.

Rand, G.M. 1985. Behavior. In *Fundamentals of Aquatic Toxicology: Methods and Applications* Rand, G.M. and Petrocelli, S.R. (eds), pp. 221–256. Hemisphere Publishing, New York.

Reynolds, W.W. 1977. Temperature as a proximate factor in orientation behavior. *Journal of the Fisheries Research Board of Canada*. 34: 734–739.

Reynolds, W.W. and Thomson, D.A. 1974. Responses of young gulf grunion, *Leuresthes sardina*, to gradients of temperature, light, turbulence and oxygen. *Copeia* 1974: 747–758.

Rice, P.J., Drews, C.D., Klubertany, T.M., Bradbury, S.P. and Coats, J.R. 1997. Acute toxicity and behavioral effects of chlorophyrophos, permethrine, phenol, strychnine and 2,4,-dinitro phenol to 30 day old Japanese medaka (*Oryzias atipes*). *Environmental Toxicology and Chemistry* 16: 696–704.

Richmonds, C. and Dutta, H. 1992. Effect of malathion on the optomotor behavior of bluegill sunfish, *Lepomis macrochirus*. *Comparative Biochemistry and Physiology* 102C: 523–526.

Sandheinrich, M.B. and Atchison, G.J. 1990. Sublethal toxicant effects on fish foraging behavior: Empirical vs mechanistic approaches. *Environmental Toxicology and Chemistry* 9: 107–120.

Sapolsky, R., Rivier, C., Yamamoto, G., Plotsky, P. and Vale, W. 1987. Interleukin-1 stimulates the secretion of hypothalamic corticotropin releasing factor. *Science* 238: 522–523.

Satchell, G.H. 1991. *Physiology and Form of Fish Circulation*, Ch. 5. Cambridge University Press, Cambridge.

Scherer, E. 1971. Effects of oxygen depletion and carbon dioxide build-up on the photic behavior of the walleye (*Stizostedion vitreum vitreum*). *Journal Fisheries Research Board of Canada* 28: 1303–1307.

Scherer, E. 1992. Behavioral responses as indicators of environmental alterations: approaches, results, developments. *Journal of Applied Ichthyology* 8: 122–130.

Scherer, E. and Harrison, S.E. 1979. The optomotor response test. In *Toxicity Tests for Freshwater Fish*. Canada Special Publication for Fisheries Aquatic Science 44. Scherer, E. (ed.), p. 179. Freshwater Institute, Winnipeg, Canada.

Scherer, E. and McNicol, R.E. 1998. Preference-avoidance responses of lake whitefish (*Coregonus clupeaformis*) to competing gradients of light and copper, lead and zinc. *Water Research* 32: 924–929.

Scherer, E,. Armstrong, F.A.J. and Nowak, S.H. 1975. Effects of mercury-contaminated diet upon walleyes, *Stizostedion vitreum vitreum* (Mitchill). Environment Canada. Fisheries and Marine Services Developmental Report No. 597.

Schreck, C.B. and Bradford, C.S. 1990. Interrenal corticosteroid production: potential regulation by the immune system in the salmonid. *Progress in Clinical and Biological Research* 342: 480–486.

Schreck, C.B. and Lorz, H.W. 1978. Stress response of coho salmon (*Oncorhynchus kisutch*) elicited by cadmium and copper and potential use of cortisol as an indicator of stress. *Journal of Fisheries Research Board of Canada* 35: 1124–1129.

Schröder, J.H. and Peters, K. 1988. Differential courtship activity of competing guppy males (*Poecilia reticulata* Peters; Pisces: Poecilidae) as an indicator for low concentrations of aquatic pollutants. *Bulletin Environmental Contamination and Toxicology* 40: 385–390.

Schumacker, P.D. and Ney, J.J. 1980. Avoidance response of rainbow trout (*Salmo gairdneri*) to single dose chlorination in a power plant discharge canal. *Water Research* 14: 651–655.

Selye, H. 1950. Stress and the general adaptation syndrome. *British Medical Journal* 1: 1383–1392.

Smith, E.H. and Bailey, H.C. 1988. Development of a system for continuous biomonitoring of a domestic water source for early warning of contaminants. In *Automated Biomonitoring: Living Sensors as Environmental Monitors* Gruber, D.S. and Diamond, J.M. (eds), pp. 182–205. Ellis Harwood Publishers Ltd, Chichester, UK.

Smith, R.F.J. and Lawrence, B.J. 1988. Effects of acute exposure to acidified water on the behavioral response of fathead minnows, *Pimephales promelas*, to alarm substance (Shreckstoff). *Environmental Toxicology and Chemistry* 7: 329–355.

Sprague, J.B. 1964a. Avoidance of copper–zinc solutions by young salmon in the laboratory. *Journal of the Water Pollution Control Federation* 36: 990–1004.

Sprague, J.B. 1964b. Lethal concentrations of copper and zinc for young Atlantic salmon. *Journal of the Fisheries Research Board of Canada* 21: 17–23.

Sprague, J.B. 1968. Avoidance reactions of rainbow trout to zinc sulfate solutions. *Water Research* 2: 367–372.

Steele, C.W. 1983. Effects of exposure to sublethal copper on the locomotor behavior of the sea catfish, *Arius felis*. *Aquatic Toxicology* 4: 83–93.

Strange, R.J., Schreck, C.B. and Golden, J.T. 1977. Corticoid stress response to handling and temperature in salmonids. *Transactions of the American Fisheries Society* 106: 213–218.

Sullivan, C.M. and Fisher, K.C. 1942. Temperature selection and effects of light and temperature on movements in fish. *American Physiology Society Federation Proceedings* 6: 213–219.

Sullivan, C.M. and Fisher, K.C. 1954. The effects of light on temperature selection in speckled trout, *Salvelinus fontinalis* (Mitchill). *Biology Bulletin of the Marine Biology Laboratory Woods Hole* 107: 278–288.

Sutterlin, A.M. 1974. Pollutants and the chemical senses of aquatic animals – perspective and review. *Chemical Senses and Flavor* 1: 167–178.

Thomas, P. and Lewis, D.H. 1987. Effects of cortisol on immunity in red drum, *Sciaenops ocellatus*. *Journal of Fish Biology* 31(Suppl. A): 123–127.

Timms, A.M., Kleerekoper, H. and Matis, J. 1972. Locomotor response of goldfish, channel catfish, and largemouth bass to a 'copper-polluted' mass of water in an open field. *Water Resources Research* 8: 1574–1580.

Waiwood, K.G. and Beamish, F.W.H. 1978. The effects of copper, pH, and hardness on the critical swimming performance of rainbow trout, *Salmo gairdneri*. *Water Research* 12: 611–619.

Warner, R.E., Peterson, K.K. and Borgman, L. 1966. Behavioral pathology in fish: a quantitative study of sublethal pesticide toxication. *Journal of Applied Ecology* 3: 223–247.

Weber, D.N. 1993. Exposure to sublethal levels of waterborne lead alters reproductive behavior patterns in fathead minnows (*Pimephales promelas*). *Neurotoxicology* 14: 347–358.

Weber, D.N., Russo, A., Seale, D.B. and Spieler, R.E. 1991. Waterborne lead affects feeding abilities and neurotransmitter levels of juvenile fathead minnows (*Pimephales promelas*). *Aquatic Toxicology* 21: 71–80.

Weeks, B.A., Anderson, D.P., DuFour, A.P., Fairbrother, A., Goven, A.J., Lahvis, G.P. and Peters, G. 1992. Immunological biomarkers to assess environmental stress. In *Biomarkers: Biochemical, Physiological, and Histological Markers of Anthropogenic Stress*. Huggett, R.J., Kimerle, R.A., Merhle, P.M. and Bergam, H.L. (eds), Ch. 5. Lewis Publishers, Boca Raton.

Werner, E.E. and Hall, D.J. 1974. Optimal foraging and the size selection of prey of the bluegill sunfish (*Lepomis macrochirus*). *Ecology* 55: 1042–1052.

Westlake, G.F. 1984. Behavioral effects of industrial chemicals in aquatic animals. In *Hazard Assessment of Chemicals: Current Developments*, Vol. III. Saxena, J. (ed.), pp. 233–250. Academic Press, Orlando.

Westlake, G.F., Kleerekoper, G.H. and Matis, J. 1974. The locomotor response of goldfish to a steep gradient of copper ions. *Water Resources Research* 10: 103–105.

Whitear, M. 1992. Solitary chemosensory cells. In *Fish Chemoreception*. Hara, T.J. (ed.), p. 103. Chapman & Hall, London.

Wiebe, J.P., Salhanick, A.I. and Myers, K.I. 1983. On the mechanism of action of lead in the testis: *in vitro* suppression of FSH receptors, cyclic AMP and steroidogenesis. *Life Sciences* 32: 1997–2005.

Woodward, D.F., Little, E.E. and Smith, L.M. 1987. Toxicity of five shale oils to fish and aquatic invertebrates. *Archives of Environmental Contamination and Toxicology* 16: 239–246.

Woodward, D.F., Goldstein, J.N. and Farag, A.M. 1997. Cutthroat trout avoidance of metals and conditions characteristic of a mining waste site: Coeur d'Alene River, Idaho. *Transactions of the American Fisheries Society* 126: 699–706.

5 Toxic responses of the reproductive system

Lisa D. Arcand-Hoy and William H. Benson

Introduction

Fish reproduction is arguably one of the most sensitive indicators of exposure to sublethal concentrations of environmental chemicals, and reproductive toxicity may simply be defined as an adverse effect on reproductive success. Alterations in reproductive output may occur in the form of altered or decreased matings or spawnings, reduction in fertility/fecundity, decreased hatching success and lowered larval survival. Typically, reproductive toxicity is directly associated with sublethal effects of chemical contaminants to feral fish populations, but indirect effects such as stress incurred as a result of chemical exposure are also of similar concern. Reproductive endpoints used to assess fish reproduction can include decreases in fertility and fecundity, egg size and number, successful hatch, developmental abnormalities, and alteration of reproductive hormones, among others. Reproductive toxicity may occur during larval, juvenile, or adult stage depending upon the species examined. Exposure at an early life stage may lead to alterations in key developmental processes (e.g. sexual differentiation), as well as increased susceptibility to chemical insult as adults. Exposure at maturity could also disrupt normal reproductive physiology. An effect on endocrine function, such as alterations in reproductive hormones which may affect mating behavior and other vital social behaviors as well as the above-mentioned egg and larval viability, may be included in the underlying factors leading to reproductive toxicity.

There are a number of ways in which chemical substances can interfere with fish reproduction. To date, a large number of reproductive studies associated with the 'endocrine disrupter' issue have examined estrogen receptor-mediated responses and adverse changes as a result of chemical interference with hypothalamic–pituitary function. Other hormones, such as thyroid hormones, are also under hypothalamic–pituitary control and may play a significant role in reproductive processes. Thyroid hormones in fish are often elevated during gonadal development and latter stage reproductive processes (e.g. gametogenesis, ovulation, spermiation). Growth hormone is also believed to play an important role in fish reproduction. Growth hormone is produced in the pituitary and is sometimes referred to as a co-gonadotropin (Scanes and Harvey, 1995). Complex communication systems also exist between endocrine and immune systems. It has been hypothesized that exposure to endocrine-disrupting chemicals may result in immunosuppression or altered immune cellular function, which, in turn, may lead to adverse reproductive effects. Other steroid hormones that regulate reproductive processes include androgens and progestins. These hormones are essential to reproductive processes in male as well as in female fish. Also, there are many non-endocrine-mediated effects (e.g. oxidative stress, bioenergetics) that could play an important role in reproduction. This chapter will serve first to outline the reproductive physiology of teleost systems, followed by an examination of reproductive toxicity in fish, and, last, it will conclude with an assessment of reproductive function, with emphasis on the use of a Japanese medaka (*Oryzias latipes*) reproduction model as described previously (Arcand-Hoy and Benson, 1998).

Hypothalamus–pituitary–gonadal axis

The hypothalamus controls the synthesis and release of hormones and chemical messengers resulting from neural stimulation from the central nervous system. Messenger substances that are synthesized in the hypothalamic nerve cells control the synthesis and release of pituitary hormones. There are a number of pituitary hormones, including adrenocorticotropin, prolactin, growth hormone, insulin-like growth factor, thyroid-stimulating hormone, and gonadotropin. Gonadotropin is the hormone principally involved in reproductive processes. The focus of this section will primarily address the role of gonadotropins in the fish reproductive cycle, although some attention will be given to other pituitary hormones that are believed to play a pivotal role in reproductive processes (i.e. thyroid hormone, growth hormone).

Gonadotropin

Gonadotropin is the catalyst of reproduction. Its role in steroid activation and regulation of reproductive processes is outlined in Figure 5.1. Two forms of gonadotropin have been isolated in fish (Kawauchi, 1989; Swanson *et al.*, 1991): GtH I and GtH II, which are analogous to mammalian follicle-stimulating hormone (FSH) and luteinizing hormone (LH), respectively (Redding and Patino, 1993). GtH I is typically involved in gametogenesis and steroidogenesis, whereas GtH II is typically involved in the final maturation stages of gametogenesis. Despite the difference with regards to the role of GtH I and GtH II, the gonadotropins are known to be responsible for stimulating the synthesis of sex steroids (i.e. androgens, estrogens, and progestins), which, in turn, act on the target tissues to regulate gametogenesis, reproduction, sexual phenotype, and behavioral characteristics.

There are a number of hormones that act as stimulating agents for the release of GtH II. These include gonadotropin-releasing hormone (GnRH), neuropeptide Y, norepinephrine, serotonin, activin/inhibin, nicotine, and cholecystokinin (Van Der Kraak *et al.*, 1997). Less is known about the stimulators for the release of GtH I (Van Der Kraak *et al.*, 1997), although it is probable that, like GtH II, the major stimulator and regulator of its release from the gonadotrophs is GnRH.

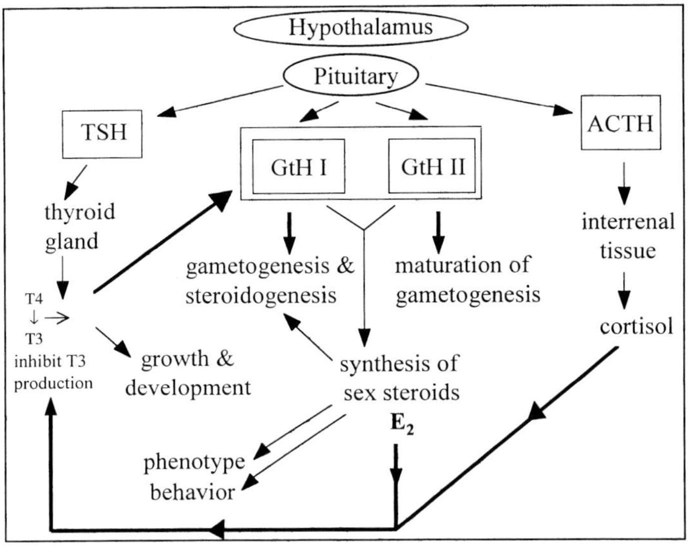

Figure 5.1 Interrelationship of immune–neuroendocrine system axes.

GnRH analogues have been used to induce ovulation in a number of fish species of economic and commercial importance (Mylonas *et al.*, 1992). GnRH is released from neuronal cell bodies via hypothalamic neurons which, in turn, stimulate the release of gonadotropins from the pituitary gonadotrophs. The most prominent inhibitor of this process is dopamine. Dopamine is able to inhibit the ability of GnRH to exert a GtH II response as well as, in some cases, reduce secretion of GtH II at the pituitary (Van Der Kraak *et al.*, 1997). The inhibitory nature of dopamine is important to endogenous feedback systems, whereby levels of GtH II can be elevated or decreased depending upon the reproductive cycle of the fish.

In the majority of fish studied, the development of oogenesis is controlled by GtH I. Circulating GtH I increases during early oocyte development and binds to receptors on the thecal and granulosa layer of the follicle. The thecal cells synthesize testosterone and allow for aromatization to result in the formation of estradiol in the granulosa layer prior to plasma secretion (Redding and Patino, 1993; Cyr and Eales, 1996). Subsequently, plasma estradiol will bind to the estrogen receptor and initiate a cascade of events resulting in the production of vitellogenin, a precursor to egg yolk protein produced in the liver. Vitellogenin is released from the liver into the blood and binds to receptors on the oocyte which incorporate the protein. As the development of the oocyte continues, levels of GtH I begin to decrease and are replaced by increasing levels of GtH II. Receptors for GtH II are found predominantly on the granulosa cells and binding stimulates the synthesis and release of progestins (Redding and Patino, 1993). Progestins play a role in gamete maturation and stimulate ovulation.

In male fish, GtH I is typically elevated throughout spermatogenesis and decreases at time of spawning, whereas GtH II is typically low throughout the growth process and is elevated at spawning (Nagahama, 1994). The gonadotropins stimulate proliferation of spermagonia as well as the synthesis of androgens required for gametogenesis and development of secondary sex characteristics. Androgen synthesis typically takes place in the Leydig cells. The type of androgen synthesized is dependent upon the species and developmental stage, but may include testosterone, 11-ketotestosterone, and/or androstenedione (Redding and Patino, 1993). To date, it is unclear as to which of these steroids is the predominant and most physiologically important androgen in teleost species; the assumption has been that 11-ketotestosterone is the predominant androgen. However, a study by Thomas *et al.* (2000) indicated that testosterone binds to the androgen receptor of Atlantic croaker (*Micropogonias undulatus*) with higher affinity than 11-ketotestosterone, and may, in fact, be the predominant androgen. Nevertheless, a decline in androgen levels and a sharp increase in progestins during spawning is primarily attributed to elevated levels of GtH II. A high level of progestins with a decline in androgen level are needed for spermiation. A more complete review of oogenesis and spermatogenesis will follow. The reader is also referred to review articles by Nagahama (1994), Nagahama *et al.* (1995) and Jobling (1995).

Thyroid hormones and reproduction

Similar to the steroid hormones, thyroid hormones are under control of the hypothalamus and pituitary. Thyroid-stimulating hormone acts on the thyroid gland and signals the synthesis and release of thyroid hormone. To date, thyroid hormone in fish has been characterized as thyroxine (T_4), which is metabolized to triiodothyronine (T_3) by means of enzymatic deiodination by iodothyronine 5′-monodeiodinase type 1 (5′-ID1). T_3 appears to be more biologically active, having a greater affinity for the receptor than T_4 (Cyr and Eales, 1996). The mechanistic action of T_3 in fish is largely unexplored, but mammalian studies suggest that T_3 binds with nuclear receptors, creating a T_3–receptor complex which, in turn, binds to a thyroid response element to initiate DNA transcription (Cyr and Eales, 1996). A major role of thyroid hormone in fish is regulation of growth and development. However, in a large majority of teleost species examined, there is an association between the thyroid hormones and reproduction. In most fish studied to date, thyroid activity increased during early gonadal development, remained elevated during the reproductive cycle, and decreased after spawning (Cyr and Eales, 1996). This relationship does not directly indicate that thyroid hormones influence reproductive function, but does suggest that there is a potential for such interactions to occur.

A number of studies have suggested that thyroid hormones may play a role in reproductive processes. More specifically, thyroid hormones are believed to play a role in oocyte growth and maturation. For example, Cyr and Eales (1996) have demonstrated that T_3 acts synergistically with gonadotropin to stimulate steroidogenesis in some female fish. These investigators reported that T_3 enhanced estradiol secretion during vitellogenesis; a gonadotropin-mediated process. Similar findings were reported by Sullivan *et al.* (1989), who showed that T_3 enhanced gonadotropin-stimulated ovulation in rainbow trout (*Salmo gairdneri*). In addition, receptors for T_3 have been found in ovarian tissue of some teleost species (Chakraborti *et al.*, 1986). Sullivan *et al.* (1989) suggested that thyroid hormones could influence the pituitary or ovary to enhance secretion of gonadotropins or GnRH. Furthermore, Cyr and Eales (1996) suggested structural similarities between thyroid-stimulating hormone and gonadotropins. A role for T_3 in male fish has also been reported whereby the inhibition of thyroid hormone resulted in an inhibition of testicular growth in developing male fish (Cyr and Eales, 1996).

Tagawa and Hirano (1991) examined the biologic significance of thyroid hormones in eggs. Female Japanese medaka were treated with thiourea in order to deplete maternal T_3 and T_4 levels and subsequent transfer of thyroid hormones to the egg and developing embryo. Analysis of T_3 and T_4 levels in the eggs after maternal thiourea treatment revealed a 90 percent decline in egg thyroid hormones over control eggs from untreated females. However, despite this decline in thyroid levels, there was no apparent decrease in or adverse effect on larval survival, size, or hatchability. It therefore appears that thyroid hormones are not necessary for early larval survival. It remains to be seen whether adverse effects associated with low levels of thyroid hormones could affect development in mature fish (Tagawa and Hirano, 1991). Continued examination of thyroid hormones in relation to reproductive function in fish exposed to compounds which interfere with thyroidal processes in the laboratory will ultimately provide valuable information concerning the relationship between reproduction and thyroid hormones.

Growth hormone and reproduction

Growth hormone has long been known for its osmoregulatory role in salmonid migration, although the mechanism has not been clearly defined. Growth hormone has been isolated from the pituitary of a number of fish species (see Degani *et al.*, 1996). Although much growth hormone research has been conducted with mammals, there is an increasing body of evidence that growth hormone plays a role in fish reproduction, and, more specifically, gonadal function. Studies by Van Der Kraak (1990) with goldfish (*Carassius auratus*) have shown that growth hormone is released from the pituitary with administration of GnRH I and GnRH II. Work by Degani *et al.* (1996) showed that fish growth and gonadal development is paralleled by an increase in growth hormone in the pituitary and plasma of female carp. For example, at maturity, there was a decrease in plasma levels of growth hormone, while pituitary levels remained unchanged. The investigators concluded that prior to maturation growth hormone plays a role in growth, but at maturity the role of growth hormone switches to that of steroid secretion. This, along with other supporting investigations, suggests that growth hormone plays a role in gonadal growth and that the role of this substance changes at maturity (Degani *et al.*, 1996).

Sakata *et al.* (1993) have reported finding high-affinity low-capacity growth hormone binding sites in rainbow trout testes. Research by Marchant and Peter

(1989) and Lin *et al.* (1993) have shown the GnRH stimulates growth hormone release in cyprinid species. However, Blaise *et al.* (1995) failed to show that GnRH played a role in the secretion of growth hormone in rainbow trout, demonstrating that a different mechanism may exist for cyprinid and salmonid species. In addition, Breton *et al.* (1993) previously demonstrated that in cyprinids dopamine influenced GnRH-induced release of gonadotropin. This is not the case with other teleosts, in particular salmonid species (Breton *et al.*, 1993).

Mammalian studies have demonstrated that growth hormone can influence estrogen receptor population in liver tissue (Scanes and Harvey, 1995). For example, work by Norstedt *et al.* (1981) showed that in ovariectomized rats hepatic estrogen receptors were reduced following hypophysectomy, whereas treatment with growth hormone was able to reverse this effect. Similar results were found in turtles (Riley *et al.*, 1987). In fish studies using hypophysectomized male and female killifish (*Fundulus heteroclitus*), treatment with growth hormone in the form of a replacement therapy resulted in an increase in testicular weight and plasma testosterone levels in male fish (Singh *et al.*, 1988). Female fish showed elevated plasma estradiol in response to growth hormone treatment (Singh *et al.*, 1988). Research by Degani *et al.* (1996) demonstrated that sex steroids stimulated the secretion of growth hormones. The mechanisms governing the action of growth hormone are not currently known.

Gonadal development

Sexual determination/differentiation

In most fish, sex determination is under genetic control; sex is determined by sex chromosomes after fertilization. In some fish, autosomes (pairs of chromosomes with sex-modifying genes) determine the sex, but this is less common (Jobling, 1995). Sexual differentiation in fish is believed to be similar to mammalian systems, whereby the presence or absence of a 'testis-determining factor' directs male or female differentiation (Jobling, 1995).

In teleosts, primordial germ cells develop exterior to the gonadal region and then undergo migration to the gonad. The development and positioning of the primordial germ cells is an important marker of sexual differentiation, which is a species- as well as sex-dependent process. In most cases, the differentiation of the gonad occurs in females before males. This is usually determined by enumerating the number of primordial germ cells entering or undergoing meiosis (Figure 5.2). For example, sexual differentiation in Japanese medaka occurs at hatching, and in female Japanese medaka germ cells are in a proliferative state. Satoh and Egami (1972) demonstrated that sexual differentiation of germs cells occurs near the time of hatching, with germ cells of female Japanese medaka entering into meiotic prophase within 24 hours of hatching. Germs cells of the male stop proliferating after hatching (Satoh and Egami, 1972). Kanamori *et al.* (1985) reported that sexual differences in the medaka were identifiable in the male medaka 10 days after fertilization; however, the actual tubular structure of

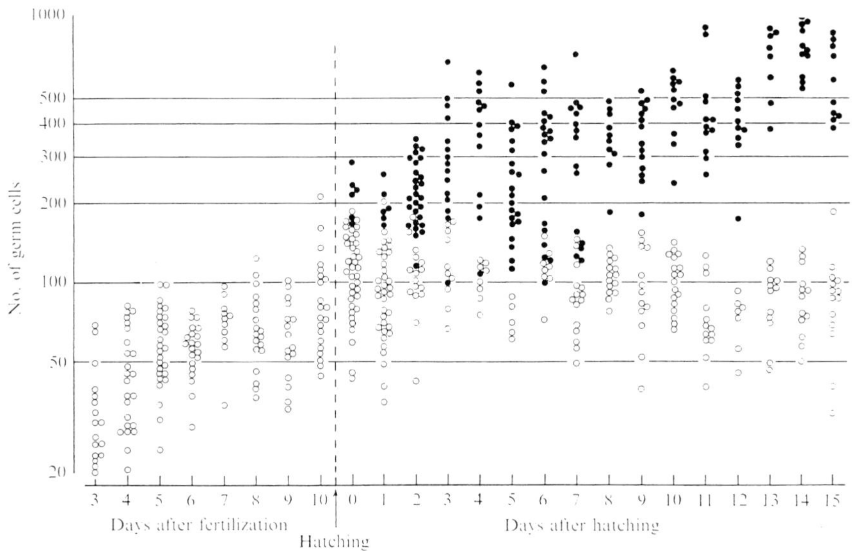

Figure 5.2 Changes in the number of germ cells in the Japanese medaka, *Oryzias latipes*, during normal embryonic development (modified from Satoh and Egami, 1972).

the testis was not visible until the fish was 15–20 mm in length, which corresponds to the time when meiosis should occur (Kanamori *et al.*, 1985).

In some species, sexual differentiation can be marked by the presence of an ovarian or testicular cavity (Nagahama, 1983). In rainbow trout, the gonads are not differentiated until 28 days post-hatch, although there are variable reports in female rainbow trout and indications of differentiation have been observed as early as 18 days post-hatch (Vanden Hurk and Sloff, 1981). The pattern of germ cell development in rainbow trout is similar to that of Japanese medaka. Le Brun *et al.* (1982) mapped out germ cell development in male and female rainbow trout. Similar to Japanese medaka, immediately after hatching, the number of germ cells were very low. However, at 5 weeks post-hatch there appeared to be germ cells undergoing meiosis and the number of germ cells markedly increased. Le Brun *et al.* (1982) reported that at 10 weeks the ovary was developed and contained meiotic oocytes. Again, similar to male Japanese medaka, the male rainbow trout remains undifferentiated until the onset of maturity (approximately 2 years of age).

Testicular development and spermatogenesis

The most common testicular structure of the teleost testis is the elongated paired tubule system attached to the dorsal wall of the organism. The main sperm duct (vas deferens) rises from this structure and leads to the opening between the urinary and anal openings known as the urogenital papilla. This lobular testicular structure is most common, although some fish species have a tubular testicular design. In the former design, the lobules are separated by connective tissue. Within the lobules are primary spermatogonia that undergo mitosis to form cysts. The cysts are formed by Sertoli cells and are the structural compartments which hold the spermatogonial cells that undergo maturation in the form of spermatogenesis. Briefly, this involves the development of primary spermatocytes which undergo a meiotic division to spermatocytes. An additional division leads to the production of spermatids which differentiate into spermatozoa. The spermatozoa, in turn, undergo a process known as spermiogenesis, in which nuclear and cytoplasmic organization occur, followed by the development of a flagellum (Nagahama, 1983). Figure 5.3 outlines the different stages of spermatogenesis in the Japanese medaka testis. As shown in Figure 5.3, the degree of spermatogenesis increases from the perimeter of the testis to the interior sperm collection ducts. When maturation is complete, the cysts rupture and the mature sperm enter the testicular lumen and travel to the connecting sperm duct.

Ovarian development and oogenesis

Typically, the teleost ovary consists of a paired cavity, where the primary function is support and containment of developing oocytes. Fish have numerous reproductive strategies and the appearance of oocytes and patterns of growth and development within the ovary reflect these different breeding styles. Terminology has been developed to express the different reproductive strategies and oocyte development. The three categories include: synchronous, group synchronous, and asynchronous. Asynchronous refers to species in which oocytes are found in all stages of development. These fish (i.e. Japanese medaka and goldfish) are capable of spawning daily for many months of the year. Synchronous refers to ovaries in

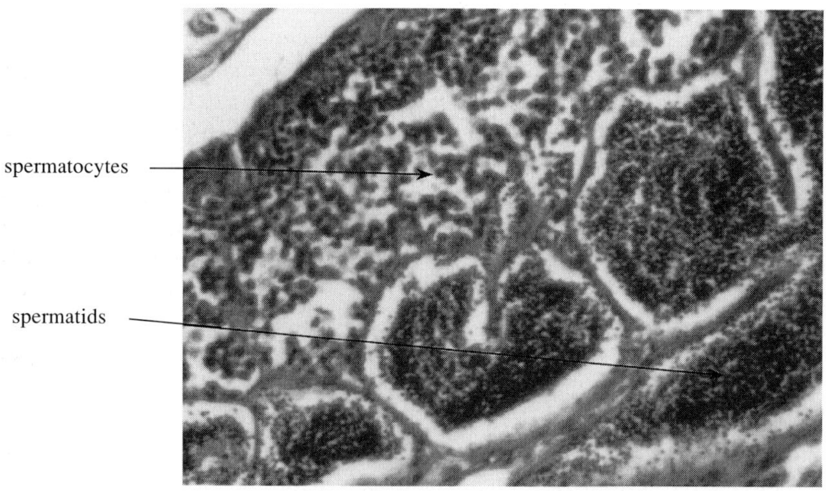

Figure 5.3 Stages of spermatogenesis in the testis of normal male Japanese medaka, *Oryzias latipes*.

which the oocytes are all in the same stage of development. This is usually reserved for species that undergo annual migrations and one-time spawnings. Group synchronous refers to ovarian oocytes that are grouped into two different stages of development. This is indicative of fish with relatively short annual breeding cycles (e.g. rainbow trout).

Oogenesis proceeds similarly to that of spermatogenesis, whereby oocytes undergo a number of mitotic divisions. The oocytes arrest at the first meiotic prophase and enter a period of growth (Nagahama, 1983). There are a number of defined stages of oogenesis. Criteria developed for rainbow trout are presented in Nagahama (1983). The egg stage is identified by oocyte size, nuclear and nucleolar appearance, and type and location of cytoplasmic inclusion (Nagahama, 1983). Figure 5.4 presents select stages of oocyte growth in female Japanese medaka, as developed by Yamamoto (1975). Ovarian follicles consist of two cell layers: the thecal and granulosa cell layers. Briefly, as oocyte growth progresses, the nucleus increases in size and becomes surrounded by nucleoli. As the oocyte continues to grow, a small dense circular body, known as a yolk nucleus, can be seen. Yolk nuclei contain a number of cellular organelles. The longer term growth period known as the vitellogenic period follows, which leads to yolk accumulation and an increase in oocyte size.

After oocyte growth, meiosis continues with the process of oocyte maturation. Gonadotropin acts at the thecal and granulosa cells and stimulates the mechanisms of steroidogenesis. Gonadotropin triggers an increase in estradiol synthesis and secretion, followed by subsequent vitellogenin production. Nagahama *et al.* (1994) proposed a two-cell model of steroidogenesis (in this case, estradiol production). Briefly, the thecal cell layer under the control of gonadotropin secretes a male steroid substrate (testosterone) that is transported to the granulosa cell layer and

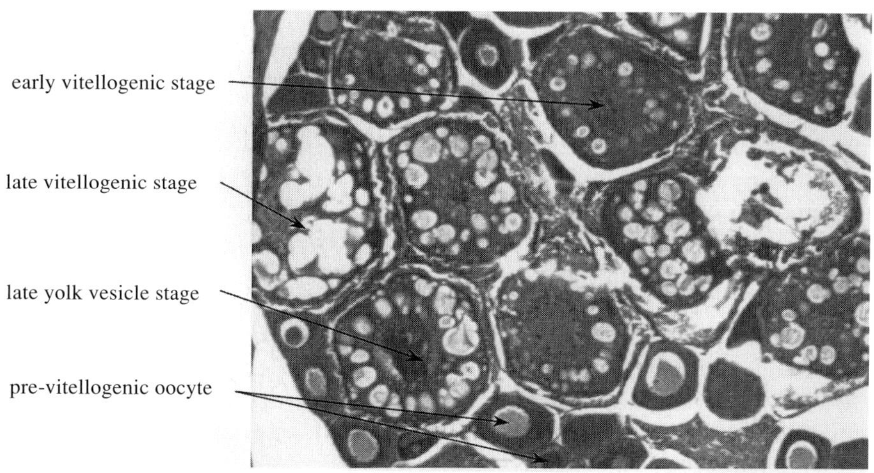

Figure 5.4 Stages of oocyte growth in the gonad of normal female Japanese medaka, *Oryzias latipes*.

converted to estradiol by the aromatizing enzyme 20β-hydroxysteroid dehydrogenase (20β-HSD). Action of gonadotropin at the follicle layer occurs by means of a receptor-mediated mechanism. During oogenesis, there is a rise in the number of gonadotropin receptors in both the thecal and granulosa layers (Kamamori and Nagahama, 1988). In addition, as the amount of gonadotropin increases, so does the level of estradiol.

Oocyte growth and maturation

Vitellogenesis: synthesis, transport, uptake, and metabolism

Vitellogenesis refers to the period of oocyte growth when the egg accumulates egg yolk nutrients for the developing embryo and larvae. In oviparous vertebrates, the synthesis of vitellogenin is controlled by endogenous estradiol, through an estrogen receptor-mediated event. The estrogen receptor is part of the nuclear receptor family, which also includes thyroid hormone, retinoic acid and vitamin D receptors (Evans, 1988). Briefly, binding of estrogen to the estrogen receptor leads to dissociation of the receptor complex. Phosphorylation and dimerization follows with subsequent binding of the dimer with the estrogen response element (ERE) of DNA, which, in turn, triggers gene transcription and translation, leading to the production of vitellogenin and other gene products. Vitellogenin, the precursor to egg yolk protein, is synthesized in the liver and transferred via the blood to developing oocytes.

Transport and uptake

Vitellogenin is synthesized in the liver and is transferred in plasma to the developing oocytes. Transport of vitellogenin in the plasma of fish appears to be free of any carrier molecules (Mommsen and Walsh, 1988). Much work has been carried out on receptor recognition and oocyte uptake in chickens, but, to date, research in fish is limited. In *Xenopus* and the chicken, vitellogenin binds to receptors on the oocyte membrane and is taken up by pinocytosis into the oocyte and is transferred in microvesicular bodies before being degraded to lipovitelin and phosvitin (see Mommsen and Walsh, 1988). The receptors appear to recognize the phosvitin region of the vitellogenin molecule. Homologous very low density lipoprotein (VLDL) receptors are found in chickens. In chickens, estradiol triggers the synthesis of VLDL by activation of the *apoVLD* gene. In fish, the production of VLDL is activated by estradiol and estrone and is positively correlated with vitellogenin. In addition, the ovary of fish and birds has been shown to be capable of vitellogenin uptake directly from the bloodstream. However, it is believed that the vitellogenin and its receptor are taken up into the oocyte by micropinocytosis. Once taken up by the oocyte, proteolytic enzymes cleave vitellogenin into lipovitelin and phosphovitins. These compounds are stored in the yolk bodies of the oocyte (Mommsen and Walsh, 1988). In addition, vitellogenin has significant ion-binding properties which may, upon entry into the oocyte, serve to bring in necessary minerals to the growing oocyte (Mommsen and Walsh, 1988). Glycogen, carotenoids, lectins, sialoglycoproteins, wax esters and sterol esters are also taken up into the oocyte at this time (Mommsen and Walsh, 1988). The role of hormones in the regulation of vitellogenin uptake is largely unknown.

Oocyte maturation

Following oocyte growth and the accumulation of egg yolk, an oocyte maturation phase occurs in which hormonally controlled meiotic divisions take place. Readers are referred to Nagahama *et al.* (1994) for a more complete review of these processes. Briefly, the first stage of meiosis involves the breakdown of the germinal vesicle and chromosome condensation. Also occurring are the formation of the first meiotic spindle and 'extrusion' of the first polar body (Nagahama *et al.*, 1994). At the second metaphase, meiosis is once again arrested just prior to ovulation. These oocytes can now be fertilized. At fertilization, the second polar body will appear.

Oocyte maturation involves a number of intricate hormonal mechanisms. Most importantly, gonadotropin, maturation-inducing hormone (MIH) and maturation-promoting hormone (MPH) have been shown to have very important and independent roles in the production of oocytes. Gonadotropin has been used in a number of *in vivo* and *in vitro* investigations to demonstrate its ability to induce maturation and ovulation. Germinal vesicle breakdown can be induced *in vitro* in response to treatment of gonadotropins (Nagahama *et al.*, 1994). However, the induction of oocyte maturation is not solely dependent upon gonadotropins. Oocytes displaced of steroidogenic cells did not respond to treatment with gonadotropin (Nagahama *et al.*, 1994). In addition to gonadotropin, MIH plays a role in oocyte maturation. A number of initiators of germinal vesicle breakdown have been found in fish. These are listed in Nagahama *et al.* (1994) as progesterone, 17a-hydroxyprogesterone, 17α,20β-dihydroxy-4-pregnen-3-one (17α,20β-DP), 17α,20β,21-trihydroxy-4-pregnen-3-one (20β-S), cortisol and deoxycorticosterone, with 17α,20β-DP and 20β-S having MIH characteristics.

17α,20β-DP has been identified as the maturation-inducing hormone of amago salmon (*Oncorhynchus rhodurus*) (Nagahama and Adachi, 1985). 17α,20β-DP was found in the medium when immature oocytes were incubated in the presence of gonadotropin. Nagahama and Adachi (1985) also found low levels of 17α,20β-DP in amago salmon vitellogenin oocytes, but higher levels were found in mature and ovulatory eggs. 17α,20β-DP has also been correlated with an increase in gonadotropin (Young *et al.*, 1983). The MIH in Japanese medaka (Fukada *et al.*, 1994) and killifish *(Fundulus heteroclitus)* (Petrino *et al.*, 1993) has also been identified as 17α,20β-DP. However, in Atlantic croaker, Trant *et al.* (1986) have identified 20β-S as the MIH. This same maturation-inducing hormone (20β-S) has also been found in the spotted sea trout (*Cynoscion nebulosus*) (Thomas and Trant, 1989). Research, to date, has not identified 20β-S in salmonids (Scott and

Canario, 1987). In Japanese medaka, 17α,20β-DP is produced in the granulosa cell in response to gonadotropin (Iwamatsu *et al.*, 1994).

Maturation-inducing hormone (MIH) production

MIH is produced in the follicle layer, with both the thecal and granulosa layers playing a role in production and metabolism of 17A,20β-DP. Nagahama *et al.* (1994) outlined a two-cell model of 17α,20β-DP synthesis. Briefly, the steroid precursor, identified as 17α-P is produced in the follicle thecal layer and metabolized to 17α,20β-DP in the granulosa layer. Gonadotropin is the trigger for the production of 17α-P. The conversion of 17α,20β-DP to 17α-P occurs via the enzyme 20β-hydroxysteroid dehydrogenase. Again, gonadotropin is the main stimulus for enhancing the activity of 20β-hydroxysteroid dehydrogenase. Kagawa *et al.* (1983) have also shown that, as oocyte maturation occurs, there is a decrease in estradiol synthesis. Oocyte maturation occurs over a very short period of time and investigators have reported that a shift from the production of estradiol to that of 17α,20β-DP occurs prior to the maturation process. In addition, 17α-P is not produced in the thecal layer until just before oocyte maturation. This two-cell type model has been demonstrated in salmonids. Japanese medaka and killifish *(Fundulus heteroclitus)* have a somewhat different mechanism as these follicular structures lack the same type of steroidogenic thecal cells (Nagahama *et al.*, 1994).

Fertilization and hatch

The chorion of the Japanese medaka oocyte is two to three layers thick, with an outer thin layer (zona radiata externa) and an inner thick layer (zona radiata interna) (Hart *et al.*, 1984). The chorion proteins are synthesized in the liver and within the developing oocyte. The hardening of the chorionic layer is essential to embryonic and larval survival. It has been hypothesized by Iwamatsu *et al.* (1995) that hardening of the chorion occurs at fertilization, when the enzyme responsible for hardening is released into the perivitelline space. During fertilization, the activated sperm enter the micropyle and attach to the plasma membrane. Most spermatozoa become activated upon contact with their aquatic environment. Sperm enter the ova through a micropyle opening on the egg surface. Research by Iwamatsu *et al.* (1997) has shown that the chorionic surface may contain specific glycoproteins that have an affinity for spermatozoa and that may play a role in terms of sperm micropyle entry.

After fertilization, the egg undergoes a number of rapid changes, including water absorption, formation of the perivitelline space and subsequent hardening of the chorion. Just before hatching, the embryo becomes very active and an enzyme (often referred to as the hatching enzyme) is released by 'hatching glands' of the embryo which serves to soften the chorion (Blaxter, 1988). Typically, successful hatch is based upon secretion of the hatching enzyme in conjunction with rapid and vigorous movements of the larval fish.

Modes of nutrition

Oviparous, ovoviparous and viviparous embryo nourishment

The reproductive biology of fish is very diverse, ranging from the production of yolk-laden eggs fertilized externally to the production of small eggs lacking yolk which are fertilized externally and require maternally supplied nutrition prior to parturition (Amoroso, 1960; Hoar, 1969; Wourms, 1981; Wourms *et al.*, 1988). Traditional terminology categorizes the different modes of reproduction as oviparous, ovoviviparous and viviparous. These terms refer to the pattern of embryo bearing and maternal contribution to embryo nourishment. Oviparous fish are those that produce eggs which develop external to the female's body and are nourished by yolk provided prior to fertilization. Ovoviviparous and viviparous fish retain developing embryos internally. Ovoviviparity and viviparity are contrasted in this terminology by the maternal nutritional contribution. Ovoviviparous females produce yolky eggs that receive no additional nourishment after fertilization, although they are retained in the female's body; viviparous females produce eggs without enough yolk for the embryo to complete development, but provide supplemental nourishment of embryos as they develop inside of the female (Balinsky, 1975).

A more detailed discussion regarding the modes of nutrition will not be included in this chapter; however, it is important to remember that chemical contaminants may elicit effects through a variety of different pathways and that the mode of reproduction may govern the degree of chemical impact. For example, those fish that retain developing embryos internally may be at greater risk to maternal–fetal transfer of chemicals than those fish that expel their eggs externally before or after fertilization. Furthermore, the composition of egg yolk nutrients may play a large role in fish with an oviparous breeding style, where no subsequent maternal nutrients are provided. Thus, when evaluating the reproductive effects of environmental chemicals, it is important to evaluate the observed effects in view of the mode of nutrition and/or mode of reproduction of the fish under investigation.

Critical developmental periods

The preceding sections have discussed various stages of development, focusing specifically on gonadal growth, egg and sperm maturation, fertilization and successful hatch. Exposure during critical stages of sexual development and differentiation could result in a number of adverse effects. Embryo and larval stages of development are sensitive to a number of natural and anthropogenic compounds present in the environment. In this chapter, developmental stages pertinent to reproduction will serve as the focus (i.e. exposure to reproductive-stage adults including maternal–fetal routes of exposure and egg-level effects, and exposure during embryo/larval sexual differentiation). The following section will discuss the potential adverse effects associated with exposure to environmental chemicals that may elicit adverse reproductive effects.

Targets for chemical toxicity

The question being asked by a large majority of investigators in toxicology is what effects does exposure to sublethal levels of chemicals have on the reproductive success of populations. Arguably, reproduction is one of the most sensitive and perhaps less tolerant indicators of sublethal effects of exposure to environmental chemicals. Reproduction is complex, involving a number of intricately timed mechanisms including release, production and action of hormonal and steroidal compounds on reproductive target tissue. Adverse reproductive effects may occur at the brain and/or gonadal level. The hypothalamus–pituitary–gonadal axis is ultimately controlled by feedback systems. For example, estrogen produced by the ovary can have a positive or negative effect on the hypothalamus, pituitary or the gonad itself, depending upon the concentration of hormone currently needed to meet the physiologic and reproductive needs of the fish. Alterations in steroid production could ultimately affect feedback pathways, or vice versa, which could lead to an impairment in reproductive processes.

With reproduction being vital to the survival of a species, it is not surprising that the ecologic significance (population level effects) of exposure to low concentrations of environmental chemicals has, to date, received the most attention. The examination of fish reproduction and the potential adverse reproductive effects posed by chemical exposure can serve as a reasonable measure of potential ecologic risks. The following sections will focus on the potential adverse reproductive effects of chemical contaminants on fish. Reproductive effects that may occur in male and female fish at different stages of development will be discussed. For example, alteration in sexual differentiation may occur at the larval or juvenile stage of some species, whereas 'egg-level effects' (e.g. oocyte atresia) may occur in adult females. The last section of the chapter will discuss the importance of developing 'measurable' endpoints of reproductive toxicity, and the importance of reproductive effects with regard to predicting potential ecologic effects.

In a review of the effects of pollution on fish reproduction, Kime (1995) examined a number of short- and long-term endpoints that serve as a measure of reproductive impairment. Indicators of long-term exposure to environmental chemicals can include examination of gonadal/somatic indices, histologic examination of ovarian tissue and the number of quality eggs produced. These endpoints also could be expanded to include male fish by examining male gonads and spermatogenesis. Short-term measures can include vitellogenin production (plasma levels in male and female fish), steroidogenesis and pituitary activity (Kime, 1995) among others.

Practical applications of these endpoints have been used in field studies of resident fish populations, e.g. a number of fish exposed to industrial effluents have shown changes in reproductive output. For example, roach (*Rutilus rutilus*) exposed to 3 percent pulp mill effluent for a complete life cycle showed reduced gonad growth (Sandstrom *et al.*, 1988). In addition, white suckers (*Catostomus commersoni*) exposed to pulp mill effluent have shown reduced plasma sex steroids, decreased egg size and gonadal somatic index, and reduced sperm motility (McMaster *et al.*, 1992). In addition, McMaster *et al.* (1991) demonstrated decreased steroid production in these white suckers, and Van der Kraak *et al.* (1992) reported a lowered response to gonadotropin and GnRH. More recently, Jobling *et al.* (1998) demonstrated a high incidence of intersexuality in wild populations of roach, resulting presumably from exposure to ambient levels of chemicals present in typical British rivers. Kime (1995) has compiled numerous references that outline the reproductive effects associated with exposure of fish to chemical contaminants. Some of this work is briefly discussed here, but the reader is referred to the review by Kime (1995).

Egg-level effects

Oocyte atresia

A number of chemical contaminants and industrial effluents have been shown to impair reproductive development and output. For example, there are a number of organochlorine pesticides that have been shown to cause oocyte atresia which may lead to several adverse effects. First and foremost, there is often a reduction in the number of eggs spawned. Second, there is potential for an increase in energy expenditure without a reproductive benefit. Kumar and Pant (1988) demonstrated that when rosy barb (*Puntius conchonius*) were exposed to aldrin and methoxyclor there was an increase in oocyte atresia. Also, in Japanese medaka exposed to γ-BHC, there was a marked increase in atretic oocytes. Kepone inhibited oviposition in Japanese medaka (Curtis and Beyers, 1978) along with a decrease in egg production and decreased hatching success (Goodman *et al.*, 1982).

McCormick *et al.* (1989) examined the effect of elevated pH on reproductive impairment in fathead minnows (*Pimephales promelas*). Upon histologic examination of ovarian tissue from stressed fish (elevated pH), there was a greater percentage of atretic oocytes than in the control population. In the study by McCormick *et al.* (1989), a decline in reproductive success of fathead minnows was found with increasing environmental stress, measured as an increase in pH. Billig *et al.* (1993) demonstrated that estrogens inhibit, whereas androgens enhance, ovarian granulosa cell apoptosis, suggesting that apoptosis may be the underlying mechanism of oocyte atresia.

As previously mentioned, oocyte maturation occurs over a very short period of time. For example, it can take fewer than 24 hours in cyprinids and up to 7 days in salmonids (Kime, 1995). Kime (1995) states that this stage is very susceptible to chemical contaminants, but also acknowledges the difficulty in studying such a narrowly defined stage. However, the implication of inhibited or failed maturation renders the oocyte incapable of fertilization and is clearly of great importance.

Alteration in egg yolk nutrients

As mentioned in previous sections, vitellogenin is produced in the liver in response to estradiol and is transported in the blood to the growing oocyte. Kime (1995) in his review of fish reproduction discussed the concentration of lipid-soluble chemicals such as organochlorine pesticides and industrial hydrocarbons in egg yolk. Such chemicals may lead to a change in the composition of yolk by interfering with naturally occurring lipids (Kime, 1995). In addition, there could be effects higher in the ovarian axis, such as alterations in mobilization and uptake of vitellogenin by oocytes. For example, Ruby *et al.* (1987) demonstrated that when rainbow trout were exposed to cyanide plasma vitellogenin levels increased and gonadal levels decreased, perhaps indicating that uptake of the protein into the oocyte was inhibited. Earlier work by Ruby *et al.* (1986) also showed that vitellogenin synthesis in rainbow trout was inhibited by cyanide during early vitellogenesis. In addition, if compounds exert hepatotoxicity, the synthesis of vitellogenin could also be altered. Pireira *et al.* (1993) showed that when winter flounder (*Pleuronectes americanius*) were exposed to cadmium, levels of vitellogenin were reduced. Singh (1989) reported that cadmium, malathion and 3-methylcholantrene reduced plasma vitellogenin in rice eel (*Monopterus albus*, Synbranchidae). The investigator suggested that reduced vitellogenin was a reflection of lowered plasma estradiol levels.

Maternal–fetal transfer

There are numerous questions and concerns pertaining to what effect alteration in vitellogenin production may have on nourishment of developing embryos and the resulting health of larval fish. For example, a correlation exists between reproductive function and plasma vitellogenin expression. Kramer *et al.* (1998) demonstrated that in fathead minnows exposed to 17β-estradiol, egg production, expressed as eggs laid per female, was significantly correlated with plasma vitellogenin. Additionally, maternal effects can arise in fish when some aspect of the environment alters the maternal contribution to yolk quality or quantity. With regard to anthropogenic agents, biologically important effects could arise through their direct incorporation into the yolk or through alteration in the egg yolking process. Lipophilic chemical contaminants have a high likelihood of being concentrated in the egg yolk. Monteverdi and Di Giulio (2000) demonstrated that oocytic accumulation of 2,3,7,8-tetrachlorodibenzo-*p*-dioxin (TCDD) and benzo(a)pyrene (BaP) by gravid killifish (*Fundulus heteroclitus*) was directly correlated with oocyte maturational status. A positive correlation was observed between oocyte maturational state (size) and both total quantity (total pmol) and concentration (pmol g^{-1} tissue) of TCDD and BaP. The maternal–fetal transfer of metals, such as copper, has also been demonstrated (Munkittrick and Dixon, 1989). Related to this, the compartmentalization of chemicals sequestered in maternal lipid stores may determine the level of exposure of ova and embryos. The manifestation of these subtle influences on reproductive biology could range from altered life histories on the affected generation to teratogenic effects.

Estrogen receptor-level effects

The fish estrogen receptor has similarities to the mammalian estrogen receptor and is a member of the steroid–thyroid–retinoic acid receptor superfamily (Evans, 1988). Research has shown an up-regulation of fish estrogen receptor mRNA and increased binding activity of radiolabeled estradiol in response to estradiol treatment (Mackay *et al.*, 1996). This up-regulation of estrogen receptor mRNA in response to estradiol treatment was demonstrated in sculpin, sea ravin, Atlantic salmon (*Salmo salar*), and winter flounder (Mackay *et al.*, 1996). Nimrod and Benson (1997) have also reported an up-regulation of estrogen receptors following exposure of channel catfish (*Ictalurus punctatus*) to estradiol and nonylphenol. In addition, Nimrod and Benson (1998) reported that the increased population of estrogen receptors in fish exposed to nonlyphenol did not result in a loss of receptor affinity, as was the case with fish exposed to estradiol. Thus, exposure to nonylphenol did not induce a desensitization mechanism that is believed to exist in association with endogenous estradiol. The significance of this finding will be further explored in the reproductive assessment section of this chapter.

During the reproductive cycle of salmonids, the amount of estrogen receptors in the liver changes. It has also been shown that an increase in the number of estrogen receptors leads to increased sensitivity to estradiol and, subsequently, the induction of vitellogenin. More recently, several investigators have demonstrated that the estrogen receptor can be activated by xenoestrogens and synthetic estrogens, in addition to endogenous estradiol.

Estradiol and male reproduction

The series of events leading to spermatogenesis and egg fertilization are governed by a series of neuroendocrine events as well as environmental and behavioral cues that also have some level of neural control. 17β-Estradiol enhances pituitary gonadotropin release and reduces androgen production by the testis (Trudeau *et al.*, 1993). Trudeau *et al.* (1993) therefore concluded that estradiol is a regulation factor in male goldfish reproduction. The investigators were able to demonstrate that when male goldfish were administered estradiol there was a decrease in testosterone and 11-ketotestosterone. It has been suggested that estradiol may directly regulate androgen production at the testicular level (Trudeau *et al.*, 1993). The significance of this negative regulation of androgen production lies in the ability for successful reproduction. The male reproductive steroids (testosterone and/or 11-ketotestosterone) are important and vital to regulation and control of expression of secondary sex characteristics, spermatogenesis, and male sexual behavior (Trudeau *et al.*, 1993). The probability that estrogen negatively affects androgen production by the testis leads one to examine further the consequences in terms of reproductive performance. Billard *et al.* (1981) demonstrated that sperm production and testicular development in rainbow trout were inhibited following exposure to 17β-estradiol. However, work by Pasmanik and Callard (1988) suggests that in male goldfish the production of estradiol is regulated during sexual development. For example, the investigators reported that aromatase levels are greatest at the time of spawning, whereas estradiol levels are greatest in sexually mature fish that are undergoing regression.

Sex differentiation

It has long been known in aquaculture that exposure to estrogens or androgens at select developmental stages can lead to the development of phenotypically altered male or female populations. For example, genetic males will possess female gonads and secondary sex characteristics, and genetic females will resemble and behave as male fish. As demonstrated by Nimrod and Benson (1998), the production of phenotypically altered sexes suggests a potential for same genetic sex matings. That is, a genetic male fish could mate with a phenotypically altered genetic male (phenotypic female), thereby producing an all genetically male population. The same holds true for genetic female–phenotypically altered genetic female matings. This type of same sex mating leading to the production of monosex tilapia populations has been exploited for aquaculture purposes (Merland, 1995). The question remains as to whether a similar scenario can exist in feral fish populations, and, if so, what are the potential ecologic implications. While it is realized that, in some fish, exposure to steroids at a critical life stage can reverse the phenotypic sex from its genetic predisposition, reproductive success of morphologically altered species has not been adequately investigated.

Environmental contaminants have been shown to exert estrogenic effects in male fish during periods of sexual differentiation. For example, Gimeno *et al.* (1996) exposed genetic male carp during a period of sexual differentiation to 4-*tert*-pentylphenol (TPP) and observed oviducts in all male fish exposed for 60 days. Gray and Metcalfe (1997) also demonstrated an intersex condition in medaka exposed to 4-nonylphenol from hatch to 3 months of age. In fish exposed to 50 $\mu g\ L^{-1}$ and 100 $\mu g\ L^{-1}$ of 4-nonylphenol, 50 percent and 86 percent of the male fish developed testis-ova, respectively. The ratio of male to female fish was 1:2, compared with 2:1 in the control population. Gimeno *et al.* (1996) also found that exposure of male carp to estradiol for 90 days produced phenotypic females with no testicular tissue.

Thyroid-level effects

Many studies pertaining to thyroid hormones have been directed toward obtaining a greater understanding of mechanistic actions. By comparison, relatively little research has been invested in examining the effect of thyroid hormones as a result of environmental chemical exposure. Ruby *et al.* (1993) examined plasma estradiol, T_3, and T_4 in rainbow trout following exposure to cyanide. These investigators demonstrated that plasma estradiol and T_3 levels were lower in cyanide-treated fish, whereas no difference was observed in plasma T_4 levels. Thus, the investigators suggested that, although cyanide inhibited the conversion from T_4 to T_3, the lowered estradiol levels may have been the result of chemical interaction along the hypothalamic–pituitary–ovarian axis. Cyr and Eales (1996) also suggested that estradiol may depress the clearance of T_4 as well as the conversion of T_4 to T_3. A study by Zhou *et al.* (1995) suggested that, in rats, methoxychlor may disrupt the T_4 to T_3 conversion by binding to 5′-ID1. Thyroid hyperplasia in salmon and herring gull populations has also been reported in select regions of the Great Lakes (Leatherland, 1992). It is suggested that this goiter condition may be a result of the inability of the organism to produce T_3. In addition, Leatherland and Sonstegard (1987) examined a stock of Lake Erie coho salmon (*Oncorhynchus kisutch*) that demonstrated lowered growth, plasma steroid hormones, fecundity, and fertilization, as well as poor development of secondary sex characteristics. This stock of salmon also had a high incidence of thyroid lesions.

Assessment of reproductive effects

Reproductive assessment: a predictor tool of population-level effects

Although the endocrine disrupter issue has received considerable notoriety, research focused on the potential for adverse health effects that are possible through disruption of the endocrine system is not new to the scientific community. However, the 'new' science and associated issues have provided additional mechanistic insight into the developmental and reproductive effects of a diverse range of environmental chemicals. Persistent and bioacummulative chemical compounds, some of which have been banned from production in the USA (e.g. PCBs and DDT) have been associated with developmental and reproductive effects in select wildlife species.

Potential reproductive-level effects in wild fish have been demonstrated in a population of white suckers exposed to bleached kraft mill effluent (BKME) along the north shore of Lake Superior (McMaster *et al.*, 1991; Munkittrick *et al.*, 1991). The exposed fish showed abnormal gonadal development, delayed age to maturity and altered steroid levels (McMaster *et al.*, 1991). Van Der Kraak *et al.* (1992) also reported reduced secretion of gonadotropin and depressed ovarian steroid synthesis in these fish. Female white suckers had a smaller number of eggs at maturity and males had a lower expression of secondary sex characteristics (McMaster *et al.*, 1991). Gonads of both male and female fish were smaller at the BKME site as well. Although females exposed to BKME had fewer and smaller eggs, there was no difference in fecundity compared with reference site white suckers. McMaster *et al.* (1991) reported no effect on hatchability, size of larvae or larvae survival. To date, the population dynamics of this white sucker population have not been investigated.

Assessment of reproductive function

There is, undoubtedly, a need to develop techniques to assess and predict the organismal cost of exposure to select chemical compounds. For example, examination of reproductive effects associated with exposure to endocrine- and reproduction-disrupting chemicals would provide information on the whole organism that could be used to predict more accurately ecologic effects. Investigations that focus on a relationship between exposure and population-level effects (e.g. reproductive success) will provide relevant information to understand further the importance and relevance of biologic indicators of exposure to chemical compounds. Furthermore, there is a need to develop approaches specifically designed to examine organismal exposure during critical developmental windows controlled by specific hormonal axes and to identify and quantify the associated adverse reproductive effects. While it is recognized that it will be difficult to relate indicators of exposure to detrimental changes in population dynamics, the development of such studies and bioassays is needed to provide relevant data regarding the ecologic significance of exposure to chemical compounds.

Reproductive toxicity guidelines

Ecologic Effects Test Guidelines have been developed by the United States Environmental Protection Agency's (US EPA) Office of Prevention, Pesticides and Toxic Substances (OPPTS) to examine the potential adverse effects of pesticides and toxic substances. Specific guidelines were designed to meet the testing mandates set forth by the Federal Insecticide, Fungicide and Rodenticide Act (FIFRA) and the Toxic Substances Control Act (TSCA). Such guidelines to assess whole organism responses include the fish life cycle toxicity test as well as the fish early life stage toxicity test (US EPA, 1996a,b). The life cycle toxicity test requires full life cycle exposure, followed by measurements of reproductive endpoints such as time of spawning, egg number, fertility and fecundity, as well as locomotion, behavioral, physiologic and pathologic effects (US EPA, 1996a). Although full life cycle exposures are often critical to simulate environmental exposures, the methodologies are time and labor intensive.

The fish early life stage toxicity test involves exposure at time of fertilization through hatch and first feeding (US EPA, 1996b). In this short-term assay, many of the reproductive parameters typically measured in adults are exchanged for successful hatch, survival, growth and developmental and behavioral abnormalities. It is important to examine early life stage endpoints, but it is also of utmost importance to examine what effect sublethal exposures at an early life stage may have on adults. This type of test method leaves room for multigenerational studies as larval fish can be grown to maturity with subsequent monitoring of the reproductive success of the second generation. This may be carried out in a more cost- and time-efficient manner that does not require full life cycle exposure, but exposure at critical windows of developmental and reproductive processes.

Reproductive model using Japanese medaka

Japanese medaka are an excellent model organism which can be induced to breed daily in response to a prescribed photoperiod, temperature and food regime. These small aquarium species require relatively simple husbandry and, in our laboratory, are fully mature within 6–8 weeks of hatch. The rapid time to maturation makes this species an attractive model organism for studying reproductive and development toxicity in F_0, as well as subsequent generations. Figure 5.5 outlines a proposed model to measure the developmental and reproductive toxicity of environmental chemicals in Japanese medaka. By examining the effects of exposure at early life and/or adult stage, the reproductive model combines the effort of the US EPA life cycle toxicity test and the early life stage toxicity test. Investigative pathways and key exposure stages are referred to in Figure 5.5.

The proposed Japanese medaka reproduction assay is based upon exposure during key developmental events and is sufficiently flexible in design to evaluate developmental and reproductive toxicity in larval and/or adult fish. Larval Japanese medaka may be exposed during a critical window of time when developmental processes are most susceptible to chemical insult (Figure 5.5; A). Experience in our laboratory indicates that 4–7 days post-hatch represents such a critical window period of development. During exposure, larval fish would be monitored daily for mortality and feeding behavior. Upon termination of exposure, larval fish are transferred to aquaria containing dilution water only and grown to maturity.

The next phase of the Japanese medaka reproduction model may involve a second exposure of the adult fish to the same chemical compound utilized during the initial larval exposure (Figure 5.5; B). Exposure at the adult stage is important because it evaluates the effect of secondary chemical insult on reproductive processes. Also, in addition to assessing reproductive output by measuring time of first spawn, fecundity, clutch size and egg hatchability, a subset of fish can be sacrificed daily and vitellogenin, as well as other diagnostic indicators of chemical exposure, may be measured. Therefore, at the end of chemical exposure and

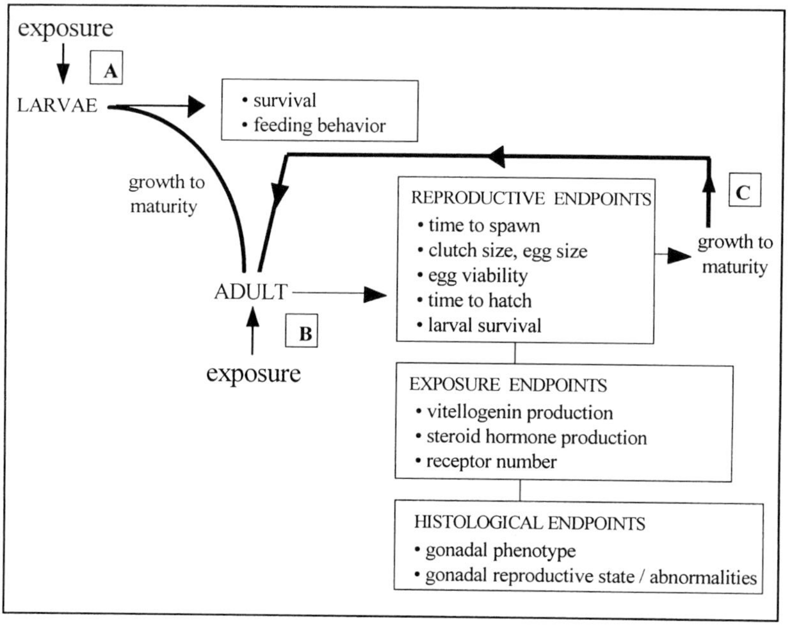

Figure 5.5 Model to assess reproductive impairment utilizing Japanese medaka, *Oryzias latipes*.

reproductive monitoring, information on adult fecundity can be related to a biologic indicator of exposure and/or effect (Figure 5.5). The Japanese medaka reproductive model will permit sufficient flexibility to examine other biologic indicators (e.g. steroid levels, receptor number) which would permit association with endpoints representative of reproductive toxicity (Figure 5.5).

The linking of diagnostic or biologic indicators to reproductive assessment endpoints is a critical step toward developing a fundamental understanding of mechanisms governing reproductive and developmental toxicity. As an example, the induction of vitellogenin in male fish has been widely used as a diagnostic indicator of exposure to estrogenic chemicals; however, little is known of the biologic consequences of such unscheduled vitellogenin induction in male fish. To begin to address this issue, Jobling *et al.* (1996) demonstrated that elevated vitellogenin levels in developing male fish were associated with decreased testicular growth. The decrease in testicular growth was only evident in developing male fish; testicular growth was not affected in mature or regressed fish. The investigators also reported that the estrogenic potency of the chemical was directly related to the inhibition of testicular growth in developing male fish. Likewise, the Japanese medaka reproduction model provides a means to measure an estrogenic response which can then be related to developmental and reproductive assessments.

As outlined in Figure 5.5, following adult exposure and measurement of reproductive parameters, a number of options for further evaluation are available to the investigator depending upon the research objectives and time and/or cost restraints of the investigation. One option would be to grow the offspring to maturity and follow the identical experimental design for the F_1 generation (Figure 5.5; C). This may or may not involve exposure at the larval stage. It is also important to note that larval exposure need not be an integral part of the proposed model. Investigators may choose to examine the relationship between biologic indicators and reproductive endpoints using only adult exposed fish.

Developmental effects measured at commencement of the reproductive assessment by means of histopathologic analysis could provide a wealth of information with regard to the nature of reproductive impairments. Thus far, data collected would include the measure of a biologic indicator (e.g. vitellogenin, steroid levels, receptor number) along with reproductive parameters such as time of first spawn, fecundity, clutch size and egg hatchability. While histology is often labor intensive, it provides an important component to the reproductive assessment data previously collected on the same population of fish. Not only does histologic information assist in determining the state of reproductive development of individual male or female fish but it also confirms the gonadal phenotype of the fish, which would have been initially sexed by examining external phenotype.

Histologic analysis of gonadal tissue has been used to explain reproductive impairment in several wildlife species. Examination of gonadal tissue from the American alligators of Lake Apopka provided valuable information regarding egg production in this species. For example, Guillette *et al.* (1994) found that female alligators from a contaminant-laden lake showed polyovular ovarian follicles and polynuclear oocytes. In addition to these histologic findings, there were extreme differences in levels of circulating sex steroids as well as reduced egg viability compared with alligators from a reference lake. Male alligators from Lake Apopka also showed poorly differentiated seminiferous tubules.

A number of studies have reported a variety of developmental effects associated with gonadal differentiation in fish following exposure to select natural and synthetic environmental estrogens. Most notably, developmental changes included phenotypic alterations, in which the sexual phenotype was different from the genetic composition, and intersex conditions, whereby gonochoristic species exist with both testicular and ovarian tissue (ova-testis). However, the majority of studies with monosex population have failed to examine the reproductive capabilities of these morphologically altered species. Should these conditions be detected in medaka exposed at the larval and/or adult stages in the proposed model, there would be a number of biochemical and physiologic endpoints that could be used to assess the reproductive potential of such fish.

Conclusions

Fish reproduction is arguably one of the most sensitive indicators of exposure to environmental chemicals. Reproductive toxicity can simply be referred to as an alteration in reproductive success. It can occur at several developmental stages (larval, juvenile, and adult) with levels of susceptibility varying among species. For the most part, larval fish undergoing gonadal differentiation and growth are highly susceptible to chemical stress. Adults also can pass the effects of chemical exposure to their young via maternal–fetal transfer of environmental chemicals or simply through a reduction in egg quality and number. Egg yolk nutrients that are essential to the developing embryo can also be altered. With widespread attention given to the endocrine disrupter issue and subsequent reproductive effects, a number of diagnostic tools are currently under development that will aid in identifying reproductive effects that, in turn, can have predictive value with respect to identifying ecologically relevant population-level effects.

References

Amoroso, E.C. 1960 Viviparity in fishes. *Symposiums of the Zoological Society of London* 1: 153–181.

Arcand-Hoy, L.D. and Benson, W.H. 1998. Fish reproduction: An ecologically relevant indicator of endocrine disruption. *Environmental Toxicology and Chemistry* 17: 49–57.

Balinsky, B.I. 1975. *An Introduction to Embryology*, 4th edn. W.B. Saunders, Philadelphia, PA.

Billard, R., Breton, B. and Richard, M. 1981. On the inhibitory effect of some steroids on spermatogenesis in adult rainbow trout (*Salmo gairdneri*). *Canadian Journal of Zoology* 59: 1479–1487.

Billig, H., Furuta, I. and Hsueh, A.J.W. 1993. Estrogens inhibit and androgens enhance ovarian granulosa cell apoptosis. *Endocrinology* 133: 2204–2212.

Blaise, O., Le Bail, P.Y. and Weil, C. 1995. Lack of gonadotropin-releasing hormone action on in vivo and in vitro growth hormone release in rainbow trout (*Oncorhynchus mykiss*). *Comparative Biochemistry and Physiology* 110C: 133–141.

Blaxter, J.H.S. 1988. Pattern and variety in development. In *Fish Physiology*, Vol. XIA, Ch. 1. Academic Press, San Diego.

Breton, B., Mikolajczyk, T. and Popek, W. 1993. The neuroendocrine control of the gonadotropin (GtH 2) secretion in teleost fish. In *Aquaculture: Fundamental and Applied Research*. Lalhore, B. and Vitiello, P. (eds), pp. 199–215. American Geophysical Union, Washington, DC.

Chakraborti, P., Maitra, G. and Bhattacharya, S. 1986. Binding of thyroid hormone isolated over nuclei from a freshwater perch *Anabas testudineus*. *General and Comparative Endocrinology* 62: 239–246.

Curtis, L.R. and Beyers, R.J. 1978. Inhibition of oviposition in the teleost *Oryzias latipes*, induced by subacute kepone exposure. *Comparative Biochemistry and Physiology* 61C: 15–16.

Cyr, D.G. and Eales, J.G. 1996. Interrelationship between thyroidal and reproductive endocrine systems in fish. *Reviews of Fish Biology* 6: 165–200.

Degani, B., Bokder, R. and Jackson, K. 1996. Growth hormone, gonad development and steroid levels in female carp. *Comparative Biochemistry and Physiology* 115C: 133–140.

Evans, R. 1988. The steroid and thyroid receptor superfamily. *Science* 240: 889–895.

Fukada, S., Sakai, N., Adachi, S. and Nagahama, Y. 1994. Steroidogenesis in the ovarian follicle of medaka (*Oryzias latipes*, a daily spawner) during oocyte maturation. *Development Growth and Differentiation* 36: 81–88.

Gimeno, S., Gerritsen, A. and Bowmer, T. 1996. Feminization of male carp. *Nature* 384: 221–222.

Goodman, L.R., Hansen, D.J. Manning, C.S. and Faas, L.F. 1982. Effects of kepone on the sheepshead minnow in an entire life-cycle toxicity test. *Archives of Environmental Contamination and Toxicology* 11: 335–342.

Gray, M.A. and Metcalfe, C.D. 1997. Induction of testis-ova in Japanese medaka (*Oryzias latipes*) exposed to p-nonylphenol. *Environmental Toxicology and Chemistry* 16: 1082–1086.

Guillette, Jr, L.J., Gross, T.S., Masson, G.R., Matter, J.M., Percival, H.F. and Woodward, A.R. 1994. Developmental abnormalities of the gonad and abnormal sex hormone concentrations in juvenile alligators from contaminated and control lakes in Florida. *Environmental Health Perspectives* 102: 680–688.

Hart, N.H., Pietri, R. and Donovan, M. 1984. The structure of the chorion and associated surface filaments in Oryzias: Evidence for the presence of extracellular tubules. *Journal of Experimental Zoology* 230: 273–296.

Hoar, W.S. 1969. Reproduction. In *Fish Physiology*, Vol. 3. Hoar, W.S. and Randall, D.J. (eds), pp. 1–72. Academic Press, New York.

Iwamatsu, T., Nakashunaa, S., Onitake, K., Matsuhisa, A. and Nagahama, Y. 1994. Regional differences in granulosa cells of preovulatory medaka follicles. *Zoological Science* 11: 77–82.

Iwamatsu, T., Shibata, Y. and Kanie, T. 1995. Changes in chorion proteins induced by the exudate released from the egg cortex at the time of fertilization in teleost, *Oryzias latipes*. *Development Growth and Differentiation* 37: 747–759.

Iwamatsu, T., Yoshizaki, N. and Shibata, Y. 1997. Changes in the chorion and sperm entry into the micropyle during fertilization in the teleostean fish, *Oryzias latipes*. *Development Growth and Differentiation* 39: 33–41.

Jobling, M. 1995. *Environmental Biology of Fishes*. Chapman & Hall, London, UK.

Jobling, S., Sheahan, D., Osbourne, J.A., Matthiessen, P. and Sumpter, J.P. 1996. Inhibition of testicular growth in rainbow trout (*Oncorhynchus mykiss*) exposed to estrogenic alkylphenolic chemicals. *Environmental Toxicology and Chemistry* 15: 194–202.

Jobling, S., Nolan, M., Tyler, C.R., Brighty, G. and Sumpter, J.P. 1998. Widespread sexual disruption in wild fish. *Environmental Science and Technology* 32: 2498–2506.

Kagawa, H., Young, G. and Nagahama, Y. 1983. Relationship between seasonal plasma estradiol-17β and testosterone levels and in vitro production by ovarian follicles of amago salmon (*Oncorhynchus rhodurus*). *Biological Reproduction* 29: 301–309.

Kamamori, A. and Nagahama, Y. 1988. Developmental changes in the properties of gonadotropin receptors in the ovarian follicles of amago salmon (*Oncorhynchus rhodurus*) to chum salmon gonadotropin during oogenesis. *General and Comparative Endocrinology* 72: 25–38.

Kanamori, A., Nagahama, Y. and Egami, N. 1985. Development of tissue architecture in the gonads of the medaka *Oryzias latipes*. *Zoological Science* 2: 695–706.

Kawauchi, H. 1989. Evolutionary aspects of pituitary hormones. *Kitasato Archives of Experimental Medicine* 62: 139–155.

Kime, D.E. 1995. The effects of pollution on reproduction in fish. 1995. *Reviews in Fish Biology and Fisheries* 5: 52–96.

Kramer, V.J., Miles-Richardson, S., Pierens, S.L. and Giesy, J.P. 1998. Reproductive impairment and induction of alkaline-lable phosphate, a biomarker of estrogen exposure, in fathead minnows (*Pimephales promelas*) exposed to waterborne 17β-estradiol. *Aquatic Toxicology* 40: 335–360.

Kumar, S. and Pant, S.C. 1988. Comparative sublethal ovarian pathology of some pesticides in the teleost, *Puntius conchonius* Hamilton. *Bulletin of Environmental Contamination and Toxicology* 41: 227–232.

Leatherland, J. 1992. Endocrine and reproductive function in Great Lake salmon. In *Chemically Induced Alterations in Sexual and Functional Development: The Wildlife/Human Connection*. Colborn, T. and Clement, C. (eds), pp. 129–145. Princeton Scientific Publishing, Princeton.

Leatherland, J.F. and Sonstegard, R.A. 1987. Comparative fecundity and egg survival in two stocks of goitered coho salmon (*Oncorhynchus kisutch* Walbaum) from Lake Erie. *Canadian Journal of Zoology* 65: 2780–2785.

Le Brun, C., Billard, R. and Jalabert, B. 1982. Changes in the number of germ cells in the gonad of the rainbow trout (Salmo gairdneri) during the first 10 post-hatching weeks. *Reproduction Nutrition Development* 22: 405–412.

Lin, X.W., Lin, H.R. and Peter, R. 1993. Growth hormone secretion in the common carp (*Cyprinus carpio* L.): in vitro interactions of gonadotropin-releasing hormone, somatostatin and the dopamine agonist apomorphene. *General and Comparative Endocrinology* 89: 62–71.

McCormick, J.H., Stokes, G.N. and Hermanutz, R.O. 1989. Oocyte atresia and reproductive success in fathead minnows (*Pimephales promelas*) exposed to acidified hardwater environments. *Archives of Environmental Contamination and Toxicology* 18: 207–214.

Mackay, M.E., Raelson, J. and Lazier, C.B. 1996. Up-regulation of estrogen receptor mRNA and estrogen receptor activity by estradiol in liver of rainbow trout and other teleostean fish. *Comparative Biochemistry and Physiology* 115C: 201–209.

McMaster, M.E., Van Der Kraak, G.J., Portt, C.B., Munkittrick, K.R., Sibley, P.K., Smith, I.R. and Dixon, D.G. 1991. Changes in hepatic mixed-function oxygenase (MFO) activity, plasma steroid levels and age at maturity of a white sucker (*Catostomus commersoni*) pollution exposed to bleached kraft mill effluent. *Aquatic Toxicology* 21: 199–218.

McMaster, M.E., Portt, C.B., Munkittrick, K.R. and Dixon, D.G. 1992. Milt characteristics, reproductive performance, and larval survival and development of white sucker exposed to bleached kraft mill effluent. *Ecotoxicology and Environmental Safety* 23: 103–117.

Marchant, T.A. and Peter, R.E. 1989. Hypothalamic peptides influencing growth hormone secretion in the goldfish, *Carassius auratus*. *Fish Physiology and Biochemistry* 7: 133–139.

Merland, C. 1995. Production of high percentage of male offspring with 17αethynlestradiol sex-reversed Oreochromis aureus. I. Estrogen sex reversal and production of F2 psuedofemales. *Aquaculture* 130: 23–34.

Mommsen, T.P. and Walsh, P.J. 1988. Vitellogenesis and oocyte assembly. In *Fish Physiology*, Vol. XI. *The Physiology of the Developing Fish.* Part A. *Eggs and Larvae*, pp. 347–406. Academic Press, San Diego.

Monteverdi, G.H. and Di Giulio, R.T. 2000. Vitellogenin-associated maternal transfer of exogenous and endogenous ligands in the estuarine fish, *Fundulus heteroclitus*. In *Proceedings of the Tenth International Symposium on Pollutant Responses in Marine Organisms.* Williamsburg, VA. April 1999. In press.

Munkittrick, K.R. and Dixon, D.G. 1989. Effects of natural exposure to copper and zinc on egg size and larval copper tolerance in white sucker (*Catostomus commersoni*). *Ecotoxicology and Environmental Safety* 18: 15–26.

Munkittrick, K.R., Portt, C.B., Van Der Kraak, G.J., Smith, I.R. and Rokosh, D.A. 1991. Impact of bleached kraft mill effluent on population characteristics, liver MFO activity, and serum steroid levels of a Lake Superior white sucker (*Catostomus commersoni*) population. *Canadian Journal of Fisheries and Aquatic Science* 48: 1371–1380.

Mylonas, C.C., Hinshaw, J.M. and Sullivan, C.V. 1992. $GnRH_a$-induced ovulation of brown trout (*Salmo trutta*) and its effects on egg quality. *Aquaculture* 10: 379–392.

Nagahama, Y. 1983. The functional morphology of the teleost gonads. In *Fish Physiology*, Vol. IXA. Hoar, W.S., Randall, D.J. and Donaldson, E.M. (eds), pp. 223–275. Academic Press, London.

Nagahama, Y. 1994. Endocrine regulation of gametogenesis in fish. *International Journal of Developmental Biology* 38: 217–229.

Nagahama, Y. and Adachi, S. 1985. Identification of maturation-inducing steroid in a teleost, the amago salmon (*Onchorhynchus rhodurus*). *Developmental Biology* 109: 428–435.

Nagahama, Y., Yoshikuni, M., Yamashita, M. and Tanaka, M. 1994. Molecular endocrinology of fish. In *Fish Physiology*, Vol. XIII. Sherwood, N.M. and Hew, C.L. (eds). Academic Press, San Diego.

Nagahama, Y., Yoshikuni, M., Yamashita, M., Tokumoto, T. and Katsu, Y. 1995. Regulation of oocyte growth and maturation in fish. In *Current Topics in Developmental Biology*, Vol. 30. Pedersen, R. and Schatten, G.P. (eds), pp. 103–145. Academic, San Diego.

Nimrod, A.C. and Benson, W.H. 1997. Xenobiotic interaction with and alteration of channel catfish estrogen receptor. *Toxicology and Applied Pharmacology* 147: 381–390.

Nimrod, A.C. and Benson, W.H. 1998. Reproduction and development of Japanese medaka following an early life stage exposure to xenoestrogens. *Aquatic Toxicology* 44: 141–156.

Norstedt, G., Wrange, O. and Gustafsson. J.-A. 1981. Multihormonal regulation of the estrogen receptor in rat liver. *Endocrinology* 108: 1190–1196.

Pasmanik, M. and Callard, G.V. 1988. Changes in brain aromatase and 5α reductase activities correlate significantly with seasonal reproductive cycles in goldfish (*Carassus auretus*). *Endocrinology* 122: 1349–1356.

Petrino, T.R., Lin, Y.W., Netherton, J.C., Powell, D.H. and Wallace, R.A. 1993. Steroidogenesis in *Fundulus heteroclitus*: Purification, characterization, and metabolism of 17α,20β-dihydroxy-4-pregnen-3-one by intact follicles and its role in oocyte maturation. *General and Comparative Endocrinology* 92: 1–15.

Pireira, J.J., Mercaldo-Allen, R., Kuropat, C., Luedke, D. and Sennefelder, G. 1993. Effect of cadmium accumulation on serum vitellogenin levels and hepatosomatic indices of winter flounder (*Pleuronectes americanus*). *Archives of Environmental Contamination and Toxicology* 24: 427–431.

Redding, M.J. and Patino, R. 1993. Reproductive physiology. In *The Physiology of Fishes*. Marine Science Series. Evans, D.H. (ed.), pp. 503–534. CRC, Boca Raton.

Riley, D., Heisermann, G.J., MacPherson, R. and Callard, I.P. 1987. Hepatic estrogen receptor in the turtle, Chrysemys picta: Partial characterization, seasonal changes and pituitary dependence. *Journal of Steroid Biochemistry* 26: 41–47.

Ruby, S.M., Idler, D.R. and So, Y.P. 1986. The effect of sublethal cyanide exposure on plasma vitellogenin in rainbow trout (*Salmo gairdneri*) during early vitellogenesis. *Archives of Environmental Contamination and Toxicology* 15: 603–607.

Ruby, S.M, Idler, D.R. and So, Y.P. 1987. Changes in plasma, liver and ovary vitellogenin in landlocked Atlantic salmon following exposure to sublethal cyanide. *Archives of Environmental Contamination and Toxicology* 16: 507–510.

Ruby, S.M., Idler, D.R. and Peng So, Y. 1993. Plasma vitellogenin, 17β-estradiol, T3 and T4 levels in sexually maturing rainbow trout Oncorhynchus mykiss following sublethal HCN exposure. *Aquatic Toxicology* 26: 91–102.

Sakata, S., Nose, T., Moriyama, S., Hirano, T. and Kawauch, H. 1993. Flounder growth hormone: Isolation, characterization and effects on juvenile rainbow trout. *General and Comparative Endocrinology* 89: 396–404.

Sandstrom, O., Neuman, E. and Karas, P. 1988. Effects of a bleached pulp mill effluent on growth and gonad function in Baltic coastal fish. *Water Science and Technology* 20: 107–118.

Satoh, N. and Egami, N. 1972. Sex differentiation of germ cells in the teleost, *Oryzias latipes*, during normal embryonic development. *Journal of Embryology and Experimental Morphology* 28: 385–395.

Scanes, C.G. and Harvey, S. 1995. Growth hormone action: reproductive function. In *Growth Hormone*. Harvey, S., Scanes, C.G. and Daughaday, W.H. (eds), Ch. 23. CRC Press, Boca Raton.

Scott, A.P. and Canario, A.V.M. 1987. Status of oocyte maturation-inducing steroids in teleosts. In *Reproductive Physiology of Fish*. Idler, D.R., Crim, L.W. and Walsh, J.M. (eds), pp. 224–234. Memorial University Press. St. John's, Canada.

Singh, H. 1989. Interaction of xenobiotics with respect to endocrine function in a protogynous teleost, Monopterus albus. *Marine Environmental Research* 28: 285–289.

Singh, H., Griffith, R.W., Takahashi, A.P., Kawauchi, H. and Stegeman, J.J. 1988. Regulation of gonadal steroidogenesis in Fundulus heteroclitus by recombinant salmon growth hormone and purified salmon prolaction. *General and Comparative Endocrinology* 72: 144–153.

Sullivan, C.G., Bernard, M.G., Hara, A. and Dickhoff, W.W. 1989. Thyroid hormones in trout reproduction: Enhancement of gonadotropin-releasing hormone analogue and partially purified salmon gonadotropin-induced ovarian maturation *in vivo* and *in vitro*. *Journal of Experimental Zoology* 25: 189–195.

Swanson, P., Suzukik, K., Kawauchi, H. and Dickhoff, W.W. 1991. Isolation and characterization of two coho salmon gonadotropins, GtH I and GtH II. *Biology of Reproduction* 44: 29–38.

Tagawa, M. and Hirano, T. 1991. Effects of thyroid hormone deficiency in eggs on early development of the medaka, *Oryzias latipes*. *Journal of Experimental Zoology* 257: 360–366.

Thomas, P. and Trant, J.M. 1989. Evidence that 17α,20β, 21-trihydroxy-4-pregnen-3-one is a maturation-inducing steroid in spotted seatrout. *Fish Physiology and Biochemistry* 7: 185–189.

Thomas, P., Loomis, T., Sperry, T., Khan, I. and Detweiler, C. 2000. Interference with the genomic and nongenomic actions of steroids in fishes by chemicals: Role of receptor binding. In *Proceedings of the Tenth International Symposium on Pollutant Responses in Marine Organisms*. Williamsburg, VA. April 1999. In press.

Trant, J.M., Thomas, P. and Shackleton, C.H.L. 1986. Identification of 17α,20β,21-trihydroxy-4-pregnen-3-one as the major ovarian steroid produced by the teleost Micropogonias undulatus during oocyte maturation. *Steroids* 47: 89–99.

Trudeau, V.L., Wade, M.G., Van Der Kraak, G. and Peter, R.E. 1993. Effects of 17β-estradiol on pituitary and testicular function in male goldfish. *Canadian Journal of Zoology* 17: 1131–1134.

US EPA (Environmental Protection Agency). 1996a. *Ecological Effects Test Guidelines.* Office of Prevention, Pesticides and Toxic Substances 850.1500. Fish life cycle toxicity. EPA 712-C-96-122. Office of Prevention, Pesticides and Toxic Substances, Washington, DC.

US EPA (Environmental Protection Agency). 1996b. *Ecological Effects Test Guidelines.* Office of Prevention, Pesticides and Toxic Substances 850.1400. Fish early-life stage toxicity test. EPA 712-C-96-121. Office of Prevention, Pesticides and Toxic Substances, Washington, DC.

Vanden Hurk, R. and Sloff, G.A. 1981. A morphological and experimental study of gonadal sex differentiation in the rainbow trout, Salmo gairdneri. *Cell Tissue Research* 218: 487–497.

Van Der Kraak, G., Rosenblum, P.M. and Peter, R.E. 1990. Growth hormone dependent potentiation of gonadotropin-stimulated steroid production by ovarian follicles of the goldfish. *General and Comparative Endocrinology* 79: 233–239.

Van Der Kraak, G.J., Munkittrick, K.R., McMaster, M.E., Portt, C.B. and Chang, J.P. 1992. Exposure to bleached kraft pulp mill effluent disrupts the pituitary–gonadal axis of white sucker at multiple sites. *Toxicology and Applied Pharmacology* 115: 224–233.

Van Der Kraak, G., Chang, J.P. and Janz, D.M. 1997. Reproduction. In *The Physiology of Fishes*, 2nd edn. Evans, D.H. (ed.), pp. 465–488. CRC, Boca Raton.

Wourms, J.P. 1981. Viviparity: The maternal–fetal relationship in fishes. *American Zoologist* 21: 473–515.

Wourms, J.P., Grove, B.D. and Lombardi, J. 1988. The maternal–embryonic relationship in viviparous fish. In *Fish Physiology*, Vol. 11B. Hoar, W.S. and Randall, D.J. (eds), pp. 1–134. Academic Press, New York.

Yamamoto, T.S. 1975. *Medaka (Killifish) Biology and Strains. Series of Stock Culture in Biological Field.* Keigaku Publishing, Tokyo.

Young, G., Crim, L.W., Kagawa, H., Kambegawa, A. and Nagahama, Y. 1983. Plasma 17α,20β-dihydroxy-4-pregnen-3-one levels during sexual maturation of amago salmon (*Oncorhynchus rhoderus*): correlation with plasma gonadotropin and *in vitro* production by ovarian follicles. *General and Comparative Endocrinology* 51: 96–105.

Zhou, L.X., Dehal, S.S., Kupfer, D., Morrell, S., McKenzie, B.A., Eccleston, Jr, E.D. and Holtzman, J.L. 1995. Cytochrome P450 catalyzed covalent binding of methoxychlor to rat hepatic, microsomal iodothyronine 5′-monodeiodinase, type I: Does exposure to methoxychlor disrupt thyroid hormone metabolism? *Archives of Biochemistry and Biophysics* 322: 390–394.

Index